PERCEPTION AND DISCOVERY

An Introduction to Scientific Inquiry

Edited by
WILLARD C. HUMPHREYS

Associate Professor of Philosophy
New College

 FREEMAN, COOPER & COMPANY

PERCEPTION AND DISCOVERY

An Introduction to Scientific Inquiry

NORWOOD RUSSELL HANSON
Late Professor of Philosophy and Fellow of Pierson College
Yale University

1736 Stockton Street, San Francisco, California 94133

Copyright © 1969 by Freeman, Cooper & Company

All rights to reproduce this book in whole or in part are reserved, with the exception of the right to use short quotations for review of the book.

Printed in the United States of America

Library of Congress Catalogue Card Number 75-95161

SBN 87735-509-6

For Trevor and Leslie

Preface

NORWOOD Russell Hanson did more in his life than three good men. It is most fortunate that this substantial part of his unpublished work has been made available through the exceptionally devoted work of his friend and former student, W. C. Humphreys. As editor of this volume he has put in far more of his time and of himself than the reader could infer from his modest comments. But the work itself bears Hanson's stamp on every page—forceful, informative, original, charming, witty and colorful.

In a general sense Hanson continues the application of the Wittgensteinian approach to the philosophy of science, as Waissman and Toulmin have also done. But he goes much further than they, exploring questions about perception and discovery in more detail, and—perhaps his greatest strength—tying in the history of science for exemplification and for its own benefit. Hanson was one of the rare thinkers in the tradition of Whewell—a man he much admired—who could really benefit from and yield benefits for both the history and philosophy of science. He founded the only combined department in this country and he was a paragon of its principles. There is a certain tension between the demands of exact historical scholarship and the more free-ranging interests of the overviewer—the philosopher of science in one of his roles. Hanson released this tension by simply working harder, so that he became an expert on such topics as the mathematics of epicycles, and the development of twentieth century physics *and* on inductive logic, etc. Philosophy of science is an immensely demanding field in itself since any comprehensive approach requires considerable scientific knowledge (*not* great research performances) as well as philosophical expertise, and it is immeasurably helped by a really good knowledge of the history of science. Throughout his work Hanson is calling on all these skills—but he had many more, from musical to engineering. He always felt that the wider a man's scope of understanding, the better his performance would be in *each* field of his practice, despite the time he would be taking away from specializing. So he was a true generalist and the rewards of achieving this impossible goal are clearly exemplified in this book, making us more acutely aware of the loss involved in his early death.

But there is something paradoxical about regretting the fate of dynamite. Its nature can be fulfilled only in a way that carries with it the risk or fact of destruction. Hanson's great red Harley, the vintage Jaguar, the ravenous Bearcat were the kinesthetic counterparts of the intellectual daring that distinguished him. Where angels feared to tread Russ would drive tanks in tandem, laughing. And half the time he'd get away with it.

The measure of the merit of a man is what he did where others failed, and in this book there are many such successes. It is quite uninteresting that there are some less-than-successes. Safety is not the name of success. The angels who fear to tread are sissies, Hanson would have said, and the price they pay for survival is insignificance; only gamblers get rich. It's a tough line, but I like it, even if it won't work with mountains. I wish there *were* angels: by now Russ would probably have them organized into a motorcycle marching band in the mornings, and mounting the heads of pins for a population count in the afternoon.

Instead of Russ we have only a good book—but it is the best kind of book to leave behind, his first textbook.

Michael Scriven

Berkeley, California
1969

Contents

Preface   vii
Editor's Prologue   3

PART I. PROVOCATIONS AND RESTRAINTS   7

  1. On Philosophizing—and Some Logical Distinctions   9
  2. Defining Conceptual Boundaries   25
  3. Measuring and Counting: More Boundaries   42

PART II. THE ACT OF SCIENTIFIC SEEING   59

  4. There Is More to Seeing than Meets the Eye   61
  5. Seeing the Same Thing   77
  6. Seeing and Seeing As   91
  7. Seeing As and Seeing That   111
  8. Seeing, Saying, and Knowing   129
  9. Spectacles behind the Eyes   149
 10. Can We See Facts?   171
 11. Facts and Seeing That   186

PART III. PERPLEXITY: THE PROCESS OF EXPERIMENTAL RESEARCH   199

 12. Waves, Particles, and Facts   201
 13. Hypotheses Facta Fingunt   220

14. Scientific Simplicity and Crucial Experiments   238
15. The Systematic Side of Science   256
16. Discovering Causes and Becauses   271
17. What Happens as a Rule   285
18. Theory-Laden Language   298
19. The Scientists' Toolbox   314
20. Laws, Truths, and Hypotheses   328
21. Principles as Platitudes   345

PART IV. PROBABILITY AND PROBABLE
REASONING IN SCIENCE   361

22. Frequencies and the Mathematics of Probability   363
23. Using and Interpreting the Probability Calculus   377
24. Elements of Statistical Technique   392
25. The Principle of Uniformity Revisited   407

Editor's Epilogue   423
Index   431

EDITOR'S PROLOGUE

Editor's Prologue

WHEN Norwood Russell Hanson died in a tragic plane crash in April, 1967, he left a number of projects unfinished. One of these was the writing of an elementary textbook in the philosophy of science based on lectures he had given in recent years at Cambridge University, Indiana (where he founded the present Department of History and Philosophy of Science), and Yale. The basic material of this text was to come from his original Cambridge lecture notes, modified and amplified and suitably updated. In addition, a number of new chapters were to be added. Of the latter, three plus a fragment of a fourth had been written in rough draft form by the time of his death.[1] The others—dealing with experimental laws, the role of notation, and methods of representation in science—were never done.

The present book has been constructed from the materials which were in completed or nearly completed form. It aims to be an introduction, suitable for use in the first year or two of a college or university student's work in science. It presupposes a wide acquaintance with neither philosophy nor science, only an interest in understanding science more fully. At the same time, it is a genuinely philosophical study which professional philosophers and scientists will find interesting and absorbing.

A number of changes have been made in putting the lectures into book form. Some of the English figures of speech and allusions have been Americanized or internationalized in order to make them intel-

---
[1] The three complete chapters form Part I of this book.

ligible to a broader audience. Some redundancies and "lecturisms" suitable for the classroom but not necessarily the written page have been omitted or altered. Footnotes have been provided (Hanson's original references were for the most part missing or incomplete); titles for chapters, for the main sections, and, indeed, for the book itself have been added; and at the beginning of each of the four main sections a reading list of books—mostly books Hanson himself used to recommend to his students—has been inserted. Several illustrations and the calculations in the Appendix to Chapter 12 have been supplied by the Editor, either from his own recollections of Hanson's lectures at Indiana University or from other sources duly noted. In all of this the aim was to preserve as far as possible the lively and exciting style which characterized Russell Hanson's work in the classroom and all of his writings.

Since the book was essentially incomplete as it stood, an Editor's Epilogue has been added to tie together some of the loose ends and provide students with a summary of some of the main points covered in each section. For any misinterpretations therein the Editor assumes full responsibility.

In terms of subject matter, one of the most unique features of the book is its extended treatment of the nature of scientific observation. Here Hanson gives, in a fuller version than has ever before appeared in print, a defense and exposition of the Wittgensteinian, ordinary language theory of perception and its ramifications for scientific observation. Opposed points of view (phenomenalism and the ocular-neural causal theory) are given fair treatment, but essentially Hanson is concerned to bring to bear on scientific practice the lessons which the ordinary language movement has to teach.

Since the ordinary language theory of perception has many points of contact with Gestalt experimental psychology, there is a good deal of discussion in Part II about Gestalt theory. Had Hanson lived he doubtless would have wished to bring the experimental references there fully up to date. The Editor has not done so chiefly because Hanson's argument is not substantially altered by recent findings of experimental psychology; it is a philosophical argument about perception anyway, not one which can be refuted or confirmed by new experimental findings (which is not to say, however, that they are altogether irrelevant).

The Editor is indebted to Professor Stephen Toulmin, Brandeis University, who has served unofficially as literary executor of Hanson's estate, for his assistance and permission in bringing the manuscript to publication. I should also like to thank Mrs. Margaret Freeman for her estimable help in copy editing; Mrs. Sally Rahi of New College, Sarasota, and the Editor's brothers, Dr. James Humphreys, Institute for Advanced Study, Princeton, and Professor Lester Humphreys, University of Massachusetts-Boston, were of great help in hunting up missing references and illustrations. Lastly, I want to thank Mrs. Fay Hanson for her cooperation in making this book possible.

Readers who are interested in pursuing other writings of Hanson's should see Volume III of *Boston Studies in the Philosophy of Science* (New York: Humanities Press, 1968), where a complete bibliography may be found. The same source contains much biographical material in the form of memorial notes from philosophers of science, historians of science, and scientists the world over. Hanson's own favorite among articles written about him is a piece called "The Bearcat Professor," by James Gilbert, associate editor, *Flying* magazine (March, 1966), p. 53.

W. C. HUMPHREYS

New College
Sarasota, Florida

Part I PROVOCATIONS AND RESTRAINTS

1. On Philosophizing—and Some Logical Distinctions
2. Defining Conceptual Boundaries
3. Measuring and Counting: More Boundaries

BIBLIOGRAPHY
PART I

Achinstein, Peter. *Concepts of Science.* Baltimore: Johns Hopkins Press, 1968.

Bridgman, P. W. *The Logic of Modern Physics.* New York: Macmillan Co., 1927.

Campbell, Norman. *An Account of the Principles of Measurement and Calculation.* New York: Longmans, Green & Co., 1928.

Churchman, C. West, and Ratoosh, P. *Measurement, Definition and Theories.* New York: John Wiley and Sons, 1959.

Cohen, M. R., and Nagel, Ernest. *Introduction to Logic and Scientific Method.* New York: Harcourt, Brace, 1934.

Ellis, Brian. *Basic Concepts of Measurement.* Cambridge: Cambridge University Press, 1966.

Hempel, C. G. *Fundamentals of Concept Formation in Empirical Science.* Chicago: University of Chicago Press, 1952.

Hospers, John. *An Introduction to Philosophical Analysis.* 2nd ed. Englewood Cliffs, N. J.: Prentice-Hall, 1967.

Kyburg, Henry E., Jr. *Philosophy of Science: A Formal Approach.* New York: Macmillan Co., 1968.

Pap, Arthur. *An Introduction to the Philosophy of Science.* New York: Free Press, 1962.

Robinson, Richard. *Definition.* London: Oxford University Press, 1950.

Smart, J. J. C. *Between Science and Philosophy.* New York: Random House, 1968.

1 | On Philosophizing—and Some Logical Distinctions

HISTORIANS of science are more than mere chroniclers. They are not content only to construct a master record of what happened and when—of discoveries, inventions, and scientific personalities, of birthdays and family connections. True, many books on the history of some science read as if the author were designing a kind of periodic table or a calendar or a genealogical tree of the events which have made the science what it is. But this is to history of science at its best as bird watching is to genetic theory.

History of science is concerned with *ideas*—with the thinking of scientists. And this is also what the philosopher of science is interested in, except in a radically different way.

Once it is admitted that doing science does require thinking, it is clear that these two related studies are immediately important to an appreciation of that thinking. Thinking evolves, and it has an internal structure. The historian explores the evolution of scientific thinking and ideas. The philosopher explores the internal structure of scientific thinking and ideas. This internal structure is only generalized from what obtains every day when any scientist is said to have an adequate grasp of a certain concept. Would we ever say this of a man who lacked *all* knowledge of the development of an idea and all knowledge of its internal structure—its logic? Hardly!

So the historian of science is not a Royal Society bookkeeper or an

A.A.A.S. librarian, kept just to settle future claims as to the priority of inventions and discoveries. He is an explorer. He seeks those factors in the intellectual environment of a given period which led to the initial formation of a certain pattern of thought. He wishes to disclose new dimensions in old concepts such as *acceleration, force, mass, charge, field, point*, etc. He does this by revealing factors which inclined men of different scientific periods to fashion these concepts one way rather than another, this way rather than that. Just as we can understand a man's career better when we know something of him—how he has behaved on similar occasions and why, what his views are on the matter which led to his action, etc.—so we shall have a better grasp of a scientific concept, e.g., H_2SO_4, when we know something of what led chemists to express themselves in this way with respect to this substance.

It has been remarked that the formula H_2SO_4 contains the whole history of mankind. As in most exaggerations, there is a kernel of truth here.

Few would deny that the sciences *have* a history, that the history of science *exists*. Philosophy of science, however, is not always granted even that minimal claim. Since this book is an exercise in this black art, we had better proceed to do in detail for philosophy of science what has been done cursorily for history of science. As before, I will begin by saying what philosophy of science is *not*, or (at least) what it need not be.

If history of science is not chronicle, then philosophy of science is not a secular religion for conscience-stricken laboratory researchers. In this decade the question "Whither science?" has been posed *ad nauseam*. Divines, demagogues, and despondent dramatists have viewed science—microphysics and biochemistry, rocketry and genetics—as the instrument of gleeful Frankensteins bent on creating the uncontrollable. And so they are led to "philosophize" about the future of our civilization under titles like "Religion and Science," "Science and Future Civilizations," "Are Scientists Human?" Doubtless, in an age of bigger and better bombs, such questions are worth discussing—they are even worth discussing carefully, which is too rarely done. But no matter how carefully they are discussed, these questions are not issues of internal importance to the teaching of science. They are concerns of a different order. They affect scientists no more than they affect other members of the community. They are matters affecting the scientist as a citizen, not as a scientist.

If there is a real case for the introduction of history and philosophy of science into undergraduate courses, it must consist of the possibility that, in some derivative sense at least, men may become better scientists as a result. It is this stronger claim that we should consider. But, however that discussion fares, the speculative, deep-purple variety of "philosophizing" to which I have alluded finds no place in philosophy of science as it will be dealt with in this book.

Philosophy of science cannot, of course, increase manual dexterity. It is not wholly unrelated, however, to the business of sharpening one's wits—the business, that is, of carefully considering the character of one's experimental problems, the logical structure of arguments and proofs, and the general nature of a science's subject matter. The details of all this will be set out. But let us first allude to another thing that philosophy of science is not, or need not be. For scientists often recoil at the sound "philosophy of science" for yet another reason.

They rightly dislike the idea of academic philosophers and collegiate historians telling them, and the world, what science is all about. If physics were beset with all the problems that professional philosophers and historians manage to find in it, then doubtless they would be handy chaps to have around the laboratory, and around every school and university concerned with the teaching of science. But here the scientist will ask, "How can book-scholars who are unlikely ever to have seen the insides of a modern physics laboratory—who have never muddled and groped through the perplexities of a research task of their own, or felt that profound unsettlement which attends every decision at the frontiers of scientific inquiry—be relied upon to know what are the conceptual problems of physics?" Well, they cannot be relied upon for that, not unless they themselves have been scientists. Indeed, to have been a scientist is an indispensable requirement for anyone concerned with writing and teaching these subjects. Unfortunately, it is not met by enough individuals who expound on the history and philosophy of science.

This revealing question gains force when one sees how unrecognizable to laboratory researchers are some of the problems which "pure" philosophers have about the natural sciences. E.g.: "How can one 'construct' concepts of electrons out of visual impressions of pointer readings?" "How can one justify the use of inductive procedures in natural science?"

"Is science possible?"

If you think the answer to this last gem is an obvious "Yes!" just because your school is crawling with science, you have not read some of the more "profound" and arresting judgments on the matter. Since the time of Kant, experimentally innocent philosophers have been industriously digging up the *presuppositions* that got buried beneath the superstructure of modern science. (Kant, incidentally, was not experimentally innocent, but neither was he grossly guilty. Besides, he had other problems.)

The search, then, is for the logical guarantee that science is built upon a rock, and not on a bog. Or, alternatively, what can science *really* help us to know? Can we ever determine matters of fact with the surety which characterizes mathematics? (Don't just sneer "Yes"; and don't just sneer "No," either.)

Suppose a Yale professor were to pride himself so much on his sobriety and rationality that he wrote an essay on sobriety. Naturally, in that work he would lament and deprecate the sentimental, gushy impressionism of some of his colleagues. But the sentimentalists will surely smile and whisper, "Isn't 'the proof' sentimental about sobriety and rationality?" Metaphysicians (2nd class) have for generations actually earned their livings by whispering (in stage whispers), "Aren't scientists unscientific about their presuppositions—about their faith in induction and their dogmatism regarding what are and are not meaningful questions, about their acceptance of principles like 'All molar physical magnitudes are linked to continuous functions,' and 'Repeat the cause of X and X will occur repeatedly'?"

Mature scientists pay little attention to these academic worries. Would a lawyer worry on being accused of uncritically presupposing the principle that important evidence may be produced by cross-questioning witnesses? He *does* presuppose it, but so what?

Thus the unabashed metaphysician can often give his argument away by confessing in his question that he knows not whereof he speaks. Such is the case also with a certain kind of epistemologist (one who theorizes about the nature of knowledge). This particular species of epistemologist, of which but few living specimens are now extant, managed to baffle himself about the data of science. Impressed by the fact that we are sometimes mistaken in our descriptions of how the world is furnished, these philosophers fancied that if science were really to succeed, the "stuff" of observations (objects, events, situations) ought to be analyzed and segregated into those components which are *strictly* supported by our sensations and those which are not—these latter being but inferred, or constructed out of what we really do have as genuine physical experiences.

"That is a galvanometer," we say—but it *might* be a wireless set, or a mousetrap. True enough.

"Ah, there is the diffraction pattern," we say. But perhaps the Christmas cheer was a little too strong.

I might declare a band of light to be almost monochromatic. But perhaps my oculist knows something about me that he isn't telling.

In short, *we could be wrong*. That, indeed, is the logic of factual statements. Nonetheless the epistemologist may point out that concerning some things we cannot be wrong; it is certain that something galvanometerish dominates my visual field. No one outside my skull can deny that diffraction-like patterns appear when I open my eyes; who should know better than I what impinges on my retina or at least of what visual imagery I am aware? That band of light may not, in fact, have wave lengths of 5890Å and 5896Å, but that I am entertaining a sodium-yellow patch is indubitable. And so it goes. All experience is experience *of*, and the incoming signals are all we really have to go on. If only scientists would come to recognize the *priority* of exclamations like "red now," "pointer-image oscillating," and "buzz, buzz"—if only they would compound these sense-experiences in a truly logical manner to "form" the material objects of the laboratory and our world—then and only then, these epistemologists suggest, science will not be the "wobbly, illogical heap of half fictions" that appear so regularly in the pages of *Nature, The Scientific American,* and *The Review of Modern Physics.*

In terms of these epistemological criteria science surely is shaky. And so is everything else. It will not surprise you therefore to learn that these criteria are seldom invoked by non-philosophers.

Later on we shall explore some of the epistemological matters much further, especially as they bear upon our notions of *fact, observation, causality, theory,* and *hypothesis*. But for now we shall simply declare with delicate dogmatism that as a general approach to our studies in philosophy of science the posture just depicted is wholly unsatisfactory.

The third "improper" question to be considered, while just as easy to puncture as the other two, is more difficult to deflate, for it is not all hot air. Let us parody it thus: How can an experimental scientist, in the reports of his research, most closely approximate to the manner of exposition of the pure mathematician or the formal logician? Can he do this at all?

Together let us answer "No!"—perhaps even "Thank heavens, no!" This verdict is written in every bit of laboratory guesswork, in every crude set of apparatus, and in every persistent perplexity which refuses to disappear simply through more deduction. Indeed the whole tradition of natural science at places like Cambridge, Göttingen, Harvard, and Moscow is expressed in the phrase "sealing wax and string." These commodities are only slightly less useful today than they were in the glorious past. Still, concerning the dispensability of axiomatization in science there have been judgments to the contrary, and passed by some very able logicians and mathematicians.

One of the latter might argue:

> It is not my intention to suggest that the ideal for laboratory research be that it might one day be carried on by chromium-plated, self-correcting, algebraically-programmed automatons. Scientific discovery will always be to some extent a groping, stumbling affair, ever requiring great ingenuity, insight, and imagination. This is because it is a step into the dark, into the uncharted unknown. And there is no way of lessening the risk incurred in taking that step. The mathematical-logical philosopher, however, is not concerned with the actual things an experimenter does, says, thinks, or feels—his inner mental life, his I.Q., or his digestion. He is concerned, rather, with the logical relations between, for example, the general statements which stand at the head (or alongside) of a given theory and the myriad specific statements which follow inferentially from them—or between an hypothesis and the evidence in support of it. It is the formal, logical structure of bodies of scientific knowledge, and not the behavior habits of any or all scientists, that interests those of us whose philosophy of science is studded with symbols, deductions, and entailments.

Now this kind of philosophizing about science is not to be despised. (Why, some of our best friends are "logical reconstructionists"!) Who will deny that many important advances in modern science were of a distinctly logical cast? I should argue strongly that this was so, that the history of scientific progress is *not* a history of increasingly refined laboratory technique but a history of changing *conceptions*. Something was looked at in a new way, the priority of some principle of nature was challenged, a set of deductions or inferences was compared with another set, ultimately to conflate the two, or to mark out differences, or even contradictions, between them. The names of those who have made such advances are familiar enough. Philosophers and logicians have rightly interested themselves in exciting systematic advances like these, and concerned themselves with the formal connections and interconnections between aspects of certain scientific theories. It is both enjoyable and intellectually profitable so to concern oneself with the sciences.

What *is* objectionable is this: The philosophy of science is often identified exclusively with just this sort of activity. Most of the important logical and philosophical aspects of the sciences can be examined without a prerequisite study of the theory of deductive systems—without even assuming any great facility in symbol manipulation, though this is, of course, a distinct advantage.

Hence, this third interpretation of "philosophy of science" is somewhat inadequate, I think, not because philosophy of science in this sense is not worth doing, or because it is incapable of interesting experimental scientists. It is inadequate because it is but a small chapter in a very large volume, a chapter too often presented as if it were all that had to be said. The danger of distortion is therefore great with the philosophers of science who spend all their time writing and rewriting this one chapter.

Apparently, then, the subject will be developed here in a different way. The questions "After science, then what?" (the consequences of science), "Is science possible?" (the assumptions of science), and "Can natural science be made into a formal discipline?" (the axiomatization of science) will not figure dominantly in our discussions. You may well ask, "Then what will?"

Let it be said once and for all that there is no *subject* to be called "philosophy of science"—not if by a "subject" is meant a subject matter, i.e., a collection of unique facts, plus a set of specially designed theories and specific rules for interpreting those theories in terms of facts, or vice versa. There is nothing to memorize, no formulae or tables to be taught. But there are lots of questions to be asked.

Now these questions are of a logical type different from those to which you may be accustomed. Here are some questions about the game of chess to illustrate differences of logical type: "How many pawns does white have?" "Why is it that the bishop cannot move along the edge of the board?" "Did Fischer make the best possible move at 15?" "Why do you speak so highly of Capablanca's game of 1925?" Note how very differently we assess the meanings of these questions. And note the different kinds of inquiry involved in giving an answer to them, and the different kinds of criteria appropriate to assessing the status of each of these answers: E.g., I can tell you how many white pawns are on the board by looking and counting them. But looking and counting are not involved in referring you to the *rule* that bishops must move diagonally. And reference to Bobby Fischer's move as the *best possible one* in the circumstances involves a subtle mixture of considerations involving matters of tactics, issues of strategy, the history of the game, and even some assessment of the abilities of Fischer and his particular opponent. Finally, thinking well of a move or a game involves many further things, some of them bordering on the aesthetic. It is in some such way as this that philosophical questions *about* and *within* science are of a logical type different from those to which you may be accustomed, as, e.g., "What is Avogadro's number?", "How does gastrulation proceed in the coelenterates—by invagination, immigration, delamination?", "What is the half-life of oxygen 17?" Questions like these will not arise here directly, though questions about these questions certainly will.

It cannot now be said precisely what it is that characterizes philosophical questions like: Are Protozoa one-celled or non-celled organisms? What are the meanings of "principle" in the expressions "principle of least action," "principle of the rectilinear propagation of light," or "principle of natural selection"? And what are the meanings of "law" in "law of nature" (e.g., Snell's law, Boyle's law, Kepler's law, Faraday's law, Mendel's law, Pauli's law)? How is the character of our observational research influenced—if at all—by the notation in which we choose to express our questions? How are "the facts" influenced by our mode of expression? What would physics today be like had we adopted Newton's formulation of the differential calculus instead of Leibniz'? Is the uncertainty principle in quantum theory a description? If so, a description of what? Observations? Facts? Limitations in measuring instruments? What? What do we mean by the word "exist" in claims like "A striped coelacanth exists," "Carbon 14 exists," "An 'organism' exists," "An anti-neutrino exists," "A contradiction in his proof exists," "A solution to this problem exists," etc.?

In short, we will here consider certain puzzles about the languages, the observations, the data, and the methods of science for the solving of which you may not before have had the time, or the interest.

A word of caution. For a scientist or science student to expect all this to make any immediate difference in his laboratory work will be to beg for disappointment. Matters of *fact* are not our direct concern—matters of logic, of ideas and reasoning, are. Do not approach our analytical program with unreal expectations. Try, rather, to treat this material as cognate to, but not immediately intimate with, your own experimental work. A scientist's attitude towards his special science may possibly be the better for it. For a good part of science consists in asking questions systematically. Anything that can make one attend more closely to the logical character of scientific questions cannot be amiss.

But what odd chapters these will be: Just a string of questions? Not quite. They will prepare for questions to be worked over in more dialectical contexts elsewhere.

The next chapter, for example, will set out some difficulties inherent in our notions of definition. A definition can do more things in general than we suppose, and less in particular than we sometimes hope. In the third chapter problems connected with measurement will be examined.

These first chapters will thus be quite broad, ranging over a wide assortment of scientific attitudes and concepts. They will be full of questions designed to stir you out of your dogmatic slumbers—or at least to complicate your dreams. Hence, the first part of this work will be framed as a challenge; we may often set out arguments with tongue in cheek (but not, hopefully, with forked tongue in both cheeks). But whether or not I am doing so is for you to decide. These first chapters are thus designed to be targets for your intellectual arrows—salt for your cerebral wounds.

The chapters of the second and third parts will be no less targets for your attack, but our tongue will not be encheeked. The objective there is to worry you, systematically, about concepts like *observation*, *facts*, experimental *data, hypotheses, theories,* crucial *experiments*, scientific *language, induction* and *deduction,* and a host of closely related topics. These chapters will be calculated not *just* to incite intellectual riot as those in the first part will be. It is hoped that there we will get some insight into the logical foundations of scientific inquiry, that we will locate methodological and philosophical brambles in uncritical views of observation and experiment, and gain a more detailed appreciation of the rules of hypothesis and theory in laboratory research.

Finally, in Part IV, we will turn to consideration of the concepts of probability and probable inference in science, weaving in threads from our earlier discussions as we go.

Let us conclude this first chapter with some logical points. These could be essential. They make all the difference between being clear-headed and being muddleheaded about the languages of science. But even so, take these observations critically; there is more to be said on each of these matters.

Distinguish a *necessary* proposition from a *contingent* proposition. If I say, "Let X be $\sqrt{4}$ and let Y be 4^2," then the proposition "$X + Y = 18$" is necessary, or necessarily true. It cannot be false. Its denial is self-contradictory. E.g., to say "$X + Y \neq 18$" is to say either that $X \neq \sqrt{4}$, or that $Y \neq 4^2$; or both—which contradicts our assumptions. Or, put another way, assigning X the value $\sqrt{4}$ and Y the value 4^2 just *is*, in a way, to assert that $X + Y = 18$. For the meaning of a claim is the entire set of its consequences. Thus part of the *meaning* of "$X = \sqrt{4}$ and $Y = 4^2$" is necessarily, that $X + Y = 18$.

A *contingent* proposition, on the other hand, can be false. Indeed, the logical possibility of its being false is, perhaps, part of what we mean when we say of some claim that it is contingent, or non-necessary. The proposition "When sucrose is heated with dilute mineral acids it takes up water and is converted into equal parts of glucose and fructose" may be denied without talking nonsense—without, that is, involving one in logical contradiction. A bona fide sample of sucrose may fail to behave in the stated way. This should make us curious, but it need not raise problems about the definitions of words or expressions. And if you would counter, "Oh, but if it does not convert into equal parts of glucose and fructose, then it just isn't sucrose"—if, that is, you make this particular behavior a defining characteristic of sucrose—then you cannot afford to skip the next chapter, where the concept of "definition" will be put under the microscope.

Clearly, most of the propositions within pure mathematics and symbolic logic are necessary (i.e., invulnerable), or analytic (i.e., with inconsistent negations), or true by definition. It is self-contradictory to

accept the axioms and rules of a symbol system, a formal game, and then deny what follows from operations on those axioms in accordance with those rules. For this reason, and others, such systems are purely formal, i.e., tell us nothing about the world, are not descriptive of the 3-D arena of experience.

Most of the propositions within natural science, however, are contingent. They are about the world, about matters of fact. They purport to describe "what is the case." There are no ultimate axioms when it comes to matters of fact: No claim is in principle unrevisable. No statement about "the external world" is self-evidently true, necessarily true, true by definition or convention. One can accept that X is a genuine sample of sucrose, heat it with mineral acids, and then consistently report that no inversion from dextro-rotatory to laevo-rotatory optical power was encountered in the resultant solution. He can do this without being accused of talking pure nonsense. He may be accused of other things, but not self-contradiction.

The following propositions are now committed to your tender mercies. What is their logical-conceptual status? (a) "The chemical atomic mass of oxygen is 16." Could this be false? Under what conditions? (b) "Force is that physical quantity which will accelerate a mass—it is equal to the product of the mass and the acceleration." Could this be false? Under what circumstances? (c) "Put a few drops of copper sulphate solution into a test tube and add a few cubic centimeters of strong caustic potash." Is this true or false? Does it even make sense to ask whether it is true or false? Why? (d) "Every event has a cause." True or false? Delineate your notion of an uncaused event. (e) "Fehling's test on glucose is better than Trommer's test." True or false? How so?

Are these propositions necessary, or contingent, or neither? Which of them can be denied meaningfully? Which cannot? And how say you of the mathematically sophisticated propositions of thermodynamics, or of genetic theory? Are they necessary or contingent—or neither or *both* (watch out for this last one!). This necessary-contingent contrast provides a good working distinction. But once placed out in the open in just this way, it seems often to lack utility in individual cases. Within elementary particle physics it would often be pointless to ask the "necessary-or-contingent" question of particular theoretical propositions. That question could obscure the fundamental function of, e.g., the principle of the conservation of energy, the exclusion principle, etc. In any case it will be a highly contextual matter whether an expression within a body of scientific language is being used in a necessary (i.e., invulnerable) or a contingent manner. Notwithstanding these subtleties, however, you will still find it worth while to separate expressions which make assertions about nature from expressions that are a mere working out of what is already implicit in the definitions, axioms, and postulates of a symbolic algorithm—and of course, keep these both distinct from "non-assertions" like commands, rules, instructions, and exhortations.

These reflections lead on to our next rough and preliminary distinction.

Let us demarcate *truth* from *consistency*, and *falsity* from *inconsistency*. All of us have been told at some time, and some of us all the time, that though our answer is correct, our mathematical reasoning is wrong—or, less frequently, that though our answer is wrong, our mathematical reasoning is correct.

When we remark of something said, "That is true," we are saying that it squares with the facts—we are agreeing that *that* is the way the world is. But to say this is not to say that all the propositions leading up to that true statement do themselves square with each other. They may not. Our argument may be unsound, our mathematics wrong, our reasoning crippled. One may conclude a lame computation by saying that the claim "At 99% of the velocity of light an electron undergoes an increase in mass of 609%" is correct. Or perhaps the mathematics was acceptable, but the answer false due perhaps to a misprint in the table from which the cm/sec equivalent of the expression "99% the velocity of light" was taken.

So that to say of a proposition "true" is not to say also of the reasoning which led up to the assertion of that proposition "consistent." And to say "inconsistent" of a bit of reasoning is not also to say "false" of the conclusion in which it terminates. The criteria for the application of these appellations are entirely different.

Naturally, we will be far more concerned here with the dichotomy "consistent-inconsistent" than with the dichotomy "true-false." [This may be a refreshing change for some.] Which brings us to a third point—one we stressed earlier in calling attention to the things some scientists do *not* do, the features of the world they do *not* notice, the experiments they do *not* consider.

The word "not" is important in strictly logical considerations. We can determine whether a claim is contingent or necessary by denying it, by asserting its negation. If the resultant sentence does not make itself disappear (by denying exactly what it asserts), then it is contingent, or non-necessary. If, however, it "draws a line through itself" as does "X is $\sqrt{25}$ and X ≠ either $+5$ or -5," then the proposition thus denied is necessarily true.

A self-contradiction, such as is set out in the last sentence, itself is but a conjunction of incompatibles, p and not-p (asserted under certain conditions the exact description of which would take us too far afield here). But that it is even possible to form a contradiction within it is an important requirement of a descriptive-informative language. Consider this a little further:

Suppose there to be a word, or a symbol, which we were prepared to apply to everything without exception. Clearly, it would be a useless term for purposes of description. When we characterize something we not only compare it with other things, we also distinguish it from other things. To speak of an animal cell as "columnar epithelium" is to distinguish it from squamous epithelium cells. A cell cannot at once be squamous and columnar. To talk of epithelium cells at all is to exclude from immediate consideration smooth muscle cells and connective-tissue cells, which latter are not epithelial cells. And this is what certain historians are up to when they call attention to what certain men in the history of science did *not* do: By perceiving what was *not* done and what was historically responsible for its not having been done, one is placed better for appreciating the historical and logical significance of whatever was done later on. Since Claudius Ptolemy did not entertain seriously enough the alternatives to geocentric planetary theory, ages of astronomers after him couched their descriptions and explanations in terms which precluded their ever hitting on the true picture of the solar system.

If a word is to have use, a logical boundary must be drawn somewhere. The word "Providence" in "Everything that happens happens through Providence" has no such boundary. This is clear enough when you ask what will count as evidence against the proposition—the answer being "nothing." The question arises, then, whether it was ever a proposition at all. And words like "actually," "really," "whizzo," and "cool" have boundaries so indistinct that it is rarely certain just what is being asserted when they are employed. After all, if the sign "Student Driver" were put on *every* motor car irrespective of the driver's skill it would convey as little as radiator caps do now. We must come to know what "Student Driver" contrasts with if we are to learn just what it means. Even to teach a child the use of a color-word like "red," it is rarely enough to show him only red things. One must also point to grass while saying "not-red," and to the sky, to the policemen's uniforms, etc.

Words as used in natural science tend towards having a well-demarcated range of application. This is due either to the boundaries a specific subject matter comes to impose upon their use or to explicit definitions which build up logical fences for many key words. Of course scientific languages do not spring forth full blown. How they are generated is an historical question of exciting complexity. But, once moderately well established, a scientific language may be expected to display clearly some network of logical boundaries.

When, in science, we explicitly use a word, an expression, a symbol, or a formula, we implicitly *exclude* a great many others. It is this which gives the words unique applications and whatever descriptive force they may possess. An expression which is applicable to everything is as vacuous as one which applies to nothing. "Three-wheeled bicycle" is vacuous; so is "thing."

Hence the word "not" is a crystallizing-out word. Its standard use is to contradict, correct, or cancel some suggestion or other. Asking what a proposition is intended to *deny* is always a good first step in finding out what it is meant to assert. A proposition which denies nothing asserts nothing.

Here is our fourth and final preliminary point.

Consider, for a moment, the general statement "No bats are oviparous," or in non-negative form "Everything that is a bat is non-oviparous," or simply, "Bats do not lay eggs."

How many examinations of how many bats and how many conclusions of the form "non-oviparous" would it take to secure this contention beyond all possible doubt? Ten? Fifty? Five hundred? Five thousand? Five million? How many? Well, if we knew all the zoology we ought to by the time we were able to raise such a question, we would probably think our conclusion beyond all *reasonable* doubt were it based on, say, twenty dissections of well-chosen specimens. (The difficult expression "well-chosen" will not detain us now.) But we should not thereby have placed our conclusion beyond all *logical* doubt. It remains logically possible that we might yet encounter a placental mammal in every respect of the order Chiroptera save that it laid eggs. We confidently identify bats at dusk, sometimes even their species, without ever having been present at the birth of a bat brood. So we might well encounter a beast which qualified as a bat to all expert observers and then learn later that it was oviparous.

From a logical point of view, then, this general proposition can never be *conclusively* established. But it can be conclusively refuted!

How? Just turn up with one egg-laying bat and you will have conclusively refuted the proposition "No bats are oviparous." *Unrestricted general propositions*, whether in positive or negative form, cannot be conclusively established by any finite number of confirmations. Yet they can be refuted by one disconfirming instance. "All birds possess convex upper surfaces on their wings" can never be finally established through observation since it ranges over an infinite class. But it can be refuted—by the discovery of one bird possessed of a wing having a concave topside, *à la* the contemporary Boeing 727.

On the other hand, there are forms of non-general propositions, e.g., "*Some* bats lay eggs," which cannot be refuted by however many observations of bats which turn out not to be egg-layers. After all, we may not be looking in the right places—remember the coelacanth! But though they can never be conclusively refuted, such claims as these can be conclusively *established*.

How? Just turn up with one egg-laying bat—perhaps in Tasmania, or Tanzania.

So by tinkering with the word "not"—by asking what it would take to refute (what would constitute grounds for denying) a proposition—we discover a fascinating asymmetry between certain forms of *general* and *non-general* propositions. A proposition of the form "All S is P" can be conclusively refuted, but never conclusively established. A proposition of the form "Some S is P" can be conclusively established, but never conclusively refuted.

To review these elementary points, then.

1. *Necessary* propositions must be distinguished from *contingent* or non-necessary propositions. It makes sense to treat the latter as vulnerable. It does not make sense, in specified contexts, to treat the former as vulnerable. E.g., the principle of the conservation of energy is not a vulnerable claim within classical physics. Yet its negation is not formally

inconsistent (or analytically false). So we must always keep the necessary-contingent dichotomy distinct from the analytic-synthetic dichotomy.

2. The search for *truth* must be demarcated from the search for *consistency*. To attack the reasoning as incorrect is not (necessarily) to attack the conclusion as false, and to endorse the conclusion as true is not necessarily to endorse the reasoning as valid.

3. In considering what meaning is to be attached to an assertion or an expression, ask what it is that is being denied or excluded. The word "not" is a kind of logical machine. It helps us to find the boundaries of an expression.

4. The assertions "All S is P" and "Some S is P" are asymmetrical in that the former can be refuted but not established beyond doubt. The latter can be established beyond doubt but not refuted.

2 | Defining Conceptual Boundaries

A USEFUL term cannot apply to everything. Some logical or conceptual boundary must appear somewhere. As we saw, the word "not" helps us to locate these boundaries; to know what is being denied is to have perceived half of what is being asserted.

Science seldom leaves us to hunt out for ourselves the logical range of an expression. In this respect scientific discourse is unlike ordinary language, the conceptual structure of which is every bit as definite as one would find in technical speech but never as explicitly set out. *Definitions* often map the area of scientific discourse for us. At the command "Define your terms!" we look to existing boundary rules, in texts or lectures, or else we formulate new ones on the spot.

However, the sorts of semantical offering which will count as definition are more numerous and varied than some scientists imagine. It is often thought that the command "Define!" denotes but one sort of operation, as do commands like "Inhale!" or "Differentiate!" Our mission here is to raise doubts about this.

Why? Why dynamite butterflies? Does anything go wrong conceptually if we think that defining is carried out in but one way rather than in many ways? The answer is "yes"—or at least "perhaps."

Let us undertake only to reveal a complexity. It is not intended that the following classification should be exhaustive, or even internally mutually exclusive—but only complicated. If we just come to see that definitions can do many more things than we suppose, and far less in any concrete situation than we may hope, we will have learned everything it is intended to set out here.

A definition is a setting out of the *meaning* of a symbol or cluster of symbols. It may be more than that. It may be less. But this is a good beginning. For it becomes immediately clear that there must be as many kinds of definition as there are ways of setting out meanings.

1. Consider first this sort of definition: " 'Rouge' means in French what 'rot' means in German." Even if we are ignorant of the meanings of both words, "rouge" and "rot," we still might be said to have learned something from this definition (though admittedly not very much). This is less a definition and more an occasion for perplexity, one which we all feel sometimes—especially when thumbing through the back pages of mathematics textbooks. We learn that two symbols are synonymous without thereby learning what either of them means. Thus, dictionaries of science will define "right line" as "straight angle," "axis origin" as "zero point," "erg" as "unit dynecentimeter," "wave velocity" as "λ," "axis cylinder" as "axon," "dextrose" as "glucose," "epinephrine" as "adrenalin," etc. These meaning-equations will not help us unless we already know the meaning of at least one of the terms. Makers of dictionaries, in their infinite wisdom, allow for this human failing by inscribing *"quod vide"* (q.v.), after every such second term. This invites us then to look up *that* term, which (we hope) will be defined either by a term we already do know or in some manner other than by the mere giving of a synonym.

2. (a) These "other manners" are several. But they fall (roughly) into two classes. One of these contains quasi-historical statements. The other contains stipulations. The first sort tells you what scientists *do in fact* mean when they use a certain symbol or expression. It reports how experts use the term in question. Such a report is therefore conveyed in a factual statement; it is a claim contingent on what is the case. It could be false. The second sort tells you what someone *proposes* to mean when he uses an expression, irrespective of what all other scientists may mean when they use it. It does not report what a word has meant or does mean. It does not report at all. It prescribes what that word will mean in some given piece of reasoning.

Obviously, most definitions offered in science textbooks and in dictionaries—and in most college lectures—are of the first variety. They set out what is most often meant by words like "displacement," "ferromagnetic," or "capacitance"; by terms like "organism," "organizer," and "organ"; by expressions like "intelligence," "learning," "society," and "inter-group hostility."

Apparently it makes sense to say of these definitions that they are *true*, factually true—contingently true. Bismuth *is* in fact said to be "diamagnetic." Four and two-tenths joules really *is* called a "calorie." The pronouncements are true and not false—as would be the definition of "displacement" as "a change of direction at a certain position." Nobody defines "displacement" in that way.

(b) But if a biologist begins a lecture by defining "protozoon" as "a non-celled organism," will we say that he has spoken falsely since (as all experts attest) Protozoa are *really* one-celled organisms? No, of course not. He assuredly has his reasons for setting out such a definition in such a way. We must follow through with him and see what he is getting at. What does his definition gain for him and us? Why should we interpret every definition in biology as being just a report on some prior ruling made by the International Commission of Zoological Nomenclature? Our lecturer in this example is inviting us to understand "non-celled organism" whenever he uses the word "protozoon." He wants us to be clear about his meanings. Invitations cannot be true or false—though, of course, they may be accepted or rejected. We turn down invitations to dinner, perhaps, but not by calling their proposers liars. Similarly, we may decline to accept our biologist-lecturer's definition of "protozoon," but not by correcting him as to the *facts*.

Mathematicians are ever alert to this distinction. The proposition "A pencil is a right cylinder whose cross-section is a regular polygon" would ordinarily be taken to define the word "pencil," i.e., give an account of what "pencil" means to most geometers. The proposition is true. Most geometers *do* mean this. But were that same proposition set out at the beginning of a paper being read to the American Mathematical Society, it would probably mean "Please understand by my use of the term 'pencil' a right cylinder whose cross-section is a regular polygon; this and only this will be meant by that word in the argument which follows." Clearly, *this* is not a true proposition; it is not a proposition at all. It is a request, a proposal, a prescription. No recitation of the facts—no reports about nature or about mathematical usage—will suffice to expose such a proposal as "false" (whatever that might mean). One's reasons for making such a proposal can be of many kinds, as we shall see.

Not many of the definitions offered in high-school and college courses in natural science show this rather arbitrary and game-like character. But some of them do. These are worth noting, if only to contrast them with myriad others which do not. Consider:

> There is, in a constant-temperature vault at Sèvres (Paris), a platinum-iridium cylinder whose mass is, by definition, 1000 gm. In the same vault there is a bar with two fine scratches on it, the distance between which is defined as the standard meter.
>
> The second of time is defined as 1/86,400 of a mean solar day.
>
> Middle C is defined by physicists as 256 vibrations per second.
>
> The chemical atomic mass of naturally-appearing oxygen is 16, by definition. Hence the mass unit is defined as being exactly 1/16 the mass of a single oxygen atom.

Think about this: What significance is to be attached to the expression "The length of the standard meter has changed," or "Chemical atomic mass is an unsuitable standard of physics"?

These standards of reference—i.e., gram, meter, second, etc.—are defined as basic to entire scientific disciplines. They are, thus, hybrid definitions. They are like the first type we considered (the factual reports of usage): Everyone agrees with them. Yet they are stipulative like the second type: The original proposals through which the standards were at first arbitrarily set seem to re-echo each time a reference to that standard is made.

So we must distinguish the *historical-authoritative* reportorial definitions offered in dictionaries, texts, and university lectures from the *stipulative-prescriptive* definitions of research scientists. Like Adam, Euclid, and Humpty-Dumpty, these scientists not only call old friends by their given names but they also christen new arrivals with the names they are to bear thenceforth.

Armed with this major distinction, we may now take on for consideration several of the *methods* scientists adopt in explaining what is meant, or, at least, what they propose to mean, by some word, term, or symbol.

1. The method of *definition by synonym* has already been hinted at. When "Chiroptera" is defined as "bats," or "phenol" is defined as "carbolic acid," when "positron" is defined as "positive electron," or "syzygy"

as "any rectilinear configuration of any three bodies," these definitions are offered on the supposition that you already know the meaning of the more familiar second expression. If you do *not,* then these definitions are only symbol-symbol definitions for you, and not a great step forward in your education. Semantically, they will serve you just as our first space explorer on Mars will be served when he learns (somehow) that, for the Martians, "urk" means the same as "glug." But if one does know the meaning of one expression in each pair, he is then invited to invest the word requiring definition with that same meaning. A new word then enters his scientific vocabulary. Its meaning has been made definite to him when he is shown that its boundaries are precisely the same as those of a conceptual territory he already knows.

2. Consider now, *definition by analysis*. One may not know what a given thing (object, process, designatum) is called. But he may understand some phrase that analyzes that thing. Thus "dodecahedron" is defined as "a solid having twelve plane faces," "disaccharide" as "two monosaccharide molecules in combination (minus one molecule of water)," "carbohydrate" as "a carbon-hydrogen-oxygen compound, other than acetic acid, lactic acid, and inositol." Words like "billion," "kilowatt," and "joule" are usually defined in this way, too.

But by far the best known form of definition by analysis is that recognized by Aristotle as division into *genus* and *differentia*. The recipe is as follows: To indicate the meaning of a word, first locate it within a larger class of words (concepts) all of which have some meaning in common. Then distinguish the word being defined from all the others in the class. Thus *Felis leo* and *Felis tigris* are two species of the same genus, *Felis*. Both lions and tigers are cats. But they are distinguished from each other by obvious marks. Indeed, the whole binomial nomenclature as employed by Linnaeus in *Species Plantarum*—and by botanists and zoologists ever since—consists in precisely this method of combining under one scientific label both the specific and generic names of some organism. So here is another debt we owe to Aristotle, who—as I hope you will realize—was a very acute man.

In the physical sciences definitions by analysis are also in evidence. Consider the classification of matter as solid or fluid according to whether or not resistance is offered to a shearing type of force; and reflect further on the classification of fluids as liquid or gas depending upon whether a free surface is displayed.

To use an ancient terminology, definition by analysis explains the nature of the whole by reference to its constituent parts. Dodecahedra are solids (with twelve flat sides), tigers are cats (with stripes), fluorine is a gas (with but one isotope), etc.

3. But scientists often define their terms in another way: by synthesis. *Definition by synthesis* explains the nature of a part by reference to the whole of which it is a constituent element. This way of defining a term (by relating to other things what it stands for) may proceed in at least two different ways.

(a) When "circle" is defined as "the plane figure covered by a moving line, one end of which is fixed," it is being causally defined. Likewise these: "A sphere is a solid generated by rotating a disc around its diameter," "A paraboloid is the solid generated by rotating a parabola around its axis of symmetry." And consider this elaborate causal definition of glycogen: "Glycogen is a white tasteless powder, soluble in water, but it forms, like starch an opalescent solution. It is insoluble in alcohol and ether. It is dextro-rotatory. With Trommer's test it gives a blue solution, but no reduction occurs on boiling . . ." [1]

(b) Words like "aorta" or "antenna," however, are not likely to be defined either by an analysis of structure or by a causal account of the way they form in the embryo and grow to size. Their definitions are more likely to be *systematic*. That is, we should probably locate the aorta and its functions inside the animal circulatory system as a whole —relating to the heart, the large branch arteries, the lungs, the liver, etc.

It is in this manner too that we are most likely to define terms like "autonomic," "Cambrian," and "reflex."

(c) Perhaps expressions such as "x-ray," "lightning," "electron," etc. are also defined systematically. But the *designata* of these words are not to be found in straightforward physical systems (like an aardvark's circulatory system or an aircraft's hydraulic system). Simple inspection will not suffice. There are important theoretical aspects to, say, the definition of an x-ray as an electromagnetic wave disturbance, or the definition of "thundercloud" as a natural electrostatic generator, or the

---

[1] William Dobson Halliburton, *The Essentials of Chemical Physiology for the Use of Students* (London, New York: Longmans, Green & Co., 1936, 13th ed.), p. 30. (The index of Halliburton's book cites this as the *definition* of "glycogen.")

definition of "electron" as "a (phase) wave packet or interference maxima (in configuration space)." It is in this broad sense of "systematic" (a sense which includes *theoretical* systems as well as physical systems) that the color red may be defined as "the color a normal person sees—i.e., is visually aware of—when light of wave length 6500–7000 Å impinges on his retina."

[Of course, some scientists get too enthusiastic. Instead of saying just that the perceptual quality intended by uses of the word "red" is correlated physiologically with light radiation of a certain wave length (e.g., 6800 A), they go on to say that the quality red *is* that wave length of light. But of the dangers of synthetic definition, more later.]

All these methods, definition by *synonym,* definition by *analysis,* and definition by *synthesis,* provide us with a word, or term, or phrase, or expression equivalent to the word explicitly requiring definition. Thus "glucose" is defined by synonym as "dextrose." It is defined by analysis as "a 6-carbon-atom sugar (a hexose)" or as "a monosaccharide." Or it may be defined synthetically as "the aldehyde of sorbitol," or as "that which, when split to CO_2 and water, is a major energy source for mammalian metabolic processes."

4. What about *this* proposition, however? "In alkaline solutions glucose reduces salts of silver, bismuth, mercury, and copper." Or this one? "Under the influence of yeast glucose is converted into alcohol and carbonic acid." These propositions could constitute at least partial definitions of "glucose." They certainly delineate part of what a chemist could *mean* by that term. Yet, unlike other definitions, these do not begin with the official formula "The word 'glucose' is defined as such and such. . . ."

Try this one: "A square has two diagonals, each of which divides it into two right-angled isosceles triangles." This is presented without the ceremonial language of definition. Nonetheless, one could learn part of what "diagonal" means from hearing just this. So far as this is so, a definition has been offered; a meaning has been made clear. "Diagonal" is thereby defined implicitly. Here now is a classical case of such *implicit definition:* "Since the snapdragon can be halved in one, and only one, plane so that the two halves are approximately mirror-images of each other, it is bilaterally symmetrical, or zygomorphic." Though the word-form here is not " 'Zygomorphic' is defined as follows . . . ," one could surely learn what "zygomorphic" meant from this proposition. To that extent it sets out the semantical content of "zygomorphic"; to that extent it defines "zygomorphic."

5. A method of definition distinct from the previous four is what might be called *denotative definition*. Here the meaning of a word or symbol is exhibited by setting out examples (verbally, of course) of the things to which it applies—of the denotata which it denotes. "The word 'ocean' means the Atlantic, the Pacific, the Indian . . . and things like them"—that is a case of denotative definition. Here are others, somewhat more viable: "The anthropoidea are the monkeys, anthropoid apes, and man." "The primary transuranic elements are neptunium, plutonium, americium, and curium."

The limitations of this sort of definition must be clear. It is of little use in mathematics: How could words like "million" or "myriagon" be defined denotatively? The technique is limited in theoretical physics, too, where terms like "infinite self-energy," "microsecond," and "structureless singularity" play important conceptual roles.

6. These five methods of definition all define a word by using *other words*. It is not always possible to do this, nor is it always advisable to try. Another method (that of *ostensive definition*), will inevitably intrude or obtrude somewhere. One would define "the standard meter" ostensively by pointing to the bar kept at Sèvres, while saying, "The distance between those two scratches *is* the standard meter."

Words denoting simple qualities like straightness and blueness probably cannot be verbally defined—which may just be a way of saying that the meaning of such words cannot be delineated discursively at all (which may raise questions concerning whether or not they are even *meaningful*: significant, yes; consequential, certainly; but possessed of semantical content? perhaps not). We originally get the point of such expressions by confronting something straight or blue while being told "That is straight" or "That is blue." A man blind from birth may speak the language of spectroscopy very well, but we will surely remark his incapacity to grasp the full meaning of color-words. Or we may allow that he can grasp such meanings since he manipulates the color-word vocabulary very well. But then we will surely remark that he lacks something usually required for communication of information about colors.

Now it may not even be the *ideal* that all definition of words should be forced into other words. How cumbersome the biologies would become were diagrams, graphs, and photographs ruled out of the class of permissible definitions—or "meaning carriers"! In anatomy definition proceeds best by pictures. Verbal accounts of the human musculature are possible, but hardly advisable—as the involved story of Nicholas Steno will later demonstrate.

Would all other methods of definition, if pressed, ultimately reduce to ostensive definition? There's a querulous crucible within which to try your philosophical mettle.

7. These are some "garden varieties" of definition. Any one of them may crop up at any time—in a text, a dictionary, a lecture, or a conversation. But there are others, too.

(a) A *regular definition* of a word or symbol is a statement of the rule which governs its employment. For example, the arithmetical symbols we call "plus" ($+$) and "times" ($\times$) are defined by the associative, commutative, and distributive laws for addition and multiplication respectively. All permissible operations with the symbols are completely specified by these three rules. They give the symbols all the meaning they have. They *are* the meaning of these symbols.

Newton's formula for gravitational attraction, viz., $F = G \dfrac{m_1 m_2}{r^2}$, may be said so to define the symbol "G" (G being the proportionality constant which gives F its proper units once m_1, m_2, and r have had units assigned). "G" has no meaning relative to the Newtonian formulation other than just what this "rule" allows. This is not to say that "G" can mean nothing else, however. In other contexts, e.g., the experiments of Cavendish and (later) Boys, the symbol receives additional semantical content.

In our ordinary discourse, too, certain words are defined primarily in this rule-giving way. How else could words like "and," "or," "if-then," "not," "all," "some" *be* defined? These "syncategorematic" concepts, as some ancient philosophers called them, encapsulate linguistic rules within single words. Consider how difficult it is for children to learn the use of the word "I"; doubtless this is because the rule for its use is very hard to gather from just hearing other people talk.

(b) Similar to regular definition is *axiomatic definition*. Euclid defines a point as "that which has no part" (a most infelicitous choice of phrase).[2] But this can hardly be said to characterize completely the meaning of "point" in the *Elements*. The meaning of the word *is* given in the axioms and rules through which Euclid circumscribes and organizes his entire subject matter. The meaning of "point" is created and controlled by those axioms. Every rule about the intersections of lines, the nature of parallels, the construction of bisectors and circles helps to give further definition (i.e., semantical sharpness) to the concept of "point."

Consider Newton's second law of motion: $F = ma$. The total meaning of "force" (F) is—so to speak—built into the whole of the classical dynamics of particles. It is not entirely captured in the terse statement of the second law. This is why the important qualification "working definition" is almost always made.

(c) Another related form is of meaning-conveyor *indirect definition*. When one considers terms like "electron," "wave function," and "gene" it becomes obvious that the total meanings of such words are not drawn *only* from the roles they play in some scientific theory. That is, we accord meaning to a word like "electron" not just in virtue of the rules which prescribe the intra-theoretical use of the term, nor merely in response to the axiomatic definition of "electron" built into the systematic parts of particle physics. Much of what we mean by this word is tied to our actual experiments and observations in spectrography and electromagnetics.

Braithwaite brought this out *via* his zip-fastener analogy.[3] The empirically testable consequences of a given theory are brought together with actual laboratory experiments; this is where the bottoms of the zip-fastener are linked. The locutions in which these testable consequences are expressed are defined by, are given their meaning in terms of, the actual pointer readings and trace patterns encountered in experiments at the bench. These meanings are transferred back up through the steps of the theory by a kind of "reverse deduction," until at last the zip-fastener gets to the "top." Here the much-traveled meaning originally

---

[2] *Euclid's Elements*, tr. by Sir Thomas L. Heath (New York: Dover Publications, Inc., 1956 (reprint)), Vol. I, p. 153.

[3] R. B. Braithwaite, *Scientific Explanation* (Cambridge: Cambridge University Press, 1959), p. 51: "A calculus designed to represent electrical theory will be constructed so that its final formulae express propositions about observable flashes of light or pointer readings of a measuring instrument; like a zip-fastener, each side will be firmly attached at one end."

accorded to the observation terms in our experimental hypotheses reaches the highly placed theoretical term "electron."

Clearly, the meaning which accrues ultimately to our theoretical term ("electron" in this case) will be dependent upon the sort of laboratory tests which figure in our initial linking together of the bottoms of the zip-fastener. There is no *a priori* reason whatever for thinking that this original bestowal of meaning or significance on a term will be one and the same, whether the tests are of a chemical, a spectrographic, or an electromagnetic variety. The valence-theorist, the cathode-ray tube technician, and the battery designer all use the term "electron" essentially, but they correlate that term with widely different observations and phenomena. These indirect definitions of theoretical terms, then, indirectly and cumulatively invest words like "electron," "wave function," and "gene" with *clusters* of meanings. It is just a mistake to think that such terms have a single hard and fast meaning once and for all. Meanings as megalithic and monolithic correlates of words will just not do for one's reflections on theoretical discourse. We will go so far as to suggest that the word "electron" is really the name of a *class* of concepts all of which are similar in theoretical ways or are the results of analogous experimental results. These concepts overlap over the *core* of the concept of electron—an entity no more capacious than is the core-concept of "game."

(d) Finally, there is the *operational definition*, of which you have all seen examples. On Professor Bridgman's account any physical concept which is to count as significant must *ultimately* be defined in terms of operations of measurement.[4] The word "ultimately" is indispensable: It allows that a physical concept need not be *directly* definable here and now in this operational way. The average velocity of a gas molecule, for instance, cannot be defined directly in terms of measurement, *cannot* as a matter of experimental principle. However, this magnitude is mathematically related (see next chapter) to the pressure and density of a gas; these are operationally definable concepts, therefore the conception of a molecule's average velocity is operationally significant.

The operationalist will ask, "What could be less useful than the definition of 'mass' as 'the resistance a body offers to changes in its posi-

---

[4] P. W. Bridgman, *The Logic of Modern Physics* (New York: Macmillan, 1927), p. 5: "In general, we mean by any concept nothing more than a set of operations; *the concept is synonymous with the corresponding set of operations.*"

tion or velocity'? What does this mean in terms of the measuring of energy outputs? Or worse yet, 'mass is the quantity of matter in a body' —Newton's own chestnut."

Compare this with Ernst Mach's definition of *mass* in terms of the concept of acceleration: Body A collides with body B—we measure (by well-understood techniques) the accelerations due to the collision and find that the ratio of the acceleration of B to the acceleration of A is constant.[5] This constant we call "the ratio of the mass of A to the mass of B." Then, by "defining" (designating arbitrarily) some body X as the standard, or unit, mass, we can assign a numerical value to the mass of A. Corresponding definitions for length and time are easily formed.

Modern science is marked by the centrality of this sort of operational definition.

Still, exaggeration must be stoutly resisted. Some scientists, and some philosophers, take operational definition to be the *only* form of definition useful and respectable in contemporary science. It must be clear now that there are many forms of defining *procedure* other than the operational one, though perhaps it is true that an operational definition of a really significant concept is always possible. But even this conjecture is vulnerable to certain analyses offered by philosophers of physics, analyses of high-order theoretical terms like "ψ" and "δ"—these identifying functions within microphysical theory whose operational justification is not at all apparent.

This must conclude our brief survey of definition. Other forms of definition equally as important as those delineated have not been mentioned. But perhaps now it will appear that the command "Define your terms!" is not a sergeant-major's order to do just a single specific thing. One is sure to have a better idea of what a science lecturer is after when one sees what sorts of definitions are being employed. Needless to say, the lecturer too will profit if he chooses definitions that will do for him precisely what they are capable of doing—no more and no less.

Not all scientists have been alert to the snares in definition. By failing to distinguish one variety from another, some scientists have encountered perplexing difficulties.

---

[5] Ernst Mach, *The Science of Mechanics*, tr. by T. J. McCormack (LaSalle, Ill.: Open Court, 1960 (reprint)), p. 266.

Nicholas Steno, a 17th Century Danish anatomist, sought to introduce rigor into the study of anatomy by emulating the manner of exposition of the great geometer Euclid. Steno's book, *Elementorum Myologiae Specimen, Seu Musculi Descriptio Geometrica,* begins with a 50-word anatomical description of fibra motrix as Definition I. That Steno did not distinguish between the kind of definition appropriate to mathematics and that appropriate to biology is evident in the ludicrously nongeometrical character of his argument: He settles for 44 definitions and five suppositions, as compared to only six lemmata and one proposition.

Suppose that when that first black swan was discovered in Australia (in the late 18th Century) a certain American zoologist behaved badly. Having just returned the page proofs of a book in which he had remarked several times that "All swans are white," he refused to accord the name "swan" to those birds which were swan-like in all respects save their color. His colleagues, of course, would have found it much more convenient to speak of *two species* of swan. But no, this zoologist-fellow persists, arguing, "It is improper to call those, or any, black birds 'swans' because all swans are white!" In this way he transforms an inductive generalization into a definitional truth. Consider this further:

Many definitions are inductive, or based upon experience. This is not so in geometry and mathematics, where definitions are just stipulations—rules of the symbol-game. Natural scientists must always risk having their publications made obsolescent while still in the press. This risk cannot be minimized, moreover, by playing fast and loose with definition—by deciding to do as Euclid did whenever nature refuses to cooperate, by arbitrarily narrowing the range of application of a theory so as to exclude all awkward counterevidence.

As we have seen, substances are often defined in terms of their so-called "causal" properties. Oxygen is, by definition, a gas that sustains combustion; hydrogen is, by definition, an element the result of whose combustion is water vapor; nitrogen is, by definition, capable of killing animals which inspire it in a pure form. These propositions all serve to explain part of what is now meant by "oxygen," "hydrogen," and "nitrogen" in the mouths of physicists, chemists, and biologists. But they are causal laws in thin disguise. They are not stipulations in any mathematical sense, e.g., as when Euclid stipulates that *the symbol "i" will be understood* in such and such a way.

One *could* define "arsenic" in this stipulative way—as a substance which has the power to poison people. One could say, "Nothing that lacks the power to poison is allowed to be called 'arsenic.'" Relative to this definition, the existence of a sample of arsenic which lacks such power is logically as impossible as the existence of a square circle or a three-wheeled bicycle. But if one *did* so define "arsenic" this would be to *change* the meaning of the word (as it is now employed) and not to preserve it at all. For such a definition would have converted the useful assertion "Arsenic is poisonous" (useful because it could be false, but usually is not and hence informs us as to the facts) into the utterly useless claim "Poisonous arsenic is poisonous." The latter is necessarily true and necessarily uninformative.

The general recipe for this confusion is as follows:

1) A law of the form "X is *a* if and only if X is *b* and *c*" is highly confirmed ($X_a \equiv_{(physical)} X_{bc}$)

2) This leads us quite naturally to offer causal definitions like "*a* is defined as *b* and *c*." ($a =_{df} bc$)

3) But suppose, because the truth of the law becomes ever more apparent, we put our definition into stipulative refrigeration. We come then to think of it, not just as a terse way of indicating a natural regularity, but as a definition in the mathematical sense—one *from which* all observed regularities *must* follow. The definition is no longer treated as the resultant expression of a natural regularity as experienced; it is now treated as its *source,* as in mathematics, where definitions often function as rules of the symbolic game. Now relative to *this* definition:

4) The law itself becomes a simple tautology: "X is *a* if and only if X is *a* ($X_a \equiv_{(logical)} X_a$)." Failure to distinguish the descriptive empirical proposition which gives rise to a causal definition from the reappraisal of that proposition after the definition has come to be regarded as all but necessary, can create the illusion that the laws of nature are deducible from pure logic *or* that nature embodies laws in the sense that they are logically necessary—our observations of regularities being but our way of detecting the formal framework of the world. This is somewhat oversimplified. Still, the danger is a real one in the physical sciences and it is instructive to view it as a confusion between two sorts of definition. But perhaps it would be more correct to say that the confusion *manifests* itself as a confusion over definition. The mathematical physicist who is unclear about the logical differences between mathe-

matical claims and physical claims will exhibit his unclarity in his definitions.

Another pitfall is this: When a scientist defines a term synthetically by *correlating* it with a set of empirical data, he occasionally oversteps and identifies the *designatum* of the defined term with the empirical data —he confounds the nature of the thing with the evidence he has for urging the existence of that thing. Physicists have said things like, "The color quality red *is* a certain wave length of light," "Heat *is* molecular motion." Psychiatrists have said things like, "The mind *is* the working of the brain." This should need no comment.

The difference between the authoritative-historical definitions we mentioned first and the stipulative definitions that were contrasted with them was signaled as partly a difference in truth-value. Authoritative definitions are assertions to the effect that, by a given expression, scientists do *in fact* mean such and such, and this assertion is true or false. Stipulative definitions are proposals to the effect that a given expression be understood *pro tem* in a special way. This proposal cannot be true or false because no proposal can. But that these two different kinds of definitions are often confused is marked by questions like: "Space isn't *really* n-dimensional, is it?" "Is it *really* possible to draw through a point more than one line parallel to a given line?" "Electrons *really* do have a simultaneous position and velocity, don't they?" "Jones defines a protozoon as a 'non-celled organism,' but is that *really* true?" "Smith places this specimen in the flagellata, but it is *really* a member of the algae, is it not?" "Robinson says that the virus vaccinia is alive, but is it *really*?"

These questions point out more about the questioner than about the actual scientific difficulties themselves. Describing happenings and clarifying meanings must always be kept sharply distinct.

A final misunderstanding: It is sometimes thought that for a discipline really to stand on its logical feet it must start out from explicit and clearly formulated definitions. We cannot play a board game until we master the rules and other preliminaries; so also with the sciences? Hence the occasional dissatisfaction with notions like *space* and *dimension* in geometry; *number, zero,* and *number successor* in arithmetic; *length, time,* and *mass* in physics; and *life* in biology. However, such concepts may be said to draw their meanings from whole scientific structures and not from any ceremonial pronouncement of the form "X is defined as . . ." Expeditions in search of the explicit definition of "space" or "mass" or "life" (without which definitions it is thought that the natural sciences must collapse) are thus based on misconception. Granted, philosophers may be the worst offenders here, but more than one physicist has felt compelled to give *the* definition of "electron," and more than one biologist has tried to serve up *the* definition of "gene." Professor Bridgman has observed: ". . . the true meaning of a term is to be found by observing what a man does with it, not what he says about it . . ." [6]

Armed, then, with this survey of scientific definitions and their possible traps, you should be prepared to wrestle with the following questions.

Definition, as outlined here, has been only of *words*. Some word or symbol is explained to us when we are told what all the experts do mean by it, or what some particular expert *proposes* to mean by it. We may get a synonym, or an analysis, or a causal account, or a rule of use, or examples—or perhaps, just a picture or a gesture towards some object. These are all ways of defining *words*.

But surely when a physiologist defines the *aorta,* he is not out just to chat about the meaning of a word. He is telling you what the aorta, the *thing* aorta, *is*. He is explaining what is its structure, its function, its importance. What medical student would take his account only as an addition to his own technical vocabulary? Something would have been learned about the world, or about the aorta anyhow. Likewise the mathematician is not playing at grammar when he defines the word "conoid." He isn't playing at anything; he is talking seriously about things, namely conoids—the solids that result from the revolution of any conic section about its axis. And definitions of gamma rays and gametes are about gamma rays and gametes, and not about the words "gamma ray" and "gamete."

Or are they? The question is for you. In your thinking about it, however, do not ignore instructive expressions like "cross-over value," "differentiation," "energy equivalent," "physical constant," etc. What would definitions of these terms be definitions of? Would they be accounts of what we *mean* when we use these words, or descriptions of some part of the world that words designate? What part of the world

---

[6] P. W. Bridgman, *op. cit.*, p. 7.

does "differentiate" designate? And "national debt," and "genius," and "homeostasis," and "weed"?

Another question: Why do we accept or reject a definition at all? Of two rival definitions, *why* should one of them carry our vote? What precisely is wrong with defining "eclipse" as "a tin of beans"? The definition is clear enough, and simple enough.

What precisely is wrong with Dr. Johnson's definition of "network" as "anything reticulated or decussated, at equal distances, with interstices between the intersections?" [7] It is precise enough and surely correct. Do not dismiss this too lightly, for it is a tricky question.

Finally, notice how most scientific reports on original research begin with a few terse definitions. This is the model for giving a final account of one's research discoveries; but is this the way discoveries are actually made? Does a chemist solve his problem by first setting out definitions clearly? Or may it not be rather that these clear definitions are the very *last* things he arrives at? Laboratory research is not geometrical demonstration. Nicholas Steno was misled. Why are scientific reports written in a manner which seems completely to reverse the order of events as they took place at the bench?

The next chapter will deal with another way of putting boundaries around scientific concepts.

---

[7] *Johnson's Dictionary: A Modern Selection,* ed. E. L. McAdam, Jr., and George Milne (New York: Pantheon Books, 1963), p. 263.

3 | Measuring and Counting: More Boundaries

IN AN ideal science every concept would have a sharp edge—a boundary. I cannot confidently imagine what such an ideal science would be like, or even whether the notion itself is wholly unobjectionable. But undeniably, complete precision in locating the boundaries of its central concepts is a goal of modern science, whether or not that goal is logically attainable.

As we suggested in the first chapter, the word "not" can help us to find the boundaries of some words and symbols. To learn that "G"— as it occurs in Newton's law of gravitation (viz., $F = G \frac{m_1 m_2}{r^2}$)—is *not* an event, and *not* a variable designating the mass of a particle, and *not* a transcendental function is of some help, though admittedly not much.

Definition, as we saw in the last chapter, carries us further in the search for logical boundaries. "G" is defined as "the proportionality constant whose role is to give F the appropriate units, once m_1, m_2, and r have been assigned units." As we saw, this will help us most (or disturb us least) when we see just what kind of definition it is.

Measurement marks boundaries still more clearly. Now we learn that G is measured experimentally as 6.664×10^{-8} cm^3/gm. $-$ sec^2.

As with most things that seem obvious in science, however, the logical character of measurement is elusive. Just how does measurement enter into science? Why is it more dominant in some disciplines than

Measuring and Counting: More Boundaries | 43

others? Why (and how) is mathematics so intimately connected with measurement and with sciences which rest on measurement?

Consider a few sample sentences:

> "This beaker has twice the capacity of that one."
> "A note and its octave are obtained when a taut string is bridged into a length and twice that length."
> "The chemical atomic mass of sulphur is twice that of oxygen."

That such assertions make sense is part of the logic of measurement. Even propositions like "The half life of carbon 11 is twice that of nitrogen 13" are quite respectable in terms of this logic.

But compare:

> "The average scientist has twice the I.Q. of the average humanist."
> "Seats in Strathcona Hall are twice as comfortable as those in Harkness Hall."

These two assertions are ludicrous. They are ludicrous, not because they are false, but because they are *not even* false. For a proposition to be at least false is a mark in its favor. What possible force can the term "twice" have in such claims?

You are probably quite sure that the earlier sentences (i.e., about beakers, octaves, and atoms) are acceptable. And you are sure that these latter sentences (i.e., about I.Q. and comfortable seats) are not acceptable. But what about the following?

> "The fluid in that beaker is twice the temperature of the fluid in this one."
> "Mercury is 13½ times more dense than water."
> "The Herbert hardness tester indicates that glass is five times harder than mild steel."

These propositions are perhaps as questionable as those about I.Q.'s and comfortable seats. What is wrong with them?

It is unnecessary to applaud the quantitative study of nature. Co-ordinating numbers with objects, processes, and their properties is at the pulse of modern science. But "number" is itself a penumbral concept. I do not refer here to the problems involved in the logical analysis of the concept *number*—the problems of logicians like Frege, Russell, Hilbert, and Brouwer. I refer, rather, to something like this:

If I say, "The number of slides in this box is 48," I am using the word "number" differently from the way I use it in "The number 48 follows 47 and precedes 49." And different from both of these is this, "The number '48' is written in three strokes of the pen." These differences are not always easy to characterize, however.

Consider our box of prepared biological slides. Besides all its other properties, one very definite property of the box is the number of slides it can contain. This property is *represented* by a number (in our third sense)—"48." Clearly, this mark is not a property of the box. It is just three lines of chalk, or a sound; in short, it is a numeral. "48," "48," "48" are three different numerals (or at least numeral tokens); but only one number is expressed by these three—(the *number* 48). So the mark "48" is a symbol for some property of the box—some property which is found to behave just as do numbers (as opposed to numerals), some property the 48th element of which follows the 47th and precedes the 49th. The expression "the number of slides the box can contain" denotes just such a property.

Now suppose I had no numerals at all; I could still determine whether two boxes had in them the same number of slides. I would remove a slide from one box and then remove one from the other. This double operation would be repeated until one, or both, of the boxes were emptied of slides. If the operation which emptied one box also emptied the other, then the boxes held the same number of slides originally. But should one of the boxes still contain slides after the other had been emptied, then that box had a larger number of slides to begin with. Indeed this constitutes a quite respectable *operational definition* of the expressions "have the same number as," "have more than," and "have less than."

This would work just as well were the counted objects different in kind. The biological slides could have been counted against the burners on a laboratory bench or against the chromosomes in a cell nucleus. Had we never heard of counting before, we could apply this process to all kinds of things, soon to discover rules by which we could abbreviate and simplify the whole process. These rules are so obvious that they are seldom noticed. Perhaps we ought therefore to set them out here.

(a) If two sets of objects, when counted against a third set, are found to have the same number as that third set, then, when counted against each other, they, likewise, will be found to have the same number. This rule enables us to determine whether two sets of objects have the same number without bringing them together at all. It suggests the possibility of portable standard collections which can be counted, first against one collection then against another, in order to tell whether these have the same number.

(b) Another rule, ancillary to the first: By starting with a single object and continually adding another single object to it, we can build up a series of collections, e.g., . , . . , . . . , ... , ... , of which some one collection will have the same number as any other collection whatever. This rule suggests the efficacy of counting collections, not against each other, but against some *single* standard collection. It suggests a way of making the standard series as little cumbersome as possible. For if we required a totally different collection for each member of the standard series (carried in small bags, say), computation would be a drawn out procedure, to say the least. A great deal of trial and error, of opening and closing little bags of pebbles (the standard "one" bag, the standard "two" bag, etc.), would have to precede our ever being able to say of two collections that they have the same number. But here we see that the earlier members of a series may be parts of the later members. We can obtain a series some member of which has the same number as any other collection you please, and yet the number of objects will be no greater than that of the largest collection we would ever want to count.

It was in this way that primitive man probably used his fingers and toes. Less primitive men used standard collections of "counters," small bead-like tokens, of which a great many could be carried at once in a single bag. (Notice how in shops we still conduct our business over *the counter*.)

We use *numerals* for exactly this same purpose. Numerals are our counting tokens. Numerals are just the distinguishable "objects" out of which we build our standard series—by adding them in turn to previous members of the series: "1", "1, 2", "1, 2, 3", "1, 2, 3, 4", and so on, indefinitely. We count other collections against these members of the standard series. In this way, we can ascertain whether or not two collections so counted have the same number. This may enable me to say that there is the same number of chromosomes per cell on a given prepared slide as there are slides in the box from which the slide came.

By a wonderful convention we describe which member of this numeral-collection series has the same number as a collection counted against it, by quoting simply the last numeral in that member. That is, we say that the collection of microscopes on the first bench encountered in the Kline laboratory has the same number as the collection "1, 2, 3, 4, 5, 6, 7, 8" by saying that the number of microscopes on the bench *is 8*. This convention, it is to be noted, represents an intellectual and scientific advance of the greatest magnitude. But what we *really* mean, when we say that the number of microscopes is 8, is that this collection of microscopes *has the same number* as the collection of numerals (taken in standard order) which ends in 8: Indeed, it has the same number as any other collection which also has the same number as the collection of numerals which ends in 8.

This is all too obvious. Or is it? Some radically different views of the nature of numbers and of counting have been held by scientists, mathematicians, and philosophers.

The phrase "distinguishable object" has figured in the foregoing. Counting can be only of the *discrete*. We cannot count the drops in a beaker of alcohol unless we first introduce a convention as to what will be regarded as "a drop." This agreed, we must then find a way of separating out these drops from one another in an invariant and uniform way.

Moreover, objects are distinguishable because *we* distinguish them. The world is not all cut up for us in advance, just waiting to be numbered; processes and events are not marked with dotted lines for ease in numerical tabulation. Counting is seldom undertaken just for its own sake, or for verbal exercise. We count when we can make significant connections between the discernible entities we are counting. We enumerate, not anything and everything, but only that which our questions, hypotheses, and theories earmark as relevant.

Counting, therefore, helps us to make our ideas precise—our ideas not only about the quantitative nature of things but also about the nature of counting itself. We come to know, implicitly at least, what can and what cannot be counted. This knowledge gets *built into* our questions about nature, so to speak. Let us now continue with this project of making what is implicit about the scientific uses of numbers quite explicit.

Measuring and Counting: More Boundaries | 47

We have considered two very obvious rules, namely (a) and (b) (pp. 44–45). They are fundamental to what we mean by "the number" of a collection. As I suggested, the obviousness of these rules must not incline us to underestimate their importance. They are essential to an understanding of measurement in science. And they are great intellectual achievements; this is clear when they are expressed thus: "Collections which equal the same collection are equal to each other" and "To any given collection whatever another element can be repeatedly added, so as ultimately to equal any collection at all." These are primitive conceptions which rivet together almost all of our contemporary mathematical induction, the first as the principle of isomorphic sets.

To these two rules we may now add a third. Attend to this one carefully. Though just as obvious as the others, it is virtually the fulcrum of any analysis of measurement.

We often wish to know the number of a collection formed by the combination of other collections. What is the number of the collection made by combining a collection of 2 objects with a collection of 3 objects? "Five," you say—well done! The first collection of 2 objects (. .) (A) is counted against the numerals "1, 2". The second collection (. . .) (B) is counted against the numerals "1, 2, 3". The collection of numerals "1, 2, 1, 2, 3" can then be counted against the standard numeral collection "1, 2, 3, 4, 5". The number of the combined collection is thus 5, by the convention cited earlier. But to get that answer we had to make this assumption:

(c) If two collections A and A have the same number, and two other collections B and B' have the same number, then the collection A'B' has the same number as the collection AB. That is, equals plus equals produce equals.

Without these obvious rules, measurement as we know it would be impossible. Without the third of these as just made explicit, we could not distinguish propositions about weights and lengths from those about I.Q.'s and degrees of comfort in lecture rooms. Nor could we distinguish either of these from propositions about temperature and density. All of these distinctions will be discussed later in the chapter.

But consider first another penumbral feature of numbers—their varied and limitless uses. Numbers occasionally serve as tags or identification marks, or they may serve to indicate the position of some thing or property in an ordinal series or scale of things and properties. Finally, and most important, numbers may be made to indicate quantitative relations between properties of things.

The use of numbers as tags is familiar enough. We number our microscopes, our benches, our specimens, our books, our cars, our rooms, our streets, our essay pages, etc. This is a most convenient way of naming things. We are thus supplied immediately with an inexhaustible store of names. But everyone recognizes that the objects so named by numbers need not stand in relation to each other as do their names: The microscope numbered 200 need not be twice as powerful as the one numbered 100. Nor is Convict Number 1000 twice as wicked as Convict Number 500. Nor are lectures given in Room 210 Strathcona twice as good as those given in Room 105.

More important to science is the ordinal use of numbers. Here the order of numerical magnitude designated is the same as that of the position held by certain things or properties on an intensive scale. Suppose that we wish to distinguish bodies with respect to their being harder or softer than certain other bodies. Our principle of order may be as follows: Diamond is harder than glass if diamond can deform (scratch) glass but not vice versa. Glass is harder than molybdenum steel if glass can deform molybdenum steel but not vice versa. And two bodies will be said to be equally hard if neither can scratch, nor be scratched by, the other. (This is, in fact, the very principle incorporated within the Vicker's diamond hardness testing machine.) We may then proceed to order bodies in a scale of hardness, if it can be shown (experimentally) that relations like the following hold: If A is harder than B, and B is harder than C, then A is harder than C. The relation of "being harder than" must be *transitive*. If it can also be shown to be asymmetrical—i.e., if A is harder than B, then B is *not* harder than A—we are free to set out an ordering of bodies on this principle. (Experiment has, of course, turned up certain exceptions to this principle as it applies to the property of hardness. Nonetheless we will return to this example, only later, as it will be more instructive then than now.)

We turn next to consider the third use of numbers. This is not to supply names, and not to order several things *à la* "greater than" and "less than." This use employs numerals as signs to indicate the quantitative relations existing between properties of single things. Clearly, the first requirement of this use of numbers is that we should at least be able to carry out operations which will allow an ordering like that found for hardness. Consider weight: We must be able to construct a scale of weights such that it will be agreed that one body is heavier than another body if, when both are placed in opposite pans of a beam balance, the pan containing the first body sinks. We must then establish experimentally that the relation "heavier than" is transitive and asymmetrical. It then remains for us to agree that two bodies are equal in weight (or are "as heavy as each other") if the body A is not heavier than B and B is not heavier than A. This means that neither pan of the balance will sink when A and B are placed in opposite pans.

Now we can construct a scale of degrees of weights, just as we did for hardness. The weights of any couple of objects can be ordered, this one being "heavier than," "lighter than," or "the same weight as" that one.

But we can do more—and this is of the utmost importance. We can now find an interpretation, in terms of some operation on bodies, for the statement that one body is *twice as* heavy as another, or that one body is three times the weight of another.

How is this interpretation possible? It is possible simply because weights can be added. *The physical process of addition is the placing of two or more weights together in the same pan of the beam balance.* Place three bodies, A, B, and C, which are equally heavy in one pan of a balance. Place another body, D, in the other pan, so that the beam balances exactly. Then D is as heavy as the three bodies A, B, and C combined— and is therefore *three times* as heavy as any one of them.

This operation can be extended to define a series of standard weights. In terms of this procedure, it is now significant to say that one object is n times as heavy or 1/nth as heavy as another. Moreover, a physical interpretation has here been given to the formal arithmetic symbol "+"—an interpretation permissible in relatively few physical contexts.

And this is the heart of the matter. Is there any physical operation which corresponds to this beam-balance procedure when it comes to hardness? Or to temperature? Or to density? I may assure myself that three specimens of mild steel are equally hard if no one of them can scratch, or be scratched by, the other two. Or perhaps my assurance lies in the fact that they are all indented to exactly the same depth by the diamond pyramid in the Vicker's hardness testing machine. But by what operation can the three of them be combined so as to offset the scratching effect of some substance said to be three times as hard as any one of them? Hardness and weight are physical properties that are different in kind, the former possessing no straightforward interpretation of "+" while the latter does.

Three equal volumes of liquid (all at the same temperature) will not, when simply combined as we combined the weights in the balance pan, form one body of liquid three times the temperature of any of the original three. This is what Aristotle was driving at when he spoke of hotness as being a qualitative but not a quantitative property. He compared it with wit and humor, things we never contrast by use of numerical degrees save in metaphorical guise—as when we judge one comedian to be twice as funny as another. In the same way three equal volumes of a liquid (all of the same density) will not, when simply combined, form one body of liquid three times the density of any of the original three.

But three equal volumes of a liquid (all of the same weight) *will*, on simple combination, form a body three times the volume and three times the weight of any one of the original three (pressure and temperature presumed to be constant). And three equal lengths of cable will, on simple combination, form a length of cable three times the length of any of the original lengths of cable. Volume, weight, and length are, thus, simply additive physical properties; hardness, temperature, and density are not.

Having done this much we have still not done enough to be sure that the numbers so assigned will have all their original arithmetic meanings. We have shown that weight is an additive property, as against hardness, temperature, and density. But it remains to be shown *by experiment* (and this must be continually stressed) that the numbers so assigned are consistent with themselves—that two different numbers never end up on the same label. It is perfectly conceivable that some physical property may meet this primary test of additivity but fail altogether when more involved manipulations of "+" become paramount.

Thus, suppose that object A is regarded as having the unit weight. We can assign weights to other objects by the process described, such that A_2 will have weight 2, A_4 will have weight 4, and A_6 weight 6. Now, can we be certain *in advance of experiment* that A_2 and A_4 will, if placed together in one pan of the beam balance, just balance A_6 placed in the other? No. It is very important to note that we cannot be certain of this until we actually perform the experiment. The emphasis of some scientists, and the usual exaggeration of that emphasis by philosophers, has sometimes been such as to minimize this. That $2 + 4 = 6$ can be demonstrated in pure arithmetic without experiment—that's a matter of formal prescription. But until we performed the proper experiments we could not be sure that the *physical* operation of addition of weights would exhibit the familiar properties of purely arithmetical addition—for physical properties are a matter of how nature is put together.

In fact, as you know very well, this physical operation of addition possesses the formal properties of arithmetical addition only in some sophisticated laboratory cases, not all. The moments of the weights of the arms of the beam balance (about the fulcrum) must be exactly equal, the two scale pans must be of equal weight, the distances between the points of suspension of the pans and the fulcrum must be equal, the center of gravity of the beam must be below the fulcrum, the metallic constitution of pans and fulcrum must be homogeneous, etc.

In other words, it takes a very well built and sensitive instrument to perform the physical operation of weighing accurately enough for us to speak of it as a physical operation of addition, the operational analogue of "+." High winds, earthquakes, strong electromagnetic force fields, etc. are incompatible with any successful resolution of this question in the case of physical properties; not so in the logico-mathematical context.

Length, time intervals, areas, angles, electrical resistance, etc. can all be measured in similar ways, even though there are special difficulties in the exact logical analysis of each of these. These properties are each additive. There are physical processes such that simply combining two objects having these properties to a certain degree produces an object having an increased degree of that property—an increase which is arithmetically predictable.

These properties are often called "extensive." This puts them into logical opposition with the properties we call "intensive" (like temperature, density, and hardness). And physical operations like weighing are called "fundamental." This puts them into logical opposition with operations that might be called "derived" (such as those used in thermometry and hydrometry); we will turn to these shortly.

Let me stress again that the treatment of certain properties in the numerical ways follows definite *experiment* and *observation* of the scientific subject matter at hand. The process of discovering that a property is measurable in the way described and the discovery of a way of measuring the property rest entirely upon experimental inquiry. Whenever a new branch of science sprouts forth, an early attempt is always made to try to find some process for measuring the novel properties being investigated. If such a process cannot be found, then other less direct methods must be contrived. Consider electricity and the striking progress of this science through the discovery (by Cavendish and Coulomb) of the law underlying the technique of measuring an electric charge, the laws discovered by Oersted and Ampere fundamental to measuring an electric current, and the laws of Ohm and Kirchhoff prerequisite for measuring electrical resistance.

Remember, many more properties definitely recognized by science are *not* measurable, in the sense now being discussed, than *are* measurable. It is remarkable that nature is even this cooperative in falling in with *our* ideas of what it ought to be.

Again, however, what is measurable depends for its discovery on the existence of experimental laws; it depends upon the facts of this experimental world. It is never wholly within our power to determine whether we will or will not measure a certain property. That is the perplexing feature of measurement which is essential in modern science; it is also the feature most likely to be hidden beneath the symbols and powerful deductive arguments of the newer research disciplines—microphysics, astrophysics, genetics, etc.

Let us consider all this in a slightly more formal way.

The bare-minimal requirements for employing numbers to measure qualitative differences are these:

(a) Given any set of n bodies, we must be able to order them serially, with respect to some ρ (heavier than, longer than) such that between any two bodies one and only one of the following hold:

$A\rho B$, $B\rho A$, $A = B$. (The relation ρ must be asymmetrical.)

(b) If $A\rho B$ and $B\rho C$ then $A\rho C$. (The relation ρ must be transitive.) These two conditions are *necessary* for all measurement. Apparently, they are *sufficient* for the measurement of intensive properties like temperature and density (this will be sharply qualified). They are not sufficient for extensive measurement, where we require some physical process of addition (symbolized by "+") which experiment proves to have the following properties:

(c) If $A + B = C$, then $B + A = C$
(d) If $A = B$, then $A + C \geqslant B$
(e) If $A = B$ and $A^1 = B^1$, then $A + A^1 = B + B^1$
(f) $(A + B) + C = A + (B + C)$

Measurement in the strict sense rests on *all* these logical conditions. It is foolish and misleading—it is "base, common, and popular"—to make statements which suggest that all six conditions hold when only the first two are in fact satisfied.

It is in this more intense light that I ask you now to reconsider the propositions with which our analysis began:

"The fluid in that beaker is twice the temperature of the fluid in this one."

"Mercury is 13½ times more dense than water."

"The Herbert hardness tester indicates that glass is five times harder than mild steel."

If what has been said is correct, these statements are radically misleading. There are no physical operations corresponding to our formal conditions (c)–(f) appropriate to any of these assertions. Granted, very frequently such things *are* said. I do not mean to suggest that anything very harmful is the immediate issue of this loose way of talking. But coming to appreciate the logic of these assertions could remove a potential obstacle from one's coming to think more clearly about measurement in science. There are problems enough about the nature of measurement in quantum mechanics and in quasar astronomy without our losing our footing on this relatively low rung of the analytical ladder.

[It is worth noting that, while we have explored only the physical process of *addition*, there are, of course, like processes for *multiplication* and *division*. Again, these represent definite experimental discoveries in each case. Contingent commitments are packed into each such determination. Operations like multiplication and division have significance in scientific numerical work just because it has been *discovered* that, with respect to a certain class of properties, a physical operation allows itself to be represented quite literally by "×" or "÷".]

However, we cannot leave things this way. It sounds as though estimations of temperature and of density are poor relations (when it comes to accuracy and clarity about what is measured) to estimations of length, mass, and time. One suggestion which must surely be spiked

54 | *Chapter 3*

is that thermometry and hydrometry are conceptually no more reliable than intelligence-testing and tea-tasting, because that clearly is not true.

As we noted, temperature and density are intensive properties and hence cannot be measured in the fundamental way appropriate to weights. They *can* be measured in a derivative way, however, which you know very well. (It is at best an open question whether intelligence-testing and tea-tasting are amenable to derived measurement in the same sense.)

What is derived measurement?

It is important to stress that, whatever derived measurement is, it must be dependent in some way on the fundamental process of measurement, as outlined. This must be the case if the numerals involved are going to succeed in telling us something significant about the properties of bodies. [The history of science confirms this conjecture. All the properties measured in the pre-scientific eras were measured by a fundamental process, e.g., weight, length, volume, area, and periods of time. Derived measurement, which we are now to consider, is a product of quite conscious scientific investigation of relatively recent date.]

Consider density, the measurement of which surely falls within the scientific period—although Archimedes did make some signal strides. If I say that a body A weighs 13½ times as much as body B, I mean that, if I put body A in one pan of a beam balance, 13 bodies of B's weight (plus another body one-half of B's weight) will be required in the other pan exactly to balance A. In other words, 13½ B's will have the same weight as A. But if I say that a given volume of a substance M has a density 13½ times the same volume of a substance W, I do not mean that 13 volumes of W (plus another ½ volume of W) will have *just* the density of M. And this is all that we mean when we say that weights are additive while densities are not, that while weight can be measured by a fundamental process, density cannot. However many volumes of water we take, all of the same density, we cannot by simple combination produce a volume with any different density. (The situation at great ocean depths is not the result of simple combination: It involves further complexities involving considerations of pressure and temperature.)

Consider these liquids: gasoline, alcohol, water, hydrochloric acid, and mercury. Suppose that we agree to call one liquid, say mercury, more dense than water, if we can find some solid body which will float on mercury, but sink in water. Experiment will then show that density, so defined, is an asymmetric and transitive relation. The liquids can, therefore, be arranged in a series of increasing density, reminiscent of our earlier ordering of substances according to their hardness. But it is clear that, since density is not an additive property, we can only measure it thus far as an intensive quality. Suppose we assign numbers 1, 2, 3, 4, 5 to designate, respectively, the positions of gasoline, alcohol, water, hydrochloric acid, and mercury. These numbers are, of course, quite arbitrary: 2, 4, 6, 8, 10 would have done as well. It surely does not follow from such an ordinal numeration that mercury is more than twice the density of alcohol. For, as we saw, the expression "more than twice the density" is without meaning in any ordinary non-technical sense.

But, of course, altogether different numbers are assigned for densities—numbers which are not chosen arbitrarily at all. The reason? Many intensive properties can be measured in a way other than simply by arranging them serially. Density is one of these. Temperature is another. Hardness, however, is not one of these—not to the present day, at any rate.

This "other way" depends on the existence of a *numerical law* holding between *other* properties of liquids, properties with which their density is invariably related. We discover experimentally (when we weigh different volumes of a liquid) that the ratio of the numbers measuring the weights and volumes of this liquid is the same, no matter how large or how small the volume measured. Thus: $W = cV$, where c is the constant for *all* samples of the same liquid, but is a different constant for other liquids. By a proper choice of our units of weight and volume we find that c is .75 for gasoline, .79 for alcohol, 1.00 for water, 1.27 for hydrochloric acid, and 13.6 for mercury (pressure and temperature constant). We also make the important discovery (and this always amazes me) that the order of these ratios is the *same* as the order of the density of liquids determined in the serial way considered earlier—where the floating of a test body was at issue. There is no *a priori* reason why this should be so, but it is, as experiment dramatically reveals. It is easy to see why scientists, ancient and modern, have so often pictured God as a professor of mathematics. For, undoubtedly, nature does cooperate with our mathematical ideas to a remarkable degree.

This ratio between the weight and volume of a liquid, to return to our main point, since it is constant for all samples of a homogeneous liquid, can therefore be taken as a measure of its density.

Still, we must remain on our guard against saying that the density of mercury, or of any other substance, is *so many times* that of water. No matter how it is measured, density is a non-additive property. It can be measured precisely, and numbers can be assigned without arbitrariness to different degrees of it—but only in virtue of there being a connection between weight and volume (two properties which are extensive).

Density, therefore, can be measured only by such a derivative method—a method which rests on the establishment of a *numerical law* between properties which are measurable by a fundamental process. The same is true of temperature, the derivative measurement of which rests on the linear expansion of some thermometric substance. Likewise buoyancy, elasticity, and mechanical efficiency.

Unless some properties were measured thus by a fundamental process, numerical laws would not be possible and derivative measurement could not be undertaken.

Please ponder the following puzzles:

1. Doubtless some will think that I have been a little tough on the concept of temperature. You will refer me to the Absolute or Kelvin scale and remark how it would often be quite appropriate to speak of a substance A as having twice the temperature of another substance B. But is the expression "twice" being used here with respect to *temperature per se* or with respect to the numerical marking of the Kelvin scale? If A's temperature is 212°F. and B's is 50°C. we may find a way of justifying the claim that A is twice as hot as B, but suppose the temperature scale of one is logarithmically ordered while the other is exponentially expanded. *That* correlation of numbers with temperature may still be precise in all the ways considered at present desirable; but the expression "twice the temperature of" will be no temptation at all. Would you say now that liquid helium is 273 times colder than ice? What could such a statement mean?

2. I have spoken of time intervals as being additive. Furthermore, I have implied that all additive properties are measured by a fundamental process. However, by what *fundamental process* could we measure the addition of time intervals? Is it not, rather, that we treat time intervals indirectly, by reference to the oscillations of a suitable pendulum, or by other such means? This is well worth worrying about.

3. If pressure, temperature, and volume all vary in "ideal" gases, the following relation holds, as you know: $\frac{P_1 V_1}{P_2 V_2} = \frac{T_1}{T_2}$
Consider the types of measurement required to establish this law. Explain their exact nature, and their interrelations.

4. The rotation of the earth is slowing down at the rate of 22 seconds per century. How can this be maintained, if the period of the earth's rotation is taken as the standard unit of time?

5. Which of the following are extensive and which are intensive properties? Which of them can be measured by a fundamental process and which only by a derived process? And which of them permits only a serial ordering without either fundamental or derived measurement:

number	heat (as opposed to temperature)	viscosity
angle	pressure	permeability
volume	color	acoustical pitch
force	shape	acoustical volume
energy	humidity	I.Q.
brightness	durability	fragrance

Part II | THE ACT OF SCIENTIFIC SEEING

4. There Is More to Seeing than Meets the Eye
5. Seeing the Same Thing
6. Seeing and Seeing As
7. Seeing As and Seeing That
8. Seeing, Saying, and Knowing
9. Spectacles behind the Eyes
10. Can We See Facts?
11. Facts and Seeing That

BIBLIOGRAPHY
PART II

Austin, J. L. *Sense and Sensibilia*. Edited by G. J. Warnock. Oxford: Oxford University Press, 1962.
Ayer, A. J. *The Foundations of Empirical Knowledge*. London: Macmillan Co., 1963.
Broad, C. D. *Scientific Thought*. New York: Harcourt, Brace, 1923.
Eddington, A. S. *Philosophy of Physical Science*. Ann Arbor: University of Michigan Press, 1958.
Hanson, Norwood Russell. *Patterns of Discovery*. Cambridge: Cambridge University Press, 1961.
Hirst, R. J. (ed.). *Perception and the External World*. New York: Macmillan Co., 1965.
Lewis, C. I. *Analysis of Knowledge and Valuation*. La Salle, Ill.: Open Court, 1946.
Mann, Ida, and Pirie, A. *The Science of Seeing*. New York: Penguin Books, 1946.
Price, H. H. *Perception*. 2nd ed. New York: Dover, 1950.
Russell, Bertrand. *Our Knowledge of the External World*. New York: Mentor Books, 1956. (Copyright 1929 by W. W. Norton.)
Strawson, P. F. *Individuals*. London: Methuen, 1959.
_____. *Introduction to Logical Theory*. New York: John Wiley and Sons, 1964. (1952).
Vernon, M. D. *A Further Study of Visual Perception*. Cambridge: Cambridge University Press, 1952.
_____. *Psychology of Perception*. New York: Penguin Books, 1962.
Whewell, William. *Philosophy of Discovery*. London: J. W. Parker and Son, 1860.
Wisdom, John. *Problems of Mind and Matter*. Cambridge: Cambridge University Press, 1963.
Wittgenstein, Ludwig. *Philosophical Investigations*. Translated by G. E. M. Anscombe. New York: Macmillan Co., 1953.
_____. *Tractatus Logico-Philosophicus*. Translated by D. F. Pears and B. F. McGuinness. New York: Humanities Press, 1961.

4 | There Is More to Seeing than Meets the Eye

IN THIS and subsequent chapters we will consider a cluster of concepts having to do with *observation*. Notions like seeing, witnessing, noticing, attending, evidence, data, facts . . . will be inspected, sometimes at close range, sometimes from far off.

Before skirting the edges of the philosophical quagmire that goes by the name "perception," notice one elementary point. The word "observe" has many uses. Two of these must be kept distinct. Consider:

When a person is hunting or searching for something, our description of what he is doing will differ in certain quite significant ways from our description of him after he has discovered what he was looking for. Thus we might say of Smith that he was hunting for the coelacanth carefully and systematically, but not that he *found* the coelacanth carefully or systematically. Astronomers will search and scan the heavens industriously and painstakingly for some planet whose existence has been predicted, but they will not espy or locate the planet industriously and painstakingly, any more than Archimedes shouted "Eureka!" industriously and painstakingly.

So too we will be said to have looked for the four-leaf clover carefully or carelessly, but not to have discovered it in these ways. Looking for the clover is a task; seeing it is an achievement. Listening is a job; hearing is a successful outcome of that job.

The word "observe" (and hence also the noun "observation") is made to perform *both in this task-performing way and in this success-indicating-way*. "Observe" sometimes serves in just the ways that "look" and "listen" serve, and it sometimes serves as "see" and "hear" serve. Thus we may speak of a man as a careful, systematic, industrious observer, or of his careful, systematic, and industrious observations. We may also, however, refer to his observations as his findings, his discoveries, what he has been successful in seeing. The word "observe," then, is sometimes used to indicate what a person is doing, and sometimes used to signal the success of what he is doing.

It is important to keep these meanings of "observation" sorted out; failure to do so can cause mischief when we ask questions about *what* it is we observe, what is its character, and how we actually proceed in our observations.

With that preliminary out of the way, let us turn to the major question of this section.

Consider two astronomers looking eastward at dawn on a clear day. One of them is a 13th Century astronomer. The other is a 20th Century astronomer. The question is: Do they both see the same thing?

A typical discussion following the posing of such a question might run as follows:

"Yes, they do see the same thing."
"No, they do not see the same thing."
"Yes, they do."
"No, they don't."
"Yes, they do."
"No, they don't."

This ought to suggest that we have here an unsettled conceptual situation. But perhaps that is putting it too strongly: Let me say, rather, that the question is a philosophical question in that there are no obviously *right* or obviously *wrong* answers to it. What we are concerned to discover is why some people should wish to say one thing and why other people should wish to dispute with them about it. It is the reasons behind the "Yes, they do" and "No, they don't" answers that will interest us. And after gaining some appreciation of this conflict in reasons, and conflict in attitudes, we must inquire which position will be most fruitful in our attempt to understand scientific research as it is carried on today.

It is only right that I announce straightaway that my inclination is to say, "No, the two astronomers do not see the same thing." This is not my way of saying that those who say otherwise speak falsely, for as we will see there are powerful arguments in favor of both positions. But some of these arguments, I feel, lead into *culs-de-sac* from which our logic can supply no escape. Others of them, I feel, open up many new paths of conceptual inquiry.

In any case, however you are inclined to speak about these matters of observation, evidence, and facts, I hope you will come to see the point of a profound remark by the great philosopher Wittgenstein: "We find certain things about seeing puzzling, because we do not find the whole business of seeing puzzling enough." [1]

We have our question; our two astronomers are looking eastward at dawn on a clear day. Does the 13th Century man see what the 20th Century man sees?

Before we can properly undertake to consider whether or not they see the same thing, we ought to consider what it is they see. For the answers to the question posed are just reflections of possible answers to the question, "What exactly do the astronomers see?"

Now there are at least three answers to this question that would incline us to say that the two astronomers see exactly the same thing. And it is to the exposition of two of these positions that the remainder of this chapter will be devoted.

One answer, which ought not perhaps to be taken very seriously despite the fact that it is the one most readily advanced, might run as follows: If both the astronomers are in possession of normal eyesight, if their organs of sight are in no way defective, if, in other words, they can see normally and are not blind, then they see the same thing because their retinas are affected in the same way. *"Seeing the same thing" is "having the same retinal reaction."* The two astronomers see the same thing if, when oriented in the same direction under the same conditions, roughly similar areas of their retinas are affected in roughly the same way. That is, the two men have the same visual impression when radiation in the form of light of roughly the same frequency produces similar

---

[1] Ludwig Wittgenstein, *Philosophical Investigations,* tr. by G.E.M. Anscombe (New York: Macmillan, 1953), p. 212. Cited hereafter as Wittgenstein, *Investigations.*

electrochemical changes in the neural receptors, the rods and the cones, which compose the surface of the retina.

And of course there is a good linguistic precedent for this position. The distinction we often wish to make between the person who has normal vision and the person who is blind is usually made by saying of the one that he can see and of the other that he cannot. The slave boy led Samson to the pillars of the temple because the slave boy could see while Samson could not see. Members of Milton's family wrote down his great poetry for him because they could see while he could not—he was blind. And in these cases of blindness due to deterioration or damage to the organs of sight, saying of someone that he cannot see is saying that the condition of his eyes is such that light radiation does not reach or does not affect the retina, or, if it does, the impulses therefrom do not reach the brain. (Notice that this last is already a qualification on the original position.)

And so we are, under certain circumstances, prompted to equate "X is blind" with "X cannot see" and "X has normal vision" with "X can see."

The connection between "seeing the sun" and "having a normal retinal reaction to the sun" is made even more intimate when we consider that if one of our astronomers is standing behind a boulder, or has his eyes closed, he will not see what the other astronomer sees, i.e., he will not have the retinal impressions the other chap has. And it is tempting to move from the consideration that, when two men do not see the same thing, it must be because of a difference in retinal reaction (due to blindness or a physical obstruction to vision like a fence or a boulder) to saying that when two men *do* see the same thing, it is just *because* their retinal reactions are the same; or, more strongly, "seeing the same thing" just is "having the same retinal reaction," no more and no less.

When certain scientists (biophysicists, biochemists, neurophysiologists, etc.) attend too carefully to the physical mechanism of sight they often end up in a position like this. They note the connections between complex events like the nuclear fission of the sun's elements, the resultant radiation of the liberated energy, its passage in the form of electromagnetic waves through space, then through our terrestrial atmosphere, through the cornea, aqueous, iris, lens, and vitreous of our astronomer's eyeballs ultimately to impinge upon the sensitive surfaces of their rods and cones, there initiating nervous impulses which, after a complicated cerebral journey, will result in the awareness of light, and perhaps even the pronouncement, "I see the sun."

Often scientists put to themselves the question, "But what is it we *really* see?" They trace the complicated path of the impulse from the sun itself to the pronouncement, "I see the sun," and they note that the only physical reaction within either of our astronomers appropriate to this particular confrontation with the sun is the physicochemical reaction in his retina. All else is a contribution to the situation by the astronomer— a contribution affected by inference from his past experience, his memory, his training, and his knowledge. And by way of this reasoning it occurs to the scientists that the seeing of the sun must be what takes place at that moment when the sun's rays impinge on the retina. To say that the astronomer sees the sun is to say that he is having a retinal reaction of a certain sort.

It is this sort of thing that inclines the neurologist, Dr. W. Russell Brain, to speak of our retinal, and other, sensations as the *indicators* by inference from which we find our way about the world.[2] What we *see* are the changes in our rods and cones. Everything above and beyond that is an intellectual construction on *our parts*. This has even induced an ophthalmologist of the stature of Dr. Ida Mann to speak of the cone-shaped cells in the macula of the eye as themselves seeing details in bright light, and of the rods as seeing approaching automobiles.[3] And the distinguished botanist Agnes Arber is similarly led to suggest that ". . . since the eye, as a mechanism, cannot receive clear impressions from different focal distances simultaneously, it can only be the mind which draws together this series of optical imprints . . ."[4]

Thus it might be argued that the question whether or not our two astronomers *see* the same thing is a question for the neurophysiologist. If their retinas are normal and they are reacting in the same way in the same circumstances, then they are seeing the same thing. Of course they may *interpret* what they see differently, they may not make the same thing out of what hits their retinas—no one is denying that—but since everything behind the retina is already brain-controlled to some extent,

---

[2] W. Russell Brain and E. B. Strauss, *Recent Advances in Neurology* (Philadelphia: P. Blackiston's Son & Co., 1929), p. 88.

[3] Ida Mann and A. Pirie, *The Science of Seeing* (New York: Penguin Books, 1946), pp. 48–9.

[4] Agnes Arber, *The Mind and the Eye* (Cambridge: Cambridge University Press, 1954), p. 116.

the *real seeing* must be what happens within the eyeball. For it is only there that we are in direct contact with the light actually sent out by the sun.

It is interesting to note that this is a kind of reversal of the ancient view that things are seen because of emanations emerging from the eyes and impinging upon outside objects. Aristotle and Galen held to this view, reckoning that all sensation was in principle a kind of touching. Even Leonardo wrote that "the eye transmits its own image through the air to all the objects which are in front of it, and receives them into itself. . . ."[5] And consider John Donne's words: "Our eye-beams twisted and did thread Our eyes upon one double string. . . . And pictures in our eyes to get Was all *our* propagation."[6]

Modern optics and neurology will no longer allow us this picture of the eyes reaching out and touching the objects seen, as if we were a kind of radar sending station. So some modern scientists simply have it the other way around; they have it that it is the objects which reach out and touch the eyes, and that is what they call "seeing."

So there is our first view of the nature of seeing. "Seeing X" is just "having a normal retinal reaction to X."

But I raised the question earlier whether this view ought to be taken at all seriously, despite its popularity in philosophical discussion among student scientists. For there are many obvious objections—I will review some of these briefly.

William Whewell, the 19th Century Cambridge philosopher of science, called attention to two possible objections in his *paradoxes of vision*.[7] The first paradox of vision for Whewell was that we see objects upright, though the images on the retina are inverted. He states his solution in this way: "We do not see the image on the retina at all, we only see by means of it."

Whewell's second paradox of vision is "that we see objects single, though there are two images on the retinas, one in each eye." His explanation of this is less satisfactory from a neurophysiological point of

---

[5] *The Notebooks of Leonardo da Vinci*, C.A. 138 r.b., comp. Edward MacCurdy (New York: Reynal and Hitchcock, 1938), Vol. II, p. 364.

[6] John Donne, "The Ecstasy," in *The Oxford Book of English Verse 1250–1918*, new edition, ed. Sir Arthur Quiller-Couch (New York: Oxford University Press, 1955), p. 233.

[7] William Whewell, "Paradoxes of Vision," in *Philosophy of Discovery* (London: J. W. Parker & Son, 1860).

view. Nonetheless, he is correct in suggesting that if seeing were just the having of retinal impressions, we ought to see double as a matter of course, and this is not the case. [In this connection it is interesting to consider how some neurologists comment on stereoscopic vision: Holmes and Horrax write that the localization of objects in the third dimension is "an intellectual operation based largely on non-visual experience, as on the testing and controlling of visual perceptions by tactile and muscular sensations. . . ."[8] W. Russell Brain concurs in this, remarking that the "visual cortex is a receptor for comparatively crude and simple visual sensations."[9] It is difficult to tell whether these neurologists are saying that seeing is not to be equated with the mere having of retinal images, or that stereoscopic vision involves more than just seeing, i.e., more than the mere having of retinal images.]

But for our purposes, of course, the way is clear. When we say that we see the sun this cannot be a comment on our retinal reaction for if it were we ought to be seeing two suns, both inverted. So to this extent at least our remarks about what we see are not remarks about the electrochemical pictures that light radiation forms on our retinas. Apropos of this, recent experiments in the psychology of perception have shown that men will see orange, and say that they see orange, when each eye is exposed individually to a color patch—a reddish colored patch for one eye and a yellowish colored patch for the other.

Another reason for resisting the equation of seeing X with having a normal retinal reaction to X is this:

The picture formed in the macula of the eye is, of course, very small. But we see things, as we say, life size. Mountains, buildings, other men, the sun—all these are seen in a way very different from the way they are recorded on the retina. And so, when we say that we see an object, we are not referring in any way to the picture of that object made within the eye.

Consider also the cases of people under hypnosis, or in a coma, or very intoxicated. The eyes of such individuals may be wide open, and functioning perfectly. But we would not say of these people that they

---

[8] Gordon Holmes and Gilbert Horrax, "Disturbances of Spatial Orientation and Visual Attention, with Loss of Stereoscopic Vision," *Archives of Neurology and Psychiatry*, Vol. 1, No. 4 (1919), p. 402.
[9] W. R. Brain and E. B. Strauss, *Recent Advances in Neurology, op. cit.*

saw the objects before them, however detailed the pictures made on their retinas by those objects. It is very like our situation when we are sleeping. Our ears are open—we have no earlids, as we do eyelids, which close when we sleep. And all the sounds around us are affecting our ear drums just as when we are awake. Yet we do not normally hear anything while we are asleep. Eyelids complicate the question because they close when we sleep, and because we do not usually see when we sleep. We therefore tend to think that open eyes are more important to the question of seeing than perhaps they are. Open eyes are certainly a necessary condition of seeing, but that they are a sufficient condition is very doubtful. The whole controversy of whether fish (who lack eyelids) ever sleep has at times gone afoul over this matter.

But of course we all know perfectly well what it is to be so taken up with a problem or a specific worry that we fail to see what is in front of our noses. We pass friends on the street without seeing them—we stare blankly, as we say. And we would say of such situations that we had not seen Jones, or Smith, even after it was proven that Jones, or Smith, was the full center in our field of vision and that our eyes were open and functioning normally.

Now it is time for a rather revolting limerick:

> There was a young man of East Bosham
> Who took out his eyeballs to wash 'em.
>   His mother said: "Jack,
>   If you don't put 'em back,
> I'll jump on the beggars and squash 'em."

And now an even less delicate reflection on that limerick. Would we wish to say that, when Jack's eyeballs were removed and detached from his brain, he was seeing anything? They might be functioning well enough, for some minutes at least, but do they see? Does Jack see? The answer here is clearly in the negative. But suppose that Jack's optic nerve were highly elastic and remained attached to both eyeball and brain. Would Jack be said to see then? Very probably. It is just here, of course, that the model of the perfect camera breaks down as an aid in the study of the eye, of sight, and of seeing. For, however marked are the similarities between cameras and eyeballs, we would all get the point of saying that cameras are blind—that they cannot see. It is perhaps the too great emphasis on the camera aspects of the eye, and the whole physicochemical chain of events having to do with sight, that has made certain scientists set it down that seeing is just what happens at the retina, *Q.E.D.* For, in truth, after the visual impulse goes beyond the retina, the camera model becomes less and less useful. But, as we have seen, it is only beyond the retina that what *we* call "seeing" takes form. All of which shows that what certain of these scientists call *seeing* is not what is usually called *seeing*.

Two more points: There are, of course, plenty of cases of blindness where the retina continues to function quite normally. And naturally we will not say, despite a person's inability to see, that he sees anyway because his retina undergoes the normal reactions to light stimuli. The people who hastily equate "seeing X" with "having a normal retinal reaction to X" are usually as well informed as we about these cases of cerebral blindness; this makes their legislation more difficult to understand. For only a philosophical prejudice which wants strictly to divide our immediate visuo-sensual reactions to physical stimuli from all that happens with respect to that reaction within our central nervous system —calling the former *seeing*, and the latter God knows what ("association," "coordination," "integration," "interpretation," "inference")— could tempt one to overlook these cases of cerebral blindness, traumatic blindness, etc., where the actual organs of sight are not at all defective.

Let us agree not to take this view of *seeing* too seriously, then. Incautious people may say of seeing that it is nothing but retinal reaction, if they please. But this is not what most of us have in mind when we speak of seeing the sun, the hills, and each other. And that must be an important consideration in this as in all philosophical inquiries. Part of learning what it is to see is coming to appreciate what is meant by the word "see" in the ordinary contexts in which it is so often used. I am unable to understand what point there would be to proceed otherwise— it would be like trying to discover what a *chap* was without considering the contexts in which we usually make reference to chaps. We might conclude such an inquiry with the pronouncement either that chaps are a kind of English undergraduate or that they are a kind of cowboy apparel—positions for which there is about as much justification, and the same kind of justification, as there is for saying that seeing X is just a normal retinal reaction to X. There is more to seeing than meets the eye.

Which brings us to a second answer to the problem of saying what it is that we see, and what it is that our astronomers might be said to see. This answer is a very much more subtle answer to the question than was the mere equation of seeing with retinal stimulation. It involves little or no reference to neurology, physiology, or psychology—though you will perhaps feel that some such reference might have been very much in order. Nor does it involve any reference to the ways in which we usually use words like "see," "observe," "sight," etc.—though I for one certainly feel that the position could have been inestimably strengthened had this reference been made occasionally. This is a strictly philosophical and logical theory about the nature of sensation, primarily visual sensation. It is called alternatively, *"the sense-datum theory"* or *"phenomenalism."*

Let's bring ourselves face to face with this elaborate answer to the question "What do we see?" by considering again our two astronomers.

We are agreed that, whatever we end up saying they see, what they see is not equivalent to what happens on their retina. So that will no longer trouble us. But consider: They say that they see *the sun*. The sun, however, is a star thousands and thousands of miles in diameter. We cannot see, at a given instant, every side and aspect of the sun. What we see is a brilliant discoid shape, so runs the argument. Speaking of what we see *as the sun* requires a long logical jump from what we are in fact visually aware of. And when we say that we see a round penny, when the coin is on a table some distance from us, what we are visually aware of is a copperish oval shape from the apprehension of which (by dint of our experience and teaching) we infer that we are looking at a round penny.

This is, of course, different from the view we have been considering —though in my opinion the two are often very badly confused. For here there is but *one* brilliant disc and but one copperish oval in question, instead of the awkward two that result when we press the crude "seeing = retinal reaction" equation to its limit. And our disc and oval are right side up and not inverted, which was the other embarrassing consequence of the earlier theory.

There is a much more important difference, however. Here the exposition of the theory requires essential terms like "awareness," "consciousness," etc. For, though the penny and I sit face to face for hours, if I am distracted by worry, or if I am in a coma, or drunk, or under hypnosis, or if there is a lesion or other malformation of my visual cortex

—if, in a word, I am blind to the presence of the penny, then, no matter whether or not my rods and cones are reacting in a wholly normal way to the light reflected from the penny, I am not having the visual sense-datum of the penny, namely the copperish oval shape, and hence I would not be said to be seeing the penny. This is a vital consideration, and it contributes strength to the theory that many critics have overlooked. For no defender of the sense-datum position—not Lord Russell, Professor Broad, Professor Price, or Professor Ayer, nor any of their philosophical ancestry or progeny—would insist that we could be having a visual sense-datum of, say, a brilliant yellow-white disc, without an accompanying awareness of the brilliant disc. Hence, for the sense-datum theorist, retinal reaction alone is not enough. If, while I am deep in thought (this happens occasionally, to all of us), the sun should peep through the fog or smog (this also happens occasionally, even in London or Los Angeles) and I do not notice this at all and can remember nothing of it afterwards when questioned, then the answer must be that I never had the sense-datum at all—I never saw the sun, indeed. In other words the role of the brain, or the mind, or the intellect, or the intelligence (call it what you will) is not eliminated with the sense-datum theory of seeing as it was in our earlier theory, however much it may be underestimated.

So where Jack's eyeballs might themselves have been said to see on our former account, here at least it is Jack that is doing the seeing, if, that is, he is mentally awake.

Now what is the nature of what he sees? What sort of experience is it to have a visual sense-datum?

We know from common experience that the world does not go dark when we shut our eyes, nor do flowers lose their scent when we catch cold. So apparently we require ways of talking about what does go dark when I shut my eyes, and what is lost when I catch cold. We need a way of speaking about what happens when one crosses one's eyes and "sees" double—for just because you cross your eyes that does not make me twins. And yet how to describe what is seen in such a case. . . .

The sense-datum theory, then, suggests that we might talk about sensations *simpliciter* by talking about *looks, appearances, glimpses, scents, flavors, sounds,* and so on. We say we see the sun. But all we really have to go on is the *look* of a brilliant yellow-white disc. We say we see a round penny, but the cash value of that in terms of sensation is just the appearance in our visual field of a copper-brown oval patch. We think we hear the telephone. But all we really hear is ding-a-ling.

Sense-datum philosophers, and some scientists, have come to hold that it is necessary to employ these idioms in order to distinguish the contributions made to our observation of common objects by our sensations from those made to it by intuition, inference, memory, conjecture, habit, imagination, and association. So in a way the motives behind speaking of a visual sense-datum rather than the seeing of a physical object are similar to the motives behind speaking of a retinal reaction rather than the seeing of such an object. Both ways of speaking are directed to distinguishing between the actual impressions the world makes on our organs of sense and the intellectual contribution that we make in our appreciation of those sense impressions. This is required, apparently, because our ordinary idioms of seeing, observing, witnessing, etc., notoriously make no such distinction. The difference between these ways of speaking is one of relative subtlety and emphasis. The retinal-reaction point of view can be shattered, or emasculated, by reference to one or two simple matters of fact. No such course of action is open to opponents of the sense-datum theory; indeed the whole point of treating it as a logical or philosophical theory, rather than a scientific one, is to stress that its creators have spared themselves no pains in systematically removing any trace of matter-of-factness from their position. And, in doing this, perhaps they have pulled the props from beneath themselves—but of that more later.

Having a visual sensation, then, is described in sense-datum terms as the getting of a momentary look, or visual appearance, of something. The sensation the astronomers must have when facing eastward at dawn is of an appearance of what they call "the sun"—an appearance that consists of a brilliant yellow-white disc. That's what they see. The sun itself is never seen.

But what is it to get this momentary look?

Initially we may note that a look of the sun, i.e., the way the sun looks, or appears, momentarily, is not an astronomical or meteorological event. Everyone, or anyone, can witness the sun's rise, but not everyone can witness the momentary look that I get of the sun. Nor can anyone else witness the momentary appearance that either of our astronomers gets of the sun—which is, in effect, to say that neither astronomer can witness what the other astronomer sees as his momentary, private picture of the sun. Just as we cannot have each other's toothaches, or each other's tickles, so we cannot share visual sense-data.

Next, this glimpse of the sun above the hill is described (in terms of the theory) as a momentary patchwork of color expanses in the astronomer's field of view. But this is no ordinary patchwork of color in the sense that tapestries, paintings, and billboards are patchworks of color. The visual appearance of the sun above the hill is not to be thought of as if it were a surface of an ordinary flat object like a tapestry or a billboard: It is simply an expanse of color, not an expanse of colored cloth or paper. It occupies the astronomer's private visual space, despite his temptation to attach it to the common objects of physical space—the sun and the hills.

And what, then, is it for the astronomer to have his visual sensation of the sun over the hills? The phenomenalists' answer is that the astronomer simply *observes* or *sees* an expanse of color patches. Indeed it is regarded as not only allowable, but illuminating, to say of people that they do not really see sunrises or hear telephones ringing; they only see color patches and hear ding-a-lings. Sense-datum theorists occasionally concede that there is a vulgar sense of "see," "hear," and "taste" in which people may say that they see sunrises, taste oranges, and hear larks. For theoretical purposes, however, we should use these verbs in a different and refined sense. We should say instead that we see color patches, taste flavors, and hear whistles.

The theorists claim that this way of talking helps with the sorts of illusions that are familiar to us all. For now we can construe in a new idiom the squinter's report that he sees two candles when there is only one to be seen. The squinter really does see two "candle-looks," and the drunk really does see one "pink elephant-appearance." Their error is simply one of extrapolation—they were not justified in supposing that two physical candles or one physical pink elephant existed in addition to the appearances of these things. We would be wrong if, on seeing the penny on the table, we supposed that the government was minting elliptical coins. But we are not wrong in insisting on the existence of something elliptical and copperish whether or not there is any physical object with which it can be correlated exactly. It is always hazardous to argue from what is in one's field of vision to what there is. But it is never hazardous to assert what it is that occupies one's field of vision—

indeed, in a queer sense of "wrong," we could never be wrong in our reports of what occupies our field of vision. How could we misidentify the objects of our private experience? (You may remember in this connection Descartes' *clear and distinct ideas*. Here was the certainty the Cartesians sought—in experiences which could not be misinterpreted.)

As for our astronomers, they say that they see the sun. But of all people they should know that there is an eight-minute delay separating the moment when a quantum of light radiation leaves the sun and the moment when that quantum impinges upon one of the light-sensitive cells of their retinas. And, of course, anything could have happened to the sun in that time interval. It could have been snuffed out like a candle—and yet the astronomer will say that he sees the sun.

Moreover, our vulgar uses of words like "see," "observe," "notice," etc. incline us to ascribe to physical objects properties which they may not, *or need* not, have. An electromagnetic wave—such as those initiated by nucleonic breakdown within the sun—is neither hot nor bright; yet these are just the properties we ascribe to the sun, the source of this radiation. But of course the hotness and brightness are just our physiological and psychological reactions to such radiation and not necessarily a valid index of the properties of the real sun.

So just a moment before dawn, when our astronomers are still in relative darkness, the real sun is already above the horizon. Only, the electromagnetic emanations have not yet reached the earth, nor our astronomers' retinas. And even when this does happen, the hotness and brightness induced are induced within us, or within astronomers. These are only indirect clues as to the nature of the real sun 93,000,000 miles distant. Hence the mental experience that the astronomers signal by their pronouncement "I see the sun" is different in time and in quality from the physical event to which they, in their vulgar way, refer. The sun is only the cause of their experience. They do not see the sun.

And so with all visual experience, and indeed all sense experience. We see only the effects of things, never the things themselves—or in Kantian terminology the things-in-themselves. Seeing a candle is, in principle, like seeing the sun. Both are experiences different in time and in quality from the sun and the candle themselves.

Lord Bertrand Russell puts the point with his usual succinctness by remarking that when the brain surgeon is operating, what he sees is not something in another person's brain, but something in his own brain—namely, the visual appearance of another person's brain.[10]

It is from such primary experiences as these, therefore, that we make our inferences to the nature of things as they really are. But sense-experience is never more than a pattern of clues about the physical world—some very reliable, it is true. We could always be mistaken in our intellectual employment of those clues. And this contrasts sharply with the sense in which we cannot be mistaken in our application of the existence and the nature of the clues themselves.

Dr. W. Russell Brain (whose camp is probably pitched midway between a retinal-reaction interpretation of *seeing* and a sense-datum interpretation) twists the point fully home by analogy with the housemaid imprisoned in the cellar.[11] Her only information about the outside world comes to her through the panel of indicators and bells hanging on the wall. Her entire knowledge of what lies beyond is the result of inference based on these indications.

Compare this with what Dr. Grey Walter of the Burden Neurological Institute says of the nervous system: "[It is] a generalized transmission system of unknown properties, a sort of 'black box' into which signals are injected and from which other signals emanate . . . within the Black Box of the cranium is some device capable of sampling incoming sensory signals . . . [it is something] like a totalizator that automatically works out the odds in favour of a . . . series of events being worthy of attention . . ."[12] So, too, our astronomers—on the indications given to them by their eyes and brains they infer that the sun is above the horizon. But theirs are inferences nonetheless and risky in a way in which the actual visual indications themselves involve no risk at all.

So while the earlier-examined position which equated "seeing X" with "having a normal retinal reaction to X" allowed us to answer the question "Do the astronomers both see the same thing?" in the affirmative

---

[10] *The Basic Writings of Bertrand Russell,* eds. Robert E. Egner and Lester E. Denonn (New York: Simon and Schuster, 1961), Part XVI, Chapter 65 "Physics and Neutral Monism," p. 607.

[11] W. Russell Brain, *Mind, Perception and Science* (London: Blackwell's, 1951), pp. 7–8.

[12] (Paraphrase by Hanson) See Chapter VII, W. Grey Walter, *The Living Brain* (New York: Norton, 1953 (1963)).

76 | *Chapter 4*

(for our astronomers are assumed both to have normal retinal reaction), the sense-datum theory obliges us to answer in the negative. For no two people can see the same thing when what is seen is construed to be a private mental event, an individual apprehension of the visual effects of some external cause. Of course, it is reasonable to assume that these private observations are largely similar for different people placed in the same circumstances. But this cannot be proved in any logically acceptable way. The sense-datum theory, then, allows us to *assume* that our 13th Century and our 20th Century astronomers see things at dawn that are mostly similar in their geometry and coloring, but on no strict account can they be said to see exactly the same thing.

In the next chapter I will try to get some missiles through the defenses of this sense-datum account of the nature of *seeing*. Indeed I hope to chop the legs from under it. I trust I will not discover that I have presented the position here with such care that it will prove there to be invulnerable.

5 | Seeing the Same Thing

IT WAS clear in the preceding chapter that if one interpreted "seeing the sun" as "having a normal retinal reaction to the sun," then our 13th and 20th Century astronomers do indeed see the same thing. But we noted a good many objections to interpreting *seeing* in this way. We found many cases where it could be established that a person's retinal reaction to the sun was altogether normal, and yet the person would not be said to be seeing the sun—because of distraction, hypnosis, intoxication, somnambulism, paresis, etc.

If, however, one interprets seeing the sun as seeing an appearance, or a look, of the sun—if saying "I see the sun" is regarded as but a crude and indefinite way of saying "I see a brilliant yellow-white disc from which I infer that the sun is up"—then the astronomers cannot possibly see the same thing. What each of them sees is private and not accessible to the other. The 13th Century man's awareness of a brilliant disc is something no one else, of whatever century, can share (for if they could it would no longer be *his* awareness, nor his visual experience). Our assumption is, of course, that the two astronomers have roughly *similar* visual sense-data when they confront the same sun under similar conditions. But this is, like most assumptions, not strictly provable; whereas it is, perhaps, strictly provable that no two observers can have the *same* visual sense-data.

That is as far as we got. We did not consider objections to this view, and to objections we must turn now. For despite the fact that it has been supported in one form or another by some very considerable scientists and philosophers of science, the view is profoundly objectionable both in its presuppositions and in its consequences.

Initially we may note that the sense-datum view outrages common sense. In itself this need not amount to very much. Common sense has been outraged before—by Copernicus, Galileo, Newton, Darwin, Pasteur—and usually for the good of mankind. But the issue in these dramatic cases was always one of *fact*. There is no dispute about the facts here. There is no appropriate experiment that would clinch the sense-datum theorist's position. By experiment and observation we can look and see whether Copernicus, Galileo, Newton, Darwin, and Pasteur were correct. But what do we look for when we are after sense-data? Agreed, the question "Are there such things as sense-data?" reads very like the questions "Are there such things as microbes?" and "Are there such things as light corpuscles?" But a *grammatical* similarity can often conceal important differences in logical structure.

Sense-data, if such things there are, are entities which—far from our having to seek them through the office of an experiment—cannot but be seen any time we see anything at all. So they are not on a footing with the entities that have been said to exist by great men in the history of science, despite the fact that common sense often opposes both. For the flouting of common sense need be of no great importance in matters of fact. Indeed, to judge by our history, it is an almost essential step for our intellectual and material progress.

But the flouting of common sense in philosophical matters, and particularly as regards the phenomenalist controversy, is of the utmost importance. Even in philosophical matters common sense may prove the loser. But the implications of this are of much greater magnitude than when common sense loses out on matters of fact. For the virtue claimed by the sense-datum theory is that it provides a more precise and more fruitful way of talking about our experience. But does it?

At dawn *I* wish to say that I see the sun: I wish to say that I feel its heat, that I am dazzled by its brightness. I would expect our astronomers to react in this way too. Is it just lack of sophistication, is it just crudity and vulgarity that make us wish to speak so?

Suppose that we, and our astronomers, witnessed the dawn and then commented on the sun's brilliance and its warmth. What would we say in these ordinary circumstances were a member of our party to deny these things? If he denied that the sun was warm and bright, if he denied even that what we see is the sun, we would of course question his seriousness, or perhaps even his sanity. Of course we see the sun, all of us. This is what we would say, and this is what we would mean. And no matter how successfully some philosopher might show such an experience to be composite, no matter how well he might point out the differences between seeing the sun and seeing yellow-white *color* patches, it is an incontestable fact that in ordinary situations we use the word "see" when we are speaking of seeing the sun. This is a typical case of our use of the word "see" and it seems as futile to counter this with the claim that we can never see the sun as it would be to say that no one ever plays chess, for what is *called* chess is really a kind of draughts played with more intricately carved tokens. Well, however that may be, that's precisely what we mean by "chess" and it is for *us* to decide what is and is not chess, however true it may be that chess is just a complex version of draughts. So too it is precisely from such situations as seeing the sun or seeing the hills that the verb "to see" draws one of its most important meanings. What sense can there be, therefore, in saying that these are not genuine *seeings,* but only pure seeing *plus* a lot of intellectual plaster in the form of inference and interpretation? Describe it how you will, show its complexities *ad nauseam* if you please, but it is this and primarily this that we call *seeing.* And in this there is about as much chance of our being in error as there is for Mrs. Jones' being in error when she has her son christened "Tom." How could she be wrong? Is it not for her to name her own child? What scientific or philosophical theory will prove her to be in the wrong?

I will be returning to this point regularly. For, while what ordinary people usually mean when they speak of electrons, or waves, or genes, or functions is of little or no importance to the understanding of such purely scientific concepts, it is of maximum importance when we are concerned to understand a general philosophical concept like *seeing.* Why this is so will become increasingly clear, I hope, in what follows.

The sense-datum theory collides not only with common sense and with what is ordinarily meant by words like "see," "hear," and "taste," it collides with itself, or rather the theory collides with its own assumptions. Thus, our reasons for holding the theory to be true—e.g., such things as that there is an eight-minute difference between what the sun does and what we perceive the sun as doing, that we can never confront a three-dimensional object on all sides at once (which makes our perception of the object different from what the object is in itself)—cannot even be *held* if the theory is true. For the reasons are themselves grounded on a very homely distinction between the way things are and the way they sometimes appear to be, the way the candle flame is in the open and the way it appears when viewed through stained glass. By heaping all of our experiences of objects into the basket labeled "mere appearances," as it must do, the theory causes the very reasons that support it to collapse. The distinction on which the reasons rest cannot even be made if the theory is true.

The observer is never aware of anything but the pictures on his private screen and the sounds of a private loudspeaker. If he speculates about what lies beyond (and surely the reasons advanced for the theory require this speculation) then nothing could count as evidence, or as reasons, one way or the other. If the theory is true, there is no ground for saying that we have real heads, or even *real* sensations.

But are things really this bad? All the theory requires is that our ordinary confidence about material objects be largely inferential, the inferences being always from sense-data. The answer is, *"Yes, things really are this bad."* When we infer from the dark clouds to the imminence of rain, or from the brightening eastern sky to the nearness of dawn, we are appealing to past experience in a manner very unlike what we are asked to believe is the case when inferring from sense-data to material objects. For what past experience could guarantee this latter inference? The theory itself gives the answer: None! And with that answer it cuts away the very reasons that were invoked in its support, reasons all of which required a contrast between sense-data and material objects. Unless a sense-datum theorist can get his audience to distinguish their perceptions from the physical causes of those perceptions, he cannot advance one cogent reason in support of his theory. But it is a consequence of his theory that the physical causes of our perceptions never *can* be known; so the contrast required cannot be made by a sense-datum theorist unless he borrows from non-sense-datum theories of perception.

Let me loft another dart at the phenomenalist balloon in the following manner:

Consider a sea captain about to leave port. He gives the command "Up anchor!" His men, however, proceed only to turn the winch, or perhaps they simply pull on the last link of the anchor chain. Have the men disobeyed the captain's orders?

If the captain insisted that they had, we would think he was ready for a shore job. For only a philosophical Captain Bligh could fail to see that, in these circumstances, pulling on the last link of the anchor chain just *is* pulling up the anchor. It would be absurd to say that pulling the last link of the chain was an altogether different thing from pulling up the anchor.

It is, perhaps, from a model something like this that both of the theories we have examined spring. When we consider the chain of events beginning with the nuclear fission within the sun and proceeding through the emission of light radiation through space, then through our terrestrial atmosphere to the eyes of our astronomers, through the cornea, aqueous, iris, lens, and vitreous of those eyes, ultimately to impinge on their retinas, we may be tempted to hunt for another physical event to answer the name of "seeing": an event in principle just like all the events preceding it. As we saw, one possibility is to locate *seeing* in the electrochemical reaction of the normal retina to light radiation. The objections to this view are in large measure avoided by the sense-datum theory, which requires that *seeing* involve a conscious awareness of what is taking up the visual field. Nonetheless, the sense-datum theory *does* treat this seeing as just another event in the chain beginning with solar nuclear fission. But, because most sense-datum theorists *are* philosophical Captain Blighs, they insist that what is *seen* is just the last link of this chain of events and never the cause of that last link. We never see the sun, nor pennies, nor trees, nor hills. All we see are brilliant yellow-white discs, copper-colored ellipsoidal patches, fuzzy green outlines, and dark rolling lines in our visual field. To my mind this is precisely like saying that anchors are never pulled up, only the last links of anchor chains.

In a way, sense-datum theorists *invent* the last link of their perceptual chain so that it will fit in with the causal-chain model they think they have inherited from natural science. What the theorists have not appreciated is that the last link in *seeing* is not something detectable (as are the events taking place in the sun, in space, on earth, and in the human eye). The last link in *seeing* is itself the detecting. Trying to treat *seeing* as something itself *seeable* is what leads to the account set out before that so offends our common sense.

The last link in visual perception is an event utterly different in kind from the other links in the chain—if, indeed, talking about links in perceptual chains is a permissible idiom at all. The last link in visual perception is not the perceiving of the final physical event in a chain of events started by the emission of sunlight. It is, rather, the perception of what started the chain of events—namely, the sun—just as getting a ship aweigh is pulling up the anchor, and not merely the ultimate link in the anchor chain.

As long as we are dominated by the causal-chain figure in perceptual matters we tend to look for the "what else" that makes seeing possible.

In other words, the major mistake of the sense-datum theory is its assimilation of the concept of *sensation* to the concept of *observation*. For the theory holds that seeing the sun consists in one's finding or intuiting a sensum, namely a brilliant yellow-white disc in one's private visual space. Having a glimpse of the sun, then, is explained in terms of having a glimpse of something else, a patchwork of colors, a white-yellow disc. But if having a glimpse of the sun entails having at least one sensation, then having a glimpse of a white-yellow disc must again involve at least one appropriate sensation, which in its turn must be analyzed into the sensing of yet an earlier sensum, and so on *ad infinitum* and *ad absurdum*.

Of course, every case of seeing is a case involving the having of at least one sensation. Indeed, this is part of the force of "seeing." But this no more means that seeing just *is* the having of a sensation than it means that bricks are houses, or letters words, or stitches very tiny suits of clothing. Visual sensations likewise are not *seeings*, though any case of seeing entails our having had at least one visual sensation. A visual sensation of a brilliant yellow-white disc is not itself a seeing of the sun, though any case of seeing the sun entails our having had at least one visual sensation of a brilliant yellow-white disc. And, of course, a strict sense-datum theory of "seeing" entails that no two scientists can see the same thing, a view which should jar us as much as hearing that two scientists could not share the same opinion, or make the same assumption, or the same inference.

A defender of the sense-datum position may deny that seeing a brilliant yellow-white disc entails having had a previous sensation *ad infinitum*. He may urge that the expression "having a sensation" is merely a vulgar way of reporting a simple intuition of a special sensible object, i.e., a brilliant yellow-white disc. And to say that one intuits such an object does not entail his being sensitively affected—as indeed people who are hypnotized, intoxicated, or psychotic may intuit such objects without prior sensitive stimulation.

But this defense does no more than to explain the having of sensations as the *not* having of any sensation. It avoids an infinite regress of the cited form by the heroic device of suggesting that sensing is a purely cognitive process which does not require its owner to be susceptible to stimuli, or to be to any degree sensitive. By treating sensation as the simple observation of special objects, viz., *sense-data,* it destroys the very concept it was professing to elucidate and it makes nonsense of the concepts of *seeing* and *observing* (since these entail the concept of sensations which are not themselves *seeings* or *observings*).

The sense-datum theorist may counter: "Whatever the natures of sensation, seeing, and observing it remains a fact that in seeing I am directly presented with patchworks of colors momentarily occupying my field of view. That sense-data are sensed is beyond question and independent of theory. In the strictest sense of 'see' what I see are two-dimensional color patches; not the sun, the hills, and the trees, but only the looks or visual appearances of the sun, hills, and trees. The squinter may not see two candles, but he sees two bright somethings; these somethings are the candle-looks, the sense-data. The theory is not inventing fictitious entities; it is only drawing attention to our immediate objects of sense, objects which we ordinarily ignore in favor of talking about common physical objects like candles, hills, trees, and the sun."

Consider again the penny on the table. It is agreed that it is round. And it is agreed that to you it looks elliptical—though there are further questions to be asked about this. Is it the case that the truth of your report that the penny looks elliptical implies that you are really scanning an object of sense which *is* elliptical? Something which, though not the penny itself, can be called the *look* or the visual appearance of the penny?

A person without a theory might feel no qualms at all in saying that the round penny looks elliptical, or even looks as if it were elliptical. But I predict that he surely would feel qualms about saying that he was seeing the elliptical look of a round penny. We seldom speak of seeing the look of the sun, of gazing at views of the hills, of catching glimpses of treetops, or of scanning the visual appearances of candle flames. For to speak in this way would be like moving from talking of eating cheese and talking of nibbling cheese, to talking of eating nibbles of cheese. "Eating nibbles" strains our understanding because "nibble" is already a noun of eating. So too "seeing looks" strains us because "look" is already a noun of seeing.

When a person ordinarily says that the tilted penny has an elliptical look he means that it looks as an elliptical but untilted penny would look. He is not saying, "I see a flat elliptical copper-colored patch" but, "I see what might be an elliptical and untilted penny."

All of which is to say that the unsophisticated sentence "The penny has an elliptical look" does not express a basic truth of perceptual experience, as the sense-datum theory assumes. It really is an expression of a fairly complex proposition. It applies to the actual look of a penny a rule or a recipe about the typical looks of untilted elliptical pennies (if such there were). In the same way, when we say that someone has a professorial appearance, we are not suggesting that there are two kinds of professorial beings, some men and some appearances of men. Similarly, there are not two kinds of elliptical objects, some pennies and some looks of pennies. A man may look rather as some professors look. And a penny may look rather as some elliptical pennies look (if such there were).

I have been denying that having a visual sensation of a brilliant yellow-white disc is a sort of observation or seeing. The sense-datum theory is often announced as if it were the report of a newly discovered class of objects, but it is only a misapprehension of the force of a familiar range of statements about how common objects are sometimes found to look. Hence all the pseudo-technical language of the sense-datum theorist with his "object of sense," "sensible object," "sensum," "sense-datum," "sense-content," "sense field," and "sensibilia"—as well as his "direct awareness" and his "acquaintance"—is just a rather ugly monument commemorating the attempt to make the concepts of sensation do the work of the concepts of observation. And this is like trying to understand what kind of *walking* arriving at one's destination is, or what sort of *drinking* quenching one's thirst is. Asking what sort of *sensation* seeing is, is asking the same sort of question—a question upon an answer to which the whole sense-datum theory rests.

⇢ Both the theories we have so far considered—the view that construed "seeing X" as "having a normal retinal reaction to X" and the view that construed "seeing X" as "having a private look at an appearance of X"—try to make us think of ourselves as imprisoned inside of our nervous systems. We are apparently just like turtles whose shells have closed all around us. All we know of the outside world are the taps on our shells and the light and dark patches which show through to the inside. Remember my reference in the previous chapter to Dr. Grey Walter, who talks of the black box of the cranium. These theories encourage us to think of ourselves as far removed from the world of physical objects. But, of course, the world of physical objects is right under our noses.

So when our astronomers say that they see the sun, they mean just what they say. They are talking neither about their retinal reactions nor about the contents of their private visual fields. They are talking about the sun, our local star 93,000,000 miles away. And so the answer to our question "Do the 13th Century and 20th Century astronomers see the same things in the east at dawn?" may be the ordinary, straightforward answer of science and of common sense, "Yes, they both see the same thing, they both see the sun."

But now our troubles really begin. Because though our astronomers may say in a single voice that 'they see the sun, only a very few questions will reveal that they do not mean the same thing by the word "sun." Nor, as I shall argue, do they see the sun in the same way. There may even be reason for answering our question as follows: "No, they do not see the same thing despite the fact that they both see the sun." So while scientists are right in saying that they see physical objects—and seeing physical objects is neither just having normal retinal reactions nor experiencing the private "looks" presented in our visual field—we cannot leave things at that. For this would lead to some very uncomplicated (and unintelligent) views as to the nature of laboratory science, views to the effect that *seeing* is just an opening of one's eyes and looking; that *observing* is simply being in a position to watch when a phenomenon takes place, and the *facts* are just the things and situations in the world that we see when our eyes are open, the things we trip over when our eyes are closed; that *hypotheses* and *theories* are just guesses differing only in degree of systematic ingenuity and riskiness. These are all too prevalent points of view, which must be challenged. And an adequate appreciation of the concepts of *seeing* and *observation* is indispensable to the making of such a challenge. To this we will turn now. But here it is appropriate to quote again the words of Wittgenstein, who said that "we find certain things about seeing puzzling because we do not find the whole business of seeing puzzling enough." [1]

What is it for two people to be looking at the same thing and yet

---
[1] Wittgenstein, *Investigations,* p. 212.

not see it in the same way? What is it for a man to have the same retinal reaction as another man, and even the same, or at least very similar, visual sense-data, and yet not see things as the other man sees them? The answer that would have been given by the two sorts of theorists we have already considered would probably run like this: The *seeing* of X is only a normal retinal reaction to X, or alternatively, the *seeing* of X is only the having of a visual sense-datum of X, a look or an appearance of X. Of course, what we make of what we see may be a very different matter. We may *interpret* differently what we see. We may bring all sorts of considerations from our individual past experience, our training, etc. to bear on what we see. But that is just an intellectual excrescence upon the business of *seeing*. It is a kind of rational plaster stuck on to the hard stones of visual sensation; it is not really relevant to an analysis of pure *seeing* and pure *observation*.

I should like now to bring forward a series of considerations which are calculated to show this two-phase account of the nature of observation in a bad light. It will become clear that the treatment of observation as a tandem operation of first *seeing* and then *thinking* is at least cumbersome, and possibly even harmful as well. Wittgenstein wrote, "If you try to reduce [the relation: physical object-sense impressions] to a *simple* formula you go wrong." [2]

What follows is intended to show that the theories about seeing we have so far considered are too simple by far to be adequate.

Consider first this well-known figure:

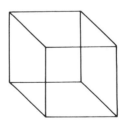

Figure 1

It will impress all normal retinas in the same way; i.e., it will etch on the retina an electrochemical copy of what is here on the page. And the

---

[2] *Ibid.,* p. 180.

sense-data each of us have relative to this figure will be roughly similar. For the figure, being drawn in the plane, will appear similarly in our private visual spaces. But do we all see the same thing?

Is it not notorious that some people will see this as a perspex cube from ahead and below, while others will see it as from ahead and above? Others will see it from directly above, as a kind of irregular polygon, while still others will just see a configuration of crisscross lines in a plane. It will be a solid transparent block to some, a hollow transparent walled box to others. Others will see a well-cut diamond shown from above, a wire-framed figure will appear to still others, and so on depending. . . . Depending on what? This will come out gradually.

All normal retinas will react to this in the same way. We have no reason for supposing our private visual impressions—our sense-data— of this figure to differ from person to person. But do we all see the same thing? And if the answer is "Yes!" how are these differences to be accounted for? By saying that they are merely different interpretations of what all normal observers would see in common? But I am not applying different interpretations when I report that I see *Figure 1* now as a box seen from below and now as a box seen from above, now as an ice cube and now as a configuration of crisscrossed sticks. I am conscious of doing no such thing. I *see Figure 1* differently, that is all. And if you insist that the seeing involves the interpretations à la bonheur, then having a different interpretation of *Figure 1* just *is* seeing it differently.

But it is questionable whether this slightly mechanical sophistication is either necessary or helpful. In any case it does an injustice to the occasionally useful word "interpretation." For there are fairly clear-cut cases where the word applies and where it does not. Thucydides presented the facts of history quite objectively. Herodotus put an interpretation on them. The Greeks discovered frictional electricity, but it took 2000 years of western science to find an interpretation for that discovery. In other words "interpretation" is a word that does not apply to everything. But now it is suggested that the reasons for variations in the reports of what is seen when various people confront *Figure 1* are the different interpretations that are part of the *seeing*. This is, in effect, to say that everyone is always interpreting, no one is ever not interpreting, particularly when up against *Figure 1*. This, however, is like saying that *all* automobile drivers are really learners and that therefore vehicles should be built with permanent "Student Driver" signs front and back. "Student

Driver" signs would then have become useless for our ordinary distinctions between beginners and expert drivers, distinctions we should nonetheless have to make whether or not "Student Driver" signs were useful in making them. This is reminiscent of a good many recommendations by philosophers, all of which get their momentary force from the straining of a common contrast. Thus it has been said that *everyone* acts from self-interest, that no one is really *free,* that no empirical proposition is ever *certain,* that such propositions are all *hypotheses,* that criminals are really sick and hence not culpable, etc. Some of these things are, of course, worth saying. But these ways of putting them are nothing if not mystifying. And it is nothing if not mystifying to hear that, although different reactions to *Figure 1* might not in themselves be due to overt acts of interpretation put on the sense impressions of, and by, particular individuals, it is nonetheless the case that these interpretations are "there" somewhere doing their work. This is just a way of putting it that, since the *seeing* cannot be different, the differences in observer reaction must be due to interpretation.

Wittgenstein considers the philosophical pronouncement that " 'I see the figure as a box' means: 'I have a particular visual experience which I have found that I always have when I interpret the figure as a box, or when I look at a box. . . .' " To this Wittgenstein replies, "But if I meant this I ought to know it. I ought to be able to refer to the experience directly, and not only indirectly. . . ."[3]

It seems to me pointless to adopt the cogs-and-wheels, wires-and-pulleys accounts of some philosophers on this issue. There is no question in *Figure 1* of putting visual grist into the intellectual mill. We do not begin with the visual sensation and only then turn our theories and interpretations loose on it. In a most important way our theories and interpretations are in the *seeing* from the outset. *Figure 1* is seen as a box from underneath or from above, or as an octagon, or as a flat, flat-line drawing, or first one of these then another—but not all of them at once. As Wittgenstein puts it: "But how is it possible to *see* an object according to an *interpretation?* The question represents it as a queer fact; as if something were being forced into a form it did not really fit. But no squeezing, no forcing took place here."[4]

---

[3] *Ibid.,* pp. 193–4.
[4] *Ibid.,* p. 200.

Other figures to which all these remarks apply are these:

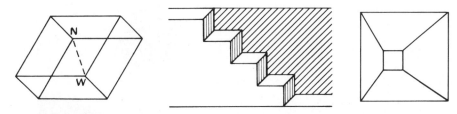

Figure 2

With each of these there are at least three possibilities open to us. We can see these figures in either a convex aspect or a concave aspect or as a flat-line drawing; or first one, then another—but not any two simultaneously. And we might ask with Wittgenstein, "Do I really see something different each time, or do I only interpret what I see in a different way?" He answers: "I am inclined to say the former. But why? To interpret is to think, to do something; seeing is a state. . . ."[5] It must be clear that the differences between our several appreciations of these figures are not easily accountable in terms of the different things we think about while looking at them. Because when I, for one, see the staircase now from above and now from below, I am not thinking anything different appropriate to each case. I am not thinking at all. And it is this that inclines me to say I am not interpreting at all. I am just seeing, now a staircase as from above, now a staircase as from below. And if the word "interpretation" is thought to be essential to making necessary distinctions here, then let us say that the interpretation is a component of the seeing, and not an operation tandem to the seeing.

Let us conclude with two well-known figures from experimental psychology:

For those of you who see (a) as a duck, let me point out Bugs Bunny; and for those of you who see only a rabbit, let me point out Donald Duck. *Figure 3* (b) is complex in the same sort of way. Some of you will see an old, rather hag-like individual here; others will see a very young, stylish woman looking away. If we call (a) the duck-rabbit, we might call (b) the wife-mother-in-law.

---

[5] *Ibid.*, p. 212.

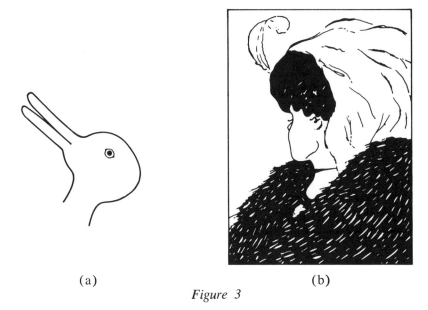

(a) (b)

Figure 3

All normal retinas would react similarly to these, would they not? The visual sense-data evoked in us by these would be likely to be similar, would they not? Would not the "looks" of (a) and (b) come to us all in much the same way, just as the "look" of a penny comes to us all in much the same way? Nonetheless some people will see a rabbit, not a duck; others will react in the opposite way.

Some will see a mother-in-law in (b), others a young wife. And it has been found that some people can *only* see these figures one way or the other; much as when we have at last solved the child's puzzle, having found the face in the branches of the tree, we can never see the tree any other way thereafter. The face is always present in the tree.

Now given two normal observers—normal retinas, relatively normal brains—given that one can only see a duck and a wife, and the other only a rabbit and a mother-in-law (in *Figure 3*), do they see the same thing?

Do the 13th Century and 20th Century astronomers see the same thing? We will pursue this exciting theme further. But in the meanwhile here is an old parlor-game question with a point: Is the animal we call the *zebra* black with white stripes, or is it white with black stripes?

6 | Seeing and Seeing As

IN THE last chapter we encountered four figures—a cube, a rhomboid, a staircase, and a tunnel—all of which displayed the phenomenon of reversible perspective. We also considered two drawings which, besides showing some variability in perspective, were marked by shifts in organization, or in aspect. These were called the "duck-rabbit" and the "wife-mother-in-law" respectively. In each case the question was asked, "Do we all see the same thing?" For there was no question here of a differing retinal reaction. The stimulus pattern is roughly the same for all onlookers. Nor is it easy to see how we could defensibly speak of our different reactions to these figures as being accompanied by different visual sensations, i.e., different sense-data. And yet, undeniably, different reports are forthcoming when we ask of people viewing these figures, "What do you see?"

We concluded the last chapter with a reiteration of our key question "Do the 13th Century and 20th Century astronomers see the same thing?" It is for the purpose of getting a better insight into the complications of this question that we will press our inquiry still further.

Let us begin with a few more variable figures: These vary, not in the perspective in which they may be perceived, but in the aspects they may present to a percipient.

Initially:

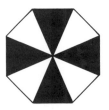

Figure 4

Some will see this as a white cross on a black ground.[1] Others will see this as a black cross on a white ground. But the difference cannot be accounted for by reference to different retinal reactions, for there need be no difference. Nor can it be accounted for by the suggestion that those mysterious entities, visual sense-data, are changing. For while I stare at such a figure, shifting from the seeing of a black cross to the seeing of a white cross, I am aware of no changes either in my retinal reaction or in my visual sense-data (whatever they are). Or if there *is* a shift in these latter I know of nothing in phenomenalism or in sense-datum theory to account for it. Indeed if I drew for you exactly what I saw when I reported "white cross on black ground," how would it differ from your drawing of what you see when you report "black cross on white ground"?

So too with Koehler's goblet:[2]

Figure 5

---

[1] Wittgenstein, *Investigations*, p. 207.
[2] See *The Mind,* eds. John Rowan Wilson and the editors of *Life* (New York: Life Science Library series, Time, Inc., 1964), p. 15.

Again, our retinas may react normally to this. But while I see a Venetian goblet, you may see two men staring at each other. Have we seen different things? Of course we have. And yet if I draw my cup for you, you may say, "By Jove, that is exactly what I saw, two men in a staring contest." Or I may myself shift my attention from the cup to the faces. Does my retinal reaction shift? Do my sense-data change? There is nothing in sense-datum theory to suggest that my sense-datum, i.e., the "look(s)" of *Figure 5,* does change. For clearly my private visual field is taken up with the same configuration of lines when I say I see a cup as it is when I say I see two faces. And yet it would be absurd to say that I saw the same thing in both cases.

In this respect *seeing* differs from *feeling,* as you would expect. For if I have had my right hand on a stove and my left hand in the refrigerator, when I plunge both hands into a basin of tepid water I will get a familiar variable reaction. Do my two hands feel the same thing? In an unimportant sense, yes; they are reacting normally to the tepid water. But it is much more natural to say that my hands feel different things: One feels the water as *hot,* the other feels it as *cold*. Different feelings, different sensations, different "sense-data" would be associated with each hand. These differences could be clearly and accurately described. But how to describe the difference between seeing a duck and seeing a rabbit in *Figure 3?* Or between any two aspects of the figures we have so far brought forward? To describe or draw such a figure in one of its aspects, say a duck or a cup, just *is* to describe or draw it in all its aspects (e.g., a rabbit or faces). Nonetheless we see different things here no less than we feel different things in the hot-cold experiment, even though the difference is not necessarily to be accounted for in terms of differing retinal reaction to stimuli, or in terms of having different pictures in the mind's eye.

We have so far considered figures with reversible perspective and figures with variable aspects. I have dwelt on these because they seem to be clear cases in which we should wish to say that we saw different things, but where we might deny that this was due either to a difference in retinal reaction or to a difference in the features of the pictures registered in our private visual fields. It is in these cases too that we should probably deny that the differences in what we see are due to differences in how we interpret what we see. For, as Wittgenstein said,

94 | *Chapter 6*

"To interpret is to think, to do something; seeing is a state."[3] Even Professor Price puts it that the perceptual act is not an *activity*. And the shifts in perspective and aspect that we have been considering might have occurred quite without thinking. Indeed, thinking hard will seldom enable one to see an aspect of a figure which he has been previously unable to notice.

I should like now to call up another group of figures that are variable in a rather less dramatic way. They are important, however, in the way that they continue to stress the *seeing as* component that has figured in all the examples so far. It is this largely overlooked component of our ordinary observations which will help us to see something more of the complexity of *observing, witnessing,* and *seeing* in scientific inquiry, and which will lead to a fuller appreciation of all that is involved in the situation wherein our two astronomers are witnessing the sun at dawn.

You may remember this one:

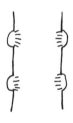

Figure 6

What is this meant to be? Your retinas and mine are similarly affected. Similar pictures of this may be assumed to be registering in our private visual fields. But do we see the same thing? I see a bear climbing up the other side of a tree. Most likely you did not see this. Did you notice, however, how the elements of this figure pulled together when you were told what I knew when drawing it? You might even say with Wittgenstein, "I see that it has not changed, and yet I see it differently. . . ."[4]

And a student once suggested this one to me:

---

[3] Wittgenstein, *Investigations,* p. 212.
[4] *Ibid.,* p. 193.

Figure 7

What do you see? A Mexican on a bicycle (seen from above)? Before I said that you might have seen just lines. But now, as Wittgenstein says, "[it] has a quite particular 'organization.'" [5]

What Wittgenstein calls here "organization" is really important, we will return to it repeatedly. We rarely see without such "organization" being operative, and yet this organization is nothing *seen* as are the lines and colors in a drawing. "Organization" is not out of the same concept-basket as are "lines" and "shapes" and "colors." This is the thin end of the wedge with which we may tumble the sense-datum account of seeing. For usually when we speak of seeing something we do so because our visual sense field is organized in certain ways. There is little in all this talk about private mental pictures that helps in any way our understanding of the organization requisite for seeing. This lacking, one might answer the question "What do you see?" with "What am I supposed to see?"—or even, "I see nothing," both of which might have been appropriate responses to the question following *Figure 7*.

Consider:

Figure 8

(a)

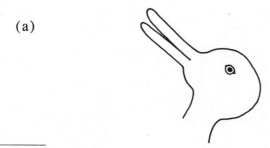

---

[5] *Ibid.*, p. 196.

in this context:
(b)

as follows:
(c)

The context clearly gives us the clue regarding which aspect of the duck-rabbit is appropriate: In such a context some people could not see the figure as a rabbit. Though in this context:

(d)

the figure may only come forward as a rabbit, e.g.:

(e)

It might even be argued, as Wittgenstein *does* argue, that the figure appearing in (e) has not the slightest similarity to the figure seen in (c), although they are congruent.[6] This flies in the face of sense-datum teaching.

Let us look further into the matter of context as it concerns aspect-vision or "seeing-as."

Of this square corner

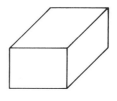

Figure 9

Wittgenstein wrote,

> You could imagine [this] appearing in several places in a book, a textbook for instance. In the relevant text something different is in question every time: here a glass cube, there an inverted open box, there a wire frame of that shape, there three boards forming a solid angle. Each time the text supplies the interpretation of the illustration.
>
> But we can also *see* the illustration now as one thing now as another. So we interpret it, and *see* it as we *interpret* it.[7]

In other words the appropriate aspect of *Figure 9* is brought out by the verbal context in which it appears, very much as one would have to talk and gesture around *Figure 3* to get an observer to see the rabbit when he had only been able to see the duck. The verbal context is, as it were, part of the illustration itself—a remark which, though it ought not to be taken literally, at least helps to show the sort of thing that brings out for a person one aspect of a visual object rather than another.

Wittgenstein also considers this triangle,

---

[6] *Ibid.*, p. 195.
[7] *Ibid.*, p. 193, Hanson's italics.

Seeing and Seeing As | 99

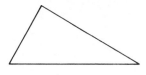

Figure 10

which he considers ". . . can be seen as a triangular hole, as a solid, as a geometrical drawing, as standing on its apex; as a mountain, as a wedge, as an arrow or a pointer, as an overturned object which is meant to stand on the shorter side of the right triangle, as a half parallelogram, and as various other things. . . . You can think now of *this,* now of *this* as you look at it, can regard it now as this, now as this, *and then you will see it now this way, now this. . . .*" [8]

Of course the context here is given in Wittgenstein's designations. For example:

". . . triangular hole . . ." does this to *Figure 10*

". . . solid . . ." does this

". . . geometrical drawing . . ." this

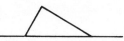

". . . standing on its base . . ."

". . . hanging from its apex . . ."

---

[8] *Ibid.,* p. 200.

"... a mountain ..."

"... a wedge ..."

and so forth.

The context that brings an appropriate aspect of a figure or an object into focus, however, need not be set out explicitly in a paragraph or in a word. Such "contexts" are very often carried around with us in our heads, having been put there by intuition, experience, and reasoning. For example, the sequence

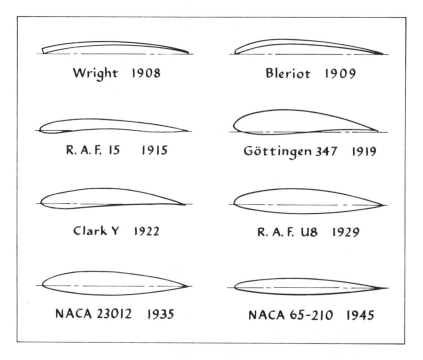

Figure 11

could mean but one thing to the aeronautical engineer. We have the same retinal reactions and the same visual sensations as he does, do we not? But we would probably not see what he does, namely the sequence of airfoil types from the earliest days of heavier-than-air flight to the wing section of the present-day airplane. To see what the aeronaut sees, we would have to know what he knows. A novice sees what the specialist sees only in the way that a person who has never seen a duck or a rabbit, nor a picture of either, sees what we see when we look at the duck-rabbit. There are aspects of *Figure 11* to which the uninformed person will remain blind. And he will remain blind to these aspects in just the way that he might be blind to the rabbit aspect of the duck-rabbit when the latter is surrounded by ducks. In both these cases he lacks a context within which he may see (in a significant way) what is before his eyes.

Try this one:

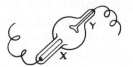

Figure 12

A trained natural scientist could only see this as one thing: an x-ray tube viewed from the cathode. Would a physicist and a non-scientist see the same thing when looking at *Figure 12?* The traditional, respectable answer to this runs: "Yes, they see the same thing, only the physicist interprets it in a way that the layman cannot." It is this "respectable" answer to the question, of course, that I have been at pains to unsettle. The answer is no more suitable here than in any of the other cases we have considered—indeed, it is positively harmful. We can agree again that the scientist and the layman have a normal retinal reaction to *Figure 12*. And we can assume that the pictures registered in their private visual fields are similar. But do they see the same thing?

Or consider just the physicist. On his first day at school years ago he had gazed in wonder at the glass and metal instruments on display in the lab. Now, after a long training in science at school and at university and in research, he returns to his old school and sees again that same x-ray tube that had so fired his imagination when he was a boy. Does he see the same thing now as he saw then? His eyes are still normal, his mental picture of the instrument is no different. But now he sees it in a very different context; he sees it in terms of electrical circuit theory, thermodynamic theory, information about the structure of metals and of glass, research into the nature of thermionic emission, optical transmission, refraction and diffraction, atomic theory, quantum theory, and relativity theory. This is a phenomenon we all know quite well. Compare the freshman's first view of his college with the senior's last view, or our first look under the hood of a newly purchased car with the same view ten exasperating years later.

"Oh yes, the physicist has learned all these things, doubtless," comes the "respectable" reply. "And they all figure in the interpretation the physicist puts upon what he sees—it is this interpretation that the layman is unable to make even though he sees exactly what the scientist sees."

But is the physicist doing any more than just seeing? As Wittgenstein says, interpreting is thinking, it is doing something. What is the physicist doing over and above what the layman does when he is seeing the x-ray tube? What do you do besides just looking and seeing when you notice the microscopes on the benches of the lab or when you see a galvanometer, or an automobile, or a close friend?

"Oh, it is just that in these familiar cases the interpretation takes up but a very short interval of time; it is all but instantaneous." So comes back the reply, and a typical philosopher's reply it is, too. It is out of the same bag of tricks that made sense-data, those final links in the perceptual causal chain, unlike all the other links, *mental, private, publicly unobservable*. These are, I feel, just dodges that philosophers have invented for the purpose of saving ideas for which they have formed a sentimental attachment. We all know very well what it is like to put an interpretation on what one sees. Artists do it, historians and journalists do it, Lysenko does it, and indeed at the frontiers of scientific research where the facts are thin and the problems thick everyone interprets what he sees. But the word *interpret* gets its bite and its use in these contexts precisely because they contrast with cases like the one where the physicist sees the x-ray tube, or where we see a bicycle. Insisting that even these last are situations involving interpretation is just another way of saying that only the apprehension of sense-data can count as *seeing,* everything in addition to that being interpretation, a saying that has been under attack since we began our inquiry.

Before a non-physicist could see *Figure 12* as a physicist sees it, before the elements of that picture will pull together, cohere, and "organize," he would have to learn a good deal of physics. It is not just that the physicist and the layman see the same thing but do not make the same thing out of it. The layman can make nothing out of it. And that is not just a figure of speech. I can make nothing out of the Arab word for *cat,* though my purely visual reaction to that word may not differ from that of an Arab child.

In the sense that I have been so far elaborating, the two do not see the same thing. To the question "What do you see?" the physicist will reply, "An x-ray tube with its cathode forward." The non-scientist may reply, "What am I supposed to see?" Both are quite appropriate answers. [In this connection it is interesting to note that very often the words "What do you see?" are used in posing the question "Can you identify the object before you?" To this question the two answers just given are comprehensive. It is not for nothing that "What do you see?" can be used in putting the "Can you identify . . . ?" question, a question which presupposes normal vision, but is calculated to test one's *knowledge*.]

Pierre Duhem puts the matter thus:

> Enter a laboratory; approach the table crowded with an assortment of apparatus, an electric cell, silk-covered copper wire, small cups of mercury, spools of wire, a mirror mounted on an iron bar; the experimenter is inserting into small openings the metal ends of ebony-headed pins; the iron oscillates, and the mirror attached to it throws a luminous band upon a celluloid scale; the forward-backward motion of this luminous spot enables the physicist to observe the minute oscillations of the iron bar. But ask him what he is doing. Will he answer 'I am studying the oscillations of an iron bar which carries a mirror'? No, he will answer that he is measuring the electric resistance of the spools. If you are astonished, if you ask him what his words mean, *what relation they have with the phenomena he has been observing and which you have noted at the same time as he,* he will answer that your question requires a long explanation and that you should take a course in electricity.[9]

The physicist, in other words, must teach his visitor everything he knows before he can show him what he sees. Not until then will his

---

[9] Pierre Duhem, *The Aim and Structure of Physical Theory,* tr. P. P. Wiener (Princeton: Princeton University Press, 1954), p. 218.

visitor be supplied with an intellectual context sufficient for throwing into relief those aspects of the cluster of objects before his eyes that the physicist sees as an indication of the electrical resistance of the spools. There is nothing wrong with his eyes. He can see in the sense that he has normal vision, i.e., he is not blind. [This is the sense in which we *can* hear, even when we do not hear the ticking of the clock behind us, or when we do not hear the street noises during sleep, even though our ears are open and our auditory organs reacting normally to every acoustical vibration. This is only to say that we are not deaf. Still, we may not hear that the oboe is out of tune, something that will strike the trained musician at our side as painfully obvious.] The visitor cannot see what the physicist sees, even though the physicist's eyes are no better than his. He cannot see what the physicist sees in much the way that he may not be able to see a rabbit but only a duck in *Figure 3*, or only a wife but not a mother-in-law. He is, in a word, blind to what the physicist sees. The elements in his visual field, though perhaps similar or identical to the elements of the physicist's visual field in color, shape, arrangement, etc., are not organized conceptually for him as they are for the physicist. And this is much the same situation as we find when both you and I gaze at *Figure 3* but I see a rabbit and you see a duck. The conceptual organization of one's visual field is the all-important factor here. It is not something visually apprehended in the way that lines and shapes and colors are visually apprehended. It is rather the *way* in which lines, shapes, and colors are visually apprehended. And in all the cases we have been examining I have been inviting you to consider a given constellation of lines and shapes (what psychologists call "a stimulus pattern") and to consider further the different ways in which this given constellation or pattern can be apprehended visually, the different sorts of conceptual organization that can be accorded to that constellation. In short, the different ways it may be seen.

Of course, the reasons why these things are seen differently are not the same for every case we have examined. A thorough examination of that, however, is a task for the experimental psychologist, a title to which I can lay no claim whatever. We have been concerned here with a conceptual inquiry, and that is the province of philosophy. We have been asking, "What *is* our concept of *seeing;* might it not be more subtle, complex, and variable than 'classical' philosophers of science would have us believe?" We have not been directly concerned with the psychological questions "How do we *arrive* at the concepts of seeing we have got, and what causes this variability in what we see?"—though of course answers to these questions would mark more clearly the boundaries of our own inquiry. As Wittgenstein would have put it, "Here the psychological is a symbol of the logical."

What all this has been leading up to is the centrality of the notion of *seeing as* within our concept of *seeing*. You see it as a duck, I see it as a rabbit; the physicist sees it as an x-ray tube, the child sees it as a kind of complicated incandescent lamp bulb; the microscopist sees it as coelenterate mesoglea, the engineer as a kind of gooey, formless stuff. And how very relevant to every case of seeing is the knowledge of him who does the looking.

Goethe said that we see only what we know. In my opinion Goethe was right in a way that "classical" philosophers of science, with all their talk of *normal perceptions, sense-data, interpretations, logical constructions,* etc., were hopelessly wrong. The point of Goethe's remark should be within our reach now. I will try to secure it by means of discussion of what it is to *see as* . . . ; my argument will be that almost everything we usually call *seeing* involves as fundamental to it what I, following Wittgenstein, have called "*seeing as.*"

I have just said that the reasons why people see things differently are not the same for all the cases we have considered. There is, however, one respect in which they do not differ, or so I shall argue. No case of seeing that we have considered is wholly independent of the knowledge of the percipient. I had to tell you what I knew about *Figure 6* before you could see it as I saw it, a bear climbing up the other side of a tree. And this bit of knowledge is intelligible only against the knowledge of what a bear is, what a tree is, and what climbing is. Almost everyone, of course, will see

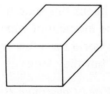

as a transparent box, or a wire-framed cube, viewed as from above, or as from below. But this need not mean that our observation of this figure is without a trace of any knowledge of the construction and properties of box-like objects and the functions of the lines used in representing such objects. On the contrary, this only goes to show that most people who are capable of experiencing the reversible perspective phenomenon of this figure, and this would of course exclude babies and dimwits as well as blind people, know enough, have learned enough, *to be able to see* this figure as a three-dimensional box, from above or from below. It is interesting to speculate as to whether a person ignorant of the existence and appearance of rabbits could see the duck-rabbit as anything but a duck. This speculation is no more inadmissible than Locke's conjecture that a man whose blindness had been cured by operation would be unable to identify and distinguish a cube from a sphere by sight alone, a conjecture admirably substantiated by Sherrington.[10] As Wittgenstein puts it, "You only see the duck and rabbit aspects if you are already conversant with the shapes of those two animals . . ."[11] Could a person who had never experienced a cup of any kind, much less one of the ornate Venetian variety, see anything but two faces in Koehler's drawing? And is nothing whatever required of us in order that the black and white crosses should alternately claim our attention? In every such case the traces of previous knowledge are to be found, and those traces figure in all the situations we have called *seeing*.

It is well known that babies, even those older than six months—the time when the retina has completely formed and a minimum of ocular coordination has been achieved—are capable of experiencing but very few of what we take to be the most ordinary visual experiences, like seeing a cloud in the sky. For all their delicate optical equipment, babies are not even in a position to be taken in by reversible perspectives or shifting aspects, much less galvanometers and x-ray tubes. They are in a "big, blooming, buzzing confusion," as William James once put it. The ophthalmologist Ida Mann likens this state to what we experience, or fail to experience perhaps, at the moment of waking when we "recapture our primitive amazement at the world for a few seconds." The usual bedroom things are before our eyes "but they look . . . bizarre and meaningless.

---

[10] See John Locke, *An Essay Concerning Human Understanding,* ed. A. S Pringle-Pattison (Oxford: Clarendon Press, 1924), Book II, pp. 75–6.

[11] Wittgenstein, *Investigations,* p. 207.

Our brains are not as awake as our eyes. . . ."[12] In such a state we could not be said to *see* anything, cubes or tubes, stairs or bears.

At this point it may be worth while to remark that some scientists and philosophers think the eye to be a kind of window in our skins. When the window is shut, as when the eyelids are closed or the cornea clouded, we cannot see. When the window is open, we can see. Normal retinal reaction on the one hand and sense-data on the other are given the title of seeing because of their apparently intimate relationship to the light coming in the window.

But the eye is not merely a transparent section of our skins. Part of it does develop embryologically from the skin, it is true; the lens and the cornea are epithelial. The retina and the optic nerve, however, are outgrowths of the brain. It could not alarm anyone, except a person with a theory to the contrary, to hear that alterations in the general state of the brain, alterations like learning what was not before known, or experiencing the heretofore inexperienced, could affect the whole character of seeing, particularly in its conceptual organization and in the singling out of "significant" aspects.

As Wittgenstein says, " 'Now he is seeing it like *this*,' 'now like *that*' would only be said of someone *capable* of making certain applications of the figure quite freely. . . . It is only if someone *can do*, has learnt, is master of, such-and-such, that it makes sense to say he has had *this* experience."[13]

Seeing a thing, therefore, is *seeing* it *as* this sort of thing, or as that sort of thing; we do not just *see* indeterminately or in general, as do infants and lunatics. And seeing a thing as *this* or *that* sort of thing presupposes a *knowledge* of *this* or *that* sort of thing. Our two astronomers would not say merely that they saw a brilliant yellow-white disc and leave it at that. What they see they see *as* the sun. And this presupposes a knowledge of what sort of thing the sun is, which digs up a nugget I buried three chapters ago. The knowledge of what the sun was in the 13th Century was very, very different from the knowledge of what the sun is now in the 20th Century. I will say no more about this now other than to suggest that the two astronomers are to the sun as you and I might be to the duck-rabbit when you see only a duck and I only a rabbit. The

---

[12] Mann and Pirie, *The Science of Seeing, op. cit.,* p. 18.
[13] Wittgenstein, *Investigations*, pp. 208–9.

difference is in our conceptual organization of the elements of our visual experience. So too the sun, hills, and trees may be seen as in quite a different relation to the medieval scientist from the relation in which they appear to the modern astronomer. This is a point I will press further in the next chapter.

Here I wish to make it quite clear that I am not denying that there are a good many cases in scientific inquiry where the data before us are wonderfully confused, and about the nature of which we may not have an inkling. It occurs to me, however, that the importance of observation in such cases is overrated and its character is not enough understood. The model for such *seeing* is what we undergo in the oculist's office where we report on the apparent distance between the point of white light and the reference line seen with the left eye: "They are coming together now. There, the point is right on the line." Or the oculist will request, "Say when you can see a green light with your left eye." How similar to situations in microscopy where, when we are confronted with a totally new and unfamiliar phenomenon, we report our visual sensations in as lustreless and phenomenal a way as possible: "It has a green tint to it in this light, and those erratic jerky movements it makes along its longitudinal axis are noteworthy. Ah, there's another one, slightly longer and thicker; there are cilia-like appendages near the narrow end, and two darkened areas at its middle." So too the physicist who expresses a given experimental situation thus: "The needle is oscillating most erratically, I wonder what's up; and see that faint streak near the neon parabola, it looks almost like a reflection of the main parabola, and there are scintillations at the periphery of the cathode scope that have never before been dominant."

I certainly do not wish to say that these are not genuine cases of seeing. If I did I would be just as far off course as those who insist that these are the *only* genuine cases of seeing. What I would urge is that these observational situations have a point to them just because they contrast with our more usual cases of seeing. The language of shapes, color patches, oscillations, and pointer-readings is the language appropriate to the unsettled experimental situation, where confusion and perhaps even conceptual muddlement dominate. And the *seeing* that figures in such situations is of the sort where the observer *does not know what he is seeing*. He will not be satisfied until he does know, until his observations cohere and are intelligible as against the general background of his already accepted and established knowledge. And it is this latter kind of seeing that is the goal of observation. For it is largely in terms of it, and seldom in terms of merely phenomenal seeing, that new inquiry will proceed.

This is part of Goethe's meaning when he says that we see only what we know. New visual phenomena are noteworthy only against our accepted knowledge of the observable world. In psychologists' language, we are *set* to see, observe, notice, or attend to certain sorts of things, but not others. The ancient Greeks failed to notice thousands of things about the world that children now regard as commonplace, but this was not due to faulty vision or lack of curiosity. Galen's followers did not see that the middle wall of the heart was usually solid and not perforated. Physicists up until 1900 failed to detect the flaw in Galileo's proof that the acceleration of a freely falling body was proportional to the time and not the distance fallen. And Darwin himself remarked of an early expedition with a colleague, "Neither of us saw a trace of the wonderful glacial phenomena all around us; we did not notice plainly scored rocks, the perched boulders, the lateral and terminal moraines. . . ."[14]

So of course it is often an essential step in the advancement of science to account for ourselves as observers in a strictly phenomenal way. Every great scientist has had to subject himself to the severities of a strict reporting of what lies in his visual field, of the shapes, lines, colors, and movements he sees.

But that is far from the end of the matter. Everyone who is forced by experimental difficulties and conceptual perplexities to observe his data as if he were in an oculist's office aims at coming to see his data in this other sense: where he knows what he is seeing, where he sees his data as it is (and not merely as it appears), where he can see that if a certain operation were performed on his data a certain other action would be observed, just as we see that if the first story of a tall building were demolished the upper stories would come crashing down.

The point is that coming to see one's data in the completely lustreless and noncommittal way that we see the objects of the oculist's test requires a highly specialized and rigorous training in science.[15] Learning

---

[14] Charles Robert Darwin, *His Life Told in an Autobiographical Chapter,* ed. Francis Darwin (London: John Murray, 1902), p. 25.

[15] See M. L. Johnson, "Seeing's Believing," *New Biology,* Vol. 15 (Oct., 1953), pp. 66–79ff.

to restrict and control one's vision in this way is a scientific accomplishment of the first magnitude, and it is far from being the birthright of every man who decides to study natural science. All of which is to say that *phenomenal* seeing is something acquired, something unusual, something different from our ordinary ways of seeing. Using *phenomenal seeing* as the typical, paradigm case of *seeing* is unjustified and misleading. Rather than our ordinary cases of seeing being logical constructions out of the research scientists' phenomenal variety of seeing, it is the latter which is a logical destruction of our ordinary kinds of seeing. It is something done in a calculated, systematic, premeditated way. But of course if *all* our seeing were carried on in this way we would collapse from exhaustion in a fortnight.

Hence I am not denying that "phenomenal" seeing is genuine seeing. I am urging that it is not the *only* genuine type of seeing, not the paradigm case of seeing, and indeed, it is only a case of seeing at all when considered against the more usual sort of seeing I have been discussing. The more usual sort of seeing is, as Goethe suggested, a seeing of what we know. It is, hence, a theory-laden operation, about which more will be said later, and hence relative in most respects to the observer's knowledge. It is this knowledge which in large measure affects what the observer will see things *as*. Wittgenstein put it this way:

> The concept of 'seeing' makes a tangled impression . . . There is not one genuine proper case of [what is seen]—the rest being just vague, something which awaits clarification. . . . What we have . . . to do is to accept the everyday language-game, and to note false accounts of the matter *as* false. . . .[16]

I will try later to explore further the theory-loaded character of *seeing*, moving from the *seeing as* component we have been discussing to what I call *seeing that*. This will bring us to the large questions having to do with the interrelations between knowledge, language, and our ordinary observation.

Here may I commend to your reflections the story that Freud tells of the visitor to the fur shop who remarked on how wonderful it was that all the pelts had two holes in them just where the animal's eyes were situated.

---

[16] Wittgenstein, *Investigations*, p. 200.

7 | Seeing As and Seeing That

WE HAVE been dealing with the concept of "seeing as. . . ." It has been argued that in most of the cases where we speak of what we see, as when we see boxes, ducks, rabbits, bears, bicycles, and x-ray tubes, we are speaking of having visual impressions which we *see as* boxes, ducks, rabbits, etc. This *seeing as* is a central component of what we ordinarily call "seeing." Several cases were noted of people having roughly the same retinal reactions to a visual stimulus X, and perhaps having the same visual impressions, or sense-data, of it as well, but where the parties concerned did not see X in the same way. One sees X as a Y, the other sees it as a Z, etc. And besides considering slightly bizarre examples of this, things like reversible figures and shifting-aspect figures, we saw that in certain important respects the seeing of airfoils, x-ray tubes, bicycles—indeed, the seeing of most of our familiar material objects—consists in part of this element I have been calling "seeing as . . ." When our seeing of physical objects lacks this "seeing as . . ." quality, e.g., in the oculist's office or when in research we encounter a visual phenomenon wholly new to our experience, we are up against the *unusual* cases, the non-typical cases. That is my contention. These cases are outstanding and important only when contrasted with our more usual varieties of seeing things. The "phenomenal" description of what the research microscopist sees is only instrumental to his goal of one day describing what is before him on the slide in physical object terms, in terms of what the thing before his eyes is seen *as*.

Very little has been said about *seeing as* itself, however, beyond suggesting that it is in some way dependent on the knowledge of the observer. Some knowledge is required even to get the reversible perspective and shifting-aspect effects in our several examples. Infants and idiots cannot react as we did to the figures of the last two chapters. A good deal of knowledge is required to see a complex arrangement of glass and metal as an x-ray tube, or to see a brilliant yellow-white disc as our "fixed" local star. Neither aboriginals nor 13th Century astronomers would see these things as physicists and astronomers today might.

But the relationship between *seeing as* and the corpus of our knowledge is not a simple one, and to the exploration of this we must now turn.

What is it to see the figures we have seen as transparent boxes, staircases, ducks, rabbits, goblets, bears, airfoil sections, and x-ray tubes? It is, as has been said, to have knowledge of certain sorts. (Robots and cameras react to light, but they cannot see. Electric eyes are blind.) It is, in short, to *see that* if certain things were done to the objects before our eyes, certain other things would probably follow. *Seeing that*, therefore, is another component of seeing, a component that is inextricably bound up with "*seeing as*. . . ." But when I say this I do not wish to make it sound as if *seeing* is composite in the way in which automobiles are composite—the *seeing as* and *seeing that* components meshing together in a mechanically perfect way to give us what we ordinarily call "seeing." This is not so. *Seeing* is just *seeing*, as *winning* is just *winning* and *finding* is just *finding*. However, one can ask logical questions about these concepts; one can ask, for example, what sort of things must have taken place for us to describe a man as finding a collar stud, or to describe him as winning a spelling contest. So too we can ask what sorts of things must have taken place for a man to be described as seeing a bicycle, or seeing an x-ray tube, or a spirochete; unless a person had had at least one visual sensation and knew what a spirochete was, he would not say that he had seen a spirochete. And seeing in this manner, as opposed to seeing in the sense-datum (or oculist's-office) manner, involves our *seeing* a visual object *as* a bicycle, or an x-ray tube, or a spirochete. Another logical element of this sort of seeing is what I shall call SEEING THAT.

For to *see*

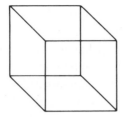

as a transparent box, or as an ice cube, or as a wire-framed block of glass is to *see that* it is six-faced, twelve-edged, and eight-cornered. It is to see that the corners of the box pictured are solid right angles. It is to see that, as a box, or cube, it would be constructed of certain rigid or semi-rigid material, and not some liquescent or gaseous material. We expect it to be as tangible as it is visible. We see that it will take up space in an exclusive way, other objects being unable to occupy the space that it is in. And we see that the box will be a locatable object. It is the sort of thing we should be able to find here, or there, or over there, but at least somewhere. Nor will it cease to exist every time we blink our eyes. Seeing this figure as a box is just to *see that* all these things will obtain of the box pictured.

This constitutes *knowledge* about boxes. Some of it is of a rather more logical nature (e.g., we should not say of anything that it was a physical object, like a box, were it not locatable in space or itself a tangible, space-occupying entity; nor should we say of any physical object that it is a cube unless it is six-faced, twelve-edged, and eight-cornered). On the other hand, that liquids and gases (*per se*) are not suitable for the formation of boxes and cubes and rigid frames is something we must learn from experience in a way rather different from the ways in which we gain our knowledge about what *objects* and *cubes* are. The words "transparent box," or "ice cube," or "wire-frame block of glass" would be inappropriate to the expressing of what is *seen as* if any or all of these further considerations were denied.

But why drag in all these reflections when we are strictly concerned with what is seen? Are these allusions to the nature of knowledge really to the point? We were talking about *seeing;* how do all these epistemological factors enter?

To this objection let me say only that if it is true that we see *things*—like cubes, ducks, bicycles, and x-ray tubes—and *call* what we see by such names, then what I have referred to as *seeing as* and *seeing that* are clearly involved in our ordinary seeing. Thus, seeing a speck in the sky as a duck involves our seeing that it will not suddenly rocket vertically upward, or come at us doing snap rolls or Emmelmann turns. We will see that it will have two rather undersized wings sawing rapidly at the air, and that when it lands in the water it will make contact feet and tail first with a goodly splash. This is, of course, a good deal more than meets the eye. And indeed, we *could* be wrong. It might turn out to be an owl, or a heron, or an airplane. But to see it as a duck, however momentarily, just is to see it in all these *seeing that* connections. Sense-datum theorists and others have wanted to say that just because *seeing as* and *seeing that* are involved in what is vulgarly called "seeing" the latter cannot properly be called *seeing* for theoretical purposes. Something pure, unadulterated by inference or intellect, or knowledge, must be the legitimate heir to the title of seeing. And this is what leads sense-datum theorists to talk so much about the cases where we go wrong in our observations, where what we see does not accord with what there is. And in this they forget or ignore the wide range of cases wherein we are right in our observations. That is, they are so concerned to discover what it is that we are right about when we are right in saying we see a duck (when only an owl is before us) that they leave unexamined all that is involved when we are right in saying we see a duck when there is a duck to be seen—a surprisingly frequent occurrence. In doting on our observational mistakes the phenomenalist portrays a world in which the senses are generally misleading and deceptive. But the world of science is not like this, or not everywhere like this. So why not attend a little more to what is involved when people say they see ducks and dinosaur tracks and are correct? This "pure visual something," whether it be the crude retinal reaction or the more subtle sense-datum, is what no one without a theory would dream of calling *seeing*, save in those relatively rare contexts where *seeing as* and *seeing that* are not possible, as with the oculist's eye-exercises or at the furthest frontier of scientific research or in the visual responses of infants and idiots. Were an ordinary man in an ordinary situation to react to his visual environment as does an infant or an idiot, we would think him to be out of his head and *not* seeing what is around him, or what he is doing. (We might also think him to be a philosopher

Seeing As and Seeing That | 115

going to extremes to prove his point, but the minute cocktail time comes he will see just what we see in the pitcher and probably get his martini poured first.)

The *seeing that* is what threads our knowledge so intimately into our everyday seeing. It is what saves us from asking "What's that?" of everything that meets our eyes. And this knowledge is *there* in the seeing and not an adjunct of it, if only for no better reason than that we never or rarely catch ourselves adjoining bits of knowledge to what we see. You see a person standing there and you are unlikely to catch yourself unawares busily tacking on to the visual impression that you have the knowledge that if he turned around you would see his back. You see him as solid, tangible, but not as bilaterally symmetrical (in the sense that he would present the same facade to you if he rotated through 180°). Of course, if he were to turn out to be shaped like one of Aristophanes' primeval men, if you were to see him rotate half a turn only to exhibit the same front view, you would say that you did not believe your eyes. But of course it isn't your eyes that would be lying. Yes, cameras and photocells do not lie. It would be the *seeing as* and the *seeing that* components of your seeing that would have got jarred up through such an experience. And needless to say, were it established that he really was Janus-faced and Janus-bodied, you would never see him again with the same eyes, and you would never see him as you saw him before, however similar his front view then to his front view now. It is just as when one cannot see a close friend in quite the same way after catching him in a lie. As Wittgenstein would have said, seeing him as solid, tangible, and with a back view as well as a front view requires no forcing, no squeezing. *It is all there in the seeing.* We see only what we know, that is what makes conjuring tricks possible. Deceptions must proceed by an exploitation of what is the normal, ordinary case. The skilled thief will dress as the broker ordinarily dresses, the skilled footballer will move his feet as he ordinarily would when about to move left—and then move right. The cigar box looks as a book ordinarily looks. And that a sleight-of-hand artist can get our minds and our eyes, i.e., our *seeing*, moving in one direction while catching us out in another direction is a clear indication of the way our SEEING usually proceeds. It is because our thoughts are so intimately a part of seeing that we must sometimes rub our eyes at illusions.

So we may see

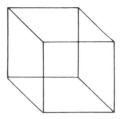

as a perspex box or as an ice cube but we may be wrong. It may be a crystal lattice or the floor plan of a rat maze. But then discovering our mistake consists in seeing that one or several of our expectations of perspex boxes or ice cubes is not being fulfilled. And hence we are *no longer at liberty to see* what we see as a perspex box or as an ice cube, just as, when we have seen the conjurer saw the young lady in half, we are no longer at liberty to see this as an ordinary case of joinery, or should I say *dis*-joinery. We cannot see the conjurer's saw as an ordinary saw, nor his actions as those of an ordinary woodsman, nor the situation as an ordinary case of sawing, if when he is finished with his work the young lady smiles and waves gaily while happily kicking her feet. And when we say we cannot believe our eyes in such a case, we indicate that seeing things as we saw them originally was just to *see that* certain things could not follow. The conjurer's trick consists in upsetting this part of our original seeing. But on a second showing of his performance would we see the same thing as before? And if what we took to be a perspex box momentarily collapses into a kind of jelly only then to snap back into its former shape, are we still at liberty to see it as a perspex box? If we persist in doing so we can only do it as the poet does it, in spite of evidence to the contrary.

Seeing something as an X is seeing that it may be expected to behave in all the ways X's do. Seeing that its behavior does not accord with what we expect of an X impedes our seeing it as an X any longer. We can no longer see porpoises as fish, nor the earth as flat, nor the heavens as bowl-shaped—all things familiar to 13th Century scientific eyes. And, save in the extraordinary cases of the oculist's tests and when we are genuinely confused and clueless about the nature of the thing before us, to see anything at all is to *see it as some X or other*. As Wittgenstein put it, ". . . what I perceive in the dawning of an aspect is not a property of the object, but an internal relation between it and other objects."[1]

At first you do not see what

is all about. You "see" it, but you see nothing. I have confronted scores of people with this sketch asking "What do you see?" In most cases, after a long hesitation, the reply comes, "I do not know," despite the fact that most people could draw just what I have drawn. Then, when I announce what the figure is meant to be, you *see* it *as* a bear climbing up the other side of a tree. And this is just to see that, were you able to get to the other side of the tree, you would see this:

Seeing

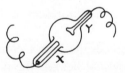

as an x-ray tube is just to see that a photographic plate placed at X will be greatly affected by x-radiation, while placing the plate at any other point will have little effect. It is to see that the target Y will get extremely hot due to its collection of the kinetic energy of the electrons shot at it from the cathode, and that since the anode has a shaft too thin for a water cooling circuit it must be made of molybdenum or tungsten. It is

---

[1] Wittgenstein, *Investigations*, p. 212.

to see that at high voltages a green fluorescence will appear around the anode. And it is to see that if a suitable crystal is placed at X a diffraction pattern will result; it is to see that with this apparatus the wave motion of the x-ray can be demonstrated.

Of course, a person could be taught to parrot "x-ray tube" every time he saw

But this is not to see the figure as an x-ray tube: He might have said "Kentucky" or "Winston" with just as much significance. He would not see that any of these other things followed.

For the physicist to see the figure as an x-ray tube, however, is just for him to see that the rest of this will follow. For him to be shown that one of these visual expectations is not fulfilled is to make it impossible for him any longer to see the figure, or the object, as an x-ray tube. It is in just such a case that we would find the guarded language noted in situations where exact identification is not possible, as in the oculist's office, or in advanced research science: E.g., ". . . well, it certainly has the shape of an x-ray tube. Look, there's the cylindrical cathode with its focusing mirror, and the anode is a quite standard one. Oh, but what is that grid of thin wires doing between the anode and the cathode? Perhaps that is what is keeping the fluorescence down. No, it's no ordinary x-ray tube . . ."

If in normal circumstances the 13th Century astronomer cannot but *see* the brilliant disc in his visual field *as* the sun, then he cannot but *see that* it is a celestial body that will behave in certain characteristic ways, ways that are explained by and serve as the foundation for his general theories about the nature of the sun. These theories are not, of course, superimposed on what he sees in the east at dawn. They do not constitute a tandem operation of interpretation distinct from the actual seeing. They are, as I have been contending, *"there"* in the seeing just as the interpretation of a piece of music is *there* in the music, not something distinct from and superimposed on the pure, unadulterated sound. Similarly we see

Seeing As and Seeing That | 119

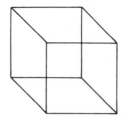

as from underneath, or as from above, or as a plane line drawing. But however it is construed, the construing, the interpreting, is in the seeing. There are not two operations, *click-click*, involved in my seeing it as an ice cube. I simply see it as an ice cube. And so the physicist sees an x-ray tube. He does not first soak up the reflected light from the glass and metal instrument and only then clamp his interpreting mechanism on to his visual impressions, thus to see them as an x-ray tube. He sees that just as you see your hands in front of you.

In the east the 13th Century astronomer sees the sun about to begin its daily journey from horizon to horizon. Just as the birds fly from tree to tree, and the clouds float from hilltop to hilltop, so the sun here at dawn is beginning its journey to the western horizon like a fiery balloon. He sees that, were he transported to heaven, he could watch the sun, the moon, and all the planets and stars circling the relatively fixed earth, the center and fulfillment of the physical universe. Seeing the sun at dawn through the geocentric spectacles of the 13th Century astronomer would be to see it in something like this way.

What the 20th Century astronomer sees, however, has an entirely different conceptual organization. Yet if he drew or painted exactly what he saw, the 13th Century astronomer would agree that that was just what *he* saw too. [And this situation is very like the one where I draw the duck that I saw, and you see in my drawing an exact image of what you saw—a rabbit.] Unlike most of us who still speak of *sunrise* and *sunset* and who treat the Nautical Almanac as a kind of textbook of ptolemaic astronomy, the 20th Century astronomer will *not* see the sun as up over the horizon ready to vault to its zenith position. The trained contemporary astronomer will probably see the horizon dipping away, or turning away, from our relatively fixed local star, the sun. After a little practice with an astronomer friend even I became able to see the sun in this way. The shift of aspect from *sunrise* to *horizon-turn*, or from seeing the sun circle the earth to seeing the earth circle the sun, is in many respects like the shift-of-aspect figures we have so far encountered: the duck-rabbit, the wife-mother-in-law, the goblet-faces. And the shift here is strictly occasioned by what we have come to know about the relation of sun to earth. So when the 20th Century astronomer *sees* what is to be seen in the east *as* the sun, he *sees that*, were he able to rocket into space, he would see the planets revolving in their eccentric orbits round the sun.

Seeing as and *seeing that* are inextricable, and they are indispensable to the seeing that counts as significant in scientific observation.

But why indispensable? *Seeing as* and *seeing that* may be inextricable, but that is a matter of fact. It is a matter of fact that we do not *see* things *as* of a certain sort without *seeing that* were such and such to be done to those things certain consequences would follow, and the fashion is to say that matters of fact are philosophically irrelevant; a fashion which, like those of the Parisian dress designers, sometimes seems unnecessarily inflexible and dictatorial. But so be it. To say that men always *do* see the world through *seeing as* and *seeing that* spectacles is one thing. To say that they *must* see the world in this way is quite another. That scientific observation is not a gaping, open-mouthed encounter with unfamiliar and unconnected flashes, sounds, and bumps, but rather a quite calculated encounter with these *as* flashes, sounds, and bumps of a particular kind, would be a splendid theme for an account of what significant seeing in science actually is. But it will not begin to secure the point that such seeing could not be otherwise. Just as the true statement that there are no five-mile-high mountains in Colorado does not prove that there could not have been. We have not begun to establish the logical position that an alternative account of the seeing which figures in scientific observation would be, not merely false, but absurd.

Only a logical argument can establish that. This, then, must be our next step.

It has been urged that it is only in relatively unusual cases that we report what we see in the language of visual impressions. There are, fortunately, great differences between the ways in which we see the sun and the moon and the ways in which we see the blobs of color and points of light in the oculist's office; nor does the research scientist see his laboratory equipment, his bicycle, and his wife in the completely baffled way that he may on occasion view what he finds on the slide of his microscope or in the window of his cloud chamber. No, in most cases of seeing we *see* what we see *as* a thing of some sort, a thing about which we could supply further information if pressed to do so. And this involves our *seeing that* other specifiable observations are possible.

For example: We see

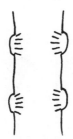

as a bear climbing up the other side of a tree. But seeing the figure in this way just is to see that, were the situation depicted turned 90° through its vertical plane,

would appear. Indeed, seeing

as a bear climbing the other side of a tree just is to see that any of a very large number of other views of the bear (from above, below, behind, etc.) could all be simultaneous views of the bear pictured

.

[It is also to see that

would not be a possible alternative to

,

nor would

;

while it would be an open question whether

or

were alternatives.] Wittgenstein asks, "Is it a case of both seeing and thinking? or an amalgam of the two, as I should almost like to say? The question is: why does one want to say this?" [2]

Notice one logical feature of *seeing that:* These locutions are always followed by clauses that could stand by themselves as complete sentences with the addition only of an initial capital and a full stop. You can *see* an ice cube, or *see* a cormorant as a duck, but you cannot *see that* an ice cube, or *see that* a duck; and this is not because of limitations on our eyesight but because of the logical and grammatical strictures of our language. Rather, you "see that" such things as that *ice cubes will melt in a hot oven;* you see that *ducks are not powerful climbing birds* or that *if*

were rotated through 90°

would appear. (The italicized words could—if not used with pictures—serve by themselves as complete sentences.)

And as the 20th Century astronomer SEES THAT *from a suitable location in space the earth would appear to describe an orbit around the sun and not vice versa,* so the 13th Century astronomer SAW THAT *from*

---

[2] *Ibid.*, p. 197.

a suitable location in space the sun would appear to describe an orbit around the earth and not vice versa.

From

which the physicist sees as an x-ray tube, he also sees that *at high voltages fluorescence will appear around the anode,* and he sees that *a regular crystal placed at X will reveal the wave nature of x-radiation.*

Now the steps between the pictures formed on the retina, or the pictures formed in one's private visual space, and the actual statements of what we see are many. Nor is pointing this out just an undertaking of psychology or neurology. It is a point of profound logical significance that the pictures on our retinas and the pictures sense-datum theorists talk about are first and foremost *pictures,* while what is called scientific knowledge is first and foremost expressed in language. There are great logical differences between pictures and language. So too there are great logical differences between what strikes our retina, or our visual sense-data, and what we know, in the sense of what is *known* to modern science.

Not all of the elements of our statements, or of our *that-clauses,* correspond to the elements of the pictures formed in the eye, or in the brain. It would show a serious misunderstanding of the uses of language to expect that they should. And yet we certainly say that we see that clouds are gathering, or we see that dawn is breaking, or we see that the bear means to climb the tree; and we mean what we say. There is much more to these "seeings," however, than what meets the eye, or even the mind's eye of the sense-datum theorist. The "more" here is a *logical* or *linguistic* "more," too.

Putting it very crudely to begin with: There is an important linguistic component to our seeing. We see things sententially, almost always. But there is nothing sentential, or linguistic, about the picture formed in our eye or in our mind's eye.

But why must we see things in this way? Why must there be a linguistic component in our seeing? There is surely a good deal in this vague contention that calls for substantiation. For we are seldom aware of the visual world clicking off before our eyes in sentence form.

Part of the answer is this: Unless there were a linguistic component to seeing, nothing that we saw, observed, witnessed, etc. would have the slightest relevance to our knowledge, scientific or otherwise. There would be no sense in speaking of *significant* observations, for nothing seen would even make sense. Before the wheels of knowledge can turn relative to a given visual experience, some assertive or propositional aspect of the experience must have been advanced.

In order to secure this matter, and at the moment it looks very insecure, I grant, I will try to display the logical gulf between pictures and language. I will try to show that retinal reaction "seeing" and sense-datum "seeing" are, at their best, *picture* accounts of seeing. Scientific knowledge, however, is a great logical remove from *picturing*. But if observations are sometimes *significant, relevant, noteworthy*, etc., they cannot be on either side of the gulf but must somehow straddle it. If we *just* saw pictures *simpliciter* there would be nothing to distinguish one picture from another, nothing to mark this observation as relevant, that one as irrelevant, this aspect significant, that one of no importance.

Seeing, then—the seeing of things like bears and bicycles, ice cubes and x-ray tubes—is what makes our visual experience important to our intellectual life, and not merely a meaningless rhapsody of visual sensations. And in maintaining this I *think* I am dissenting from the views of Professor Price as expressed in his book *Perception*, where he remarks that "The perceptual act . . . is not an activity. There is in it no element of fussiness, no wondering nor questioning. One does not have to take trouble over it—it is a blessed relief from the labour of discursive thought." [3] So it may be. But seeing is not nearly so independent from the fruits of discursive thought as Price's remarks suggest. For if this were so how would our observations ever manage to get on the same logical track as our knowledge? This must happen if observations are ever to be spoken of as relevant, or even irrelevant.

Let me make it clear, however, that the *knowledge* of which I am speaking is largely the knowledge of what there is, as expressed in the language of the textbook, the experimental report, or the lecture. I am not talking of the knowledge of how to do things. Thus we may say of a

---

[3] H. H. Price, *Perception* (London: Methuen & Co., 1964), pp. 152–3.

man that he knows how to ride a bicycle while agreeing that he could not possibly express such knowledge in language (just as most of us know how to whistle but would sound foolish if we tried to give verbal instructions about it). Nor do I recite verbal instructions to myself all the time I am repairing a clock, though I should feel no qualms in saying that I know how to repair clocks.

If, however, I were to try to instruct someone else in the gentle art of clock repairing, I should probably have to do so in language. Grunts and gestures would be as inappropriate to that task as they are to most science lectures at a university. I should have somehow to translate my knowledge of how to repair clocks into the *knowing that* idiom: I should be saying that *if the balancer is oscillating too quickly the hairspring should be given more freedom*, or that *if the driving spring wheel is wearing away the hand pinion a lantern-pinion should be substituted*, etc. And in saying such things I may be expressing myself sententially for the first time.

It is of course in this latter sense of knowledge, the linguistic sense, that I am construing *scientific knowledge*. For unless what a man knows can be expressed in language, in books or journals or lectures, he cannot very well communicate his knowledge to others. This is not, of course, to say that scientists rely little on knowing how. The "feel" of things, the "look" of a situation, are often the factors that incline a man's research one way or another. But for just this reason these imponderables are not likely to affect the general *corpus* of scientific knowledge, and hence the way the world will appear to the general scientific eye. This is the reason why surgery is so much an art, why the great anatomists were as much artists as scientists, and why the exchange of ideas between mathematical physicists is so often an almost aesthetic interaction. But without minimizing the importance of this scientific *knowing how*, particularly as it distinguishes individual scientists, I will restrict the expression "scientific knowledge" to the linguistically formed knowledge that is the substructure of the literature and teaching in a scientific education. As Wittgenstein puts it, ". . . knowing it only means: being able to describe it. . . ." [4]

Certainly, before our visual impressions can be of any relevance, or even be recognized as irrelevant, to scientific knowledge in this sense,

---

[4] Wittgenstein, *Investigations,* p. 185.

they must be observations, seeings. They must include *seeing as* and *seeing that*. For, surely, the ground floor of the language of science, the level closest to mere visual sensation, is nothing if not a series of statements. Statements are true or false. Pictures, retinal or mental, are neither true nor false. The language of science is a language of statements. Our retinas, however, are not impressed by statements, or even anything logically like statements.

Yet, *what we see* is certainly enough to determine the truth or the falsity of statements like "The sun is above the horizon," "The cube is transparent," and "There is a bear on the other side of that tree." So our visual sensations must be cast in the form of language before they can even be considered in terms of what we know to be true. Until a visual sensation can be so considered, it is not observation, it is not what we usually call *seeing*. It is more like the oculist's phenomena, or the buzzing confusion we encounter immediately on recovery from unconsciousness, or the vista of the professor who stares out of his college window lost in thoughts about ancient Greece. Or, as Wittgenstein remarks, "I looked at the flower, but was thinking of something else and was not conscious of its colour . . . [I] looked at it without seeing it." [5]

It is this linguistic requirement, made of anything that is to be relevant to our knowledge (scientific or otherwise), that makes *seeing that* a logically indispensable element in significant seeing, in observation. Our everyday knowledge is not a montage of sticks, stones, color patches, and noises. It is, rather, a system, more or less, of propositions (or utterances having a propositional force in a given context). Scientific texts are not picture books. Scientists are not cameras. [Notwithstanding Isherwood's play entitled "I am a Camera."]

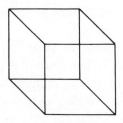

---
[5] *Ibid.*, p. 211.

asserts nothing. It is neither true nor false. Neither

nor

tells us anything about the world.

may be inaccurate, but it is not a lie. (What would evidence to the contrary be like?) When a first-year biologist presents the laboratory supervisor with a clumsy graphic representation of what his microscope reveals, the latter may accuse the freshman of carelessness, or inattention to detail, or of misrepresentation, but not of prevarication.

This is the first essential step in trying to show the discrepancies between pictures and language. In what follows we will try to wedge these two types of representation apart for good, showing thereby the great gap between retinal or mental pictures and our knowledge of what there is in the world. This will also show that seeing or observation in science, if it is to be relevant or significant as we so often have occasion to describe it, must be composite, or, as Wittgenstein might have said, ". . . half visual experience, half thought. . . ." [6]

---

[6] *Ibid.*, p. 197.

8 | Seeing, Saying, and Knowing

CONSIDER arithmetic. There are at least three possible ways in which one might consider the statement that $x^n \cdot x^n = x^{2n}$. We might consider it in terms of the historical context and conditions in which this notation and this particular arithmetic truth became established. This would involve us in talk about culture, civilization, intellectual history, and great mathematicians.

Or we could consider the statement psychologically. That is, we could consider how difficult it was for little Basil to learn to manipulate this particular sort of notation, but how he finally caught on when Professor Snodgrass explained it to him with toy blocks. And we would note the increased facility Basil showed in his future calculating due to the psychological aid he had received through the mastery of this statement.

Finally, we could consider the statement *formally*, as a relation between concepts or, perhaps, as a way of making clear the *meaning* of the concept of *exponential multiplication*. And to treat $x^n \cdot x^n = x^{2n}$ in this analytic way does not oblige one to raise either historical or psychological questions. The conceptual importance of this arithmetic proposition can be entirely and adequately grasped without considering either when or how the proposition came to be known to men or to a single man.

So too we may observe that there is something logically queer about the statement "He knows slowly that $x^n \cdot x^n = x^{2n}$" and we may note this without going hip-deep into theories of knowledge, neurology, and the history of thought. That there is something queer about the expression "knows slowly" is a commentary on the *concept* of knowledge, and not on the neurophysiological, behavioral, or historical basis of knowledge.

Likewise we would recoil in horror if a surgeon remarked that, while he had explored the patient's cranium thoroughly, locating the pineal and the pituitary as he went, he had had no success whatever in locating the patient's mind. And this would be a horror based on the realization that the surgeon had thoroughly misunderstood the logic of the word "mind," however skillful he might be with scalpel and sticky tape.

And if a zoologist were to weep over his bench lamenting on having found the pigeon's cornea, macula, and retina *but not its vision*, we should think he was not quite clear about the nature of his inquiry, however thoroughly he would recite his lessons on the embryology and neurology of the pigeon's eye.

It is in this way that I have been addressing myself to the concepts of seeing, noticing, observing, etc. I do not need to be convinced that the physiological and psychological *reasons why* we have these concepts are altogether interesting and exciting. And I dare say that some of my remarks may have set you thinking along these lines. But please be clear that *I* was not thinking along these lines, however much I might wish I had the ability and the training to do so. I have been concerned strictly with the conceptual issue: What do we usually *mean* when we say that we see X? What is the nature of the *concept* of seeing, or observing? And this question can be broached without taking on any psychological or neurophysiological burdens beyond the few simple sketches I have drawn from time to time. I have imagined myself to be talking logically, not psychologically or physiologically, even though these latter idioms may be, as Wittgenstein observed, *symbols* of the logical.

Thus, I have been trying to argue that, just as there is something logically askew in expressions like "know slowly," "looking at minds," and "discovering the vision," so there is something logically askew about the accounts of seeing that make it all sound as if we soak up light radiation like blotters and then clamp our interpretations on our visual impulses, *click-clack*—but oh, so very quickly. This is not anything like what I *mean* when I say I see a bicycle, or a duck, or a bear, or an x-ray tube. If there were something corresponding to our good old word "interpretation" going on in me while I see, then I ought to be able to consider such an act by myself as I can when I interpret history, art, or an ambiguous remark. I should not have to invent such an act just in order to set into motion my philosophical theories about seeing.

"But it's all unconscious, it's all instantaneous." *It's all blarney!* That's not what *I* mean by *interpretation* nor, I conjecture, what *most* people unencumbered by philosophical theories mean by it.

Never mind. Say that seeing involves unconscious interpretation if you must. But then notice the comprehensive difference between conscious acts of interpretation—from which the word draws most of its force—and *unconscious interpretation*, a curiously timeless, colorless, ineffable, intangible, invisible, unlocatable, indiscoverable entity. Talk about ghosts in machines!

Well, my quite modest task has only been to show that the ordinary logic of *seeing* does not square with the phenomenalist's theories about what that logic *must* be. I have been urging that *seeing* is a theory-laden activity. And for this purpose excursions into genetic psychology and neurophysiology *are not necessary* (which is not to say that they are never helpful). One can talk philosophy *by way* of a factual discipline without thereby allowing his case to stand or fall on the success of that discipline.

If the notions of *picture* and *language* can be split apart we will have displayed a logical gulf between retinal or mental pictures on the one hand and scientific knowledge on the other. The status of what are called "significant observations" or "the seeing of relevant factors" will have been once and for all removed from the area in which retinal or mental pictures reside. For *significance* and *relevance* in science are notions that depend on knowledge and upon language. Things, events, and observations are not intrinsically significant or intrinsically relevant. They are so only against the background of what we already know about things, events, observations.

To establish this we will digress slightly from our main theme: namely, the question whether a 13th Century and a 20th Century astronomer see the same thing in the east at dawn. Now we will consider how it is that visual pictures (whether retinal or mental), like most pictures, photographs, and diagrams, are not by themselves relevant to scientific knowledge without the mediation, indeed the participation, of linguistic forms in the actual visual perception. What we see *is* very often relevant to our knowledge because of the linguistic threads woven through every observation. I will try to secure this position by a *reductio ad absurdum* argument which will run as follows: If this were not so, if seeing were a purely visual phenomenon untainted by any of the effects of language, then nothing that we saw with our eyes would ever be relevant to what we know about the world, and nothing that we know about the world could even have significance with respect to what we say we see. A hopelessly intolerable situation, you will agree. In such a state of affairs our visual life really would be a rhapsody of blobs, shapes, and points of light, by definition unintelligible. And our knowledge of the visual world would be as we now construe the blind man's to be. Our vision would be without understanding, our knowledge without light. If this is absurd it will show again how different is seeing from the mere formation of retinal and mental pictures, and how central to the connection between vision and knowledge is the concept of *seeing that*.

Now, some very considerable Cambridge philosophers, e.g., Lord Bertrand Russell, Professor Ludwig Wittgenstein in the 1910's and 1920's, and Professor John Wisdom about 30 years ago, thought that the connections between the drawing of pictures and maps on the one hand and speaking and writing on the other were very intimate. At one time or another each of these thinkers tried to explain the nature of language by stressing its likeness with the features of mirror reflections, photographs, paintings, and maps. They were even led to such poetic extravagances as that "language is the mirror of reality." It will be instructive to see where and why these attempts proved, as I think, inadequate. For if one can point up the logical gap between pictures of things and descriptions of things, the difference between visual pictures and scientific knowledge will have become more apparent. It is clear that the so-called "picture theory of language" would encourage a sense-datum, phenomenalistic, or picture account of seeing, an account once offered by each of the philosophers mentioned.

In a pretty obvious way the elements of mirror reflections, pictures, and maps correspond to the elements of the things reflected, pictured, or mapped. Thus a mirror reflection of a picture of

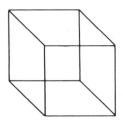

or of

will not differ with respect to the type or quantity of elements represented, e.g., corners, lines, claws, paws, etc. A drawing of an x-ray tube will show one anode target and one cathode source just as the tube itself has one of each of these elements. And a map of Cambridge, England, will show the Zoological Laboratories, King's College Chapel, and the Sports Grounds in roughly the same numbers, and the same relative position, as they would be found in Cambridge itself. Let us agree to call mirror reflections, pictures, and maps *signs* or copies. They are not, of course, signs in the sense that cumulo-nimbus clouds are a sign of rain, or smoke a sign of fire, or a skin rash the sign of a kidney ailment. That is, mirror reflections and pictures are not *symptoms;* they merely stand in a certain relation to things reflected or pictured which, for want of a better word, I shall call a *signifying* or *copying* relation. Each element of the sign or copy, then, represents some element of the signified, i.e., the original. And the arrangement of the elements of the copy shows the arrangement of the elements in the original. Thus,

and

represent a box, a bear, an x-ray tube, and the sun at dawn, just because of this correspondence.

There is, however, no differentiation of function among the elements of the copy. No one blob of color, or line, or point, or patch of light plays a role specifically different from that played by any other blob of color, line, point, or patch. All the elements of the copy, though they may signify different elements of the original, bear exactly the same relation to those elements: that of *representing*.

Finally, both the copy and the original must be of the same logical type. Reflections, pictures (retinal, sense-datum, or artistic) may be of boxes, bears, x-ray tubes, or the dawn; but they are never, and could never be, of the weight of boxes or the growling of bears, or of the silence of an x-ray tube or the warmth of the morning sun.

To sum up these features of the "copy-original" relation:

1. Each element in the copy represents some element in the original, though not necessarily *vice versa*. Maps, e.g., do not reproduce every detail of the area they map. Yet, nothing is put on a map which is absent in the area mapped.

2. The elements of the copy show no diversity of role. Color patches just show color patches. Landscapes are not punctuated with "→" or "=" or ">", and certainly not with "if-then," "is identical with," or "is greater than."

3. The arrangement of elements in the copy will show the arrangement of elements in the signified. If a picture is to be of a bear, the lines and patches in the picture must be arranged and organized just as the bear is: head on top, tail behind.

4. The copy must be of the same logical type as the signified. We can reflect or picture a bear, but not the noises it makes.

However, something is askew already. Pictures and maps are two-dimensional, while that which they depict is very often three-dimensional. Two-dimensional signs of three-dimensional things signified! Mirror images are, in an optical and a physiological sense, three-dimensional, but in another sense they too are two-dimensional. That is, we are restricted with respect to the elements of mirror images just as we are restricted by the elements of pictures and maps. We cannot be embraced by a picture or a mirror image of a bear, though real bears have been known to show such affection. So reflections and paintings of bears and x-ray tubes are not all dimensional replicas of bears and x-ray tubes in the ways that solid models of bears and x-ray tubes are. The copy-original relation as exemplified in reflections and pictures, therefore, is less than a perfect correspondence in that a dimension somehow gets lost in the process. And this must be patched up or accounted for, if it is to serve

Seeing, Saying, and Knowing | 135

as the inspiration for a theory of the nature of language. For it would be nice at least to begin with a complete one-to-one correspondence between the copy and the original.

There are two ways of making the correspondence more thoroughgoing. One is to take as our model of the copy-original relation, not things like mirror reflections and portrait paintings, but solid models and sculpture. Three-dimensional signs of three-dimensional things signified! The correspondence between the real bear and the museum bear is in this respect more thoroughgoing than the correspondence between the real bear and the oil-painting bear. But this alteration holds little promise as a clue about language. For though the child may treat his word-blocks as if they themselves were words, we old folks do not use such solid tokens in order to communicate. The sculptor may speak to us in stone, but this is surely a metaphorical use of "speak." Could the sculptor tell a lie in stone? Could he contradict himself? No. So if the sign-signified or copy-original relation is to be made more complete it is not by fattening up signs into three dimensions, for that is a blind alley when it comes to the study of language.

The other way of making the correspondence complete is *not* to tack a dimension onto the sign or copy, but to slash one off of the thing signified. At first this may seem as fruitless as the other alternative, for if there is anything that may be said about our world it is that it is indisputably three-dimensional, hard, bulky, ponderous, and not anything like a flatland of stage scenery populated by a lot of playing-card objects.

And yet this is the course each of the philosophers mentioned chose at one time or another, and for reasons that ought by now to be familiar to you, reasons which consort very well with their attitudes to language, to knowledge, to experience, and to perception. (A significant policy in one of these areas of inquiry is likely to affect all the others considerably.)

Yet, we *say* that the world is three-dimensional—but is there not something a little vulgar about this? What is it that we are *really* visually aware of? Does our eye take in the solidity of objects? Is our private visual space a kind of warehouse for the world's furniture, wherein we can visually inspect objects on all sides simultaneously, intuiting, as it were, their solid dimensions as well as our familiar one-view-at-a-time apprehension of them? No. The pictures in our eyes, and our minds, are of course FLAT. Or, if it is urged that they are three-dimensional, this cannot even be the optical three-dimensionality of mirror reflections. We can no more be embraced by the retinal or mental picture of a bear than we can by a reflection of a bear. But what is there to the visual world beyond the pictures we get of it through our eyes and our visual cortex? All the rest is mere inference from the experiences of other senses, like touch and perhaps hearing as well. And this goes for stereoscopic vision too. Consider what the neurologists Holmes & Horrax say about this: [The visual cortex is a receptor for comparatively crude and simple visual sensations, and localization of objects in the third dimension] is ". . . an intellectual operation based largely on non-visual experience, as on the testing and controlling of visual perceptions by tactile and muscular sensations from all parts of the body." [1]

So in just the way that a picture theory of language supports the sense-datum account of seeing, the sense-datum account of seeing helps us on towards the acceptance of a picture theory of language. For all that we ever *really* see consists in flat pictures in our private visual spaces (after all the intellectual masking tape has been removed from what we vulgarly call *seeing*). And that is all we can ever really apprehend of the visual world anyhow. So the correspondence between signs, like pictures and drawings, and things signified—bears, boxes, x-ray tubes—is academically quite tidy. Pictures and drawings are two-dimensional, and the sights, or views, or mental pictures we have of bears and boxes are also two-dimensional. No dimensions are lost here.

As Wisdom observes, "When a mirror mirrors a scene, then (1) for each coloured patch in the reflection there is a *coloured patch in the scene* and vice versa, and (2) the form of the scene is also the form of the reflection—if coloured patches in the scene run red, green, blue then coloured patches in the reflection run red, green, blue. . . ." [2] The scenes presented by the physical world, then, consist in color patches, expanses of color in two dimensions. The correspondence is complete; all we have is a picture picturing a picture, the paradigm of the copy-original relation. The complications of the third dimension are only intellectual importations anyhow and should never have troubled us had we but seen the visual world as the sense-datum philosopher does.

---

[1] Holmes and Horrax, *op. cit.*, p. 385.
[2] John Wisdom, "Logical Constructions, I," *Mind*, Vol. 41 (1932), p. 202. Hanson's italics.

I will not bother to attack this "twist" to the copy-original relation; in any case its weaknesses are fairly clear, particularly in the way it would dismiss stereoscopic vision as an intellectual addition to *pure* seeing, making the Cyclops' depthless vision into the standard case of seeing. It is enough to say that if this were a correct account of the relation between the pictures of the world we draw and the pictures of the world we have, then the correspondence between the sign and the signified would be complete. The way would then be open to us to stress the logical similarities between *language* and *pictures:* If using a language can be shown to be in all its essentials a kind of picturing, then (logically) language will be related to the visual world just as pictures are. Language will then be a kind of logical reflection of the world as seen; it will be a perfect logical schematic of the private pictures within our individual visual spaces, in the best tradition of *sign* and *signified, copy* and *original*.

And all our problems about how scientific knowledge and the language in which it is expressed are related will be at an end. For if language just mirrors what we "see," then scientific knowledge is merely a kind of logical cinematographic record of the observations made by great scientists. Having a visual impression just is to enrich one's language and one's knowledge by that much. And all this talk about *seeing as* and *seeing that* and all our fuss about our two astronomers (by now transfixed before a sun which must surely have got stuck on the horizon to have remained there through all these chapters) is down the drain.

But *is* language just a kind of picturing? I think it is not. There are fundamental logical differences between the character and uses of language and the character and uses of pictures. I mean to point up these differences now and split asunder language and pictures; or, rather, to show that the way (or ways) in which language is related to the world is not the way (or ways) in which signs are related to what they signify. If this can be done, and I think it can, then *scientific knowledge* will have been split asunder from the having of those visual experiences that might be called retinal pictures or mental pictures. And if this is done, then the great gap between what we know and what we experience as retinal or sense-datum pictures will be apparent. It will be clear, moreover, that if this were all there were to it, then nothing we know would be, *could* be, relevant to what we apprehend visually. Nor could what we apprehend visually affect in any way what we know.

Nor would expressions like "significant observation" or "seeing what is relevant" have any clearly specifiable use. There *is* a gap between picture-having (retinal or mental) and knowing, and the complicated concept of *seeing* is what spans it. I almost wish to say, as Wittgenstein was also inclined to say, that for such reasons seeing is half visual experience, half thought. But that would not be right, for reasons advanced earlier. It is all just seeing. We do not do two things—react visually, then think—and I invite any theorist who feels otherwise to show me the double operation in what we ordinarily call *seeing*. We just see—ducks or rabbits, bears, bicycles, x-ray tubes. And yet, indisputably, *seeing,* with its *seeing as* components, does bridge the gap between the pure camera in use and the pure electronic brain in use. *Seeing* is just what we might call *appreciating the visual world as intelligible*.

But I am ahead of myself here. I must first show how philosophers and others have tried to assimilate language-using to picturing, and indeed why they have thought it necessary to do so. Then I must make it clear how I think this assimilation to be a mistake. Language-using and picturing are two logically indistinct operations. But, if this is so, scientific knowledge (essentially language-using) is logically distinct from the having of normal retinal responses or visual sense-data (both of which are, essentially, picturing). And the bridge I mean to indicate between the two is just the bridge we have always known it to be: *seeing,* the seeing of *physical objects as* objects of a certain sort, as objects *that* may be characterized in certain sorts of ways. If I can do all this, then I am free to suggest that the differences in seeing we have so far been discussing in such detail are to be sought, not in the retinal or mental pictures of which philosophers and scientists have spoken so much (for there need not be such differences there), but in the knowledge and the language which can affect the ways in which we see, or observe, or witness the world at work.

Of signs and what they signify we said earlier that each element in the copy represents an element in the original, one-for-one. The elements in the copy, however, are not variable in their functions. They just *represent* elements in the original. The arrangement of the copy's elements, moreover, will show the arrangement of the original's elements, and the copy and the original must be of the same logical type.

Now, looked at in a certain way our language might very well be thought to approximate to these requirements of the sign-signified or copy-original relation.

Consider:

and the sentence "The bear is on the tree."

If the picture is true to life, then just as it contains a bear-element and a tree-element, so in the original there are a bear and a tree. Likewise with the sentence: If it is a true sentence, then just as it contains the word "bear" and the word "tree," so the situation it describes contains a bear and a tree. The picture has conjoined

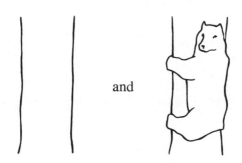

to form a representation of the relation of bear and tree in the schema "The _____ is on the _____." Just as the picture has two elements related exactly as are the two elements of the original situation, so the sentence has two word-elements in a verbal relation that signifies the actual relation of the actual bear and actual tree in the actual situation described.

And if these are true copies they can contain nothing that the original lacks. Thus if the picture were elaborated, putting tusks or a dorsal fin on the beast pictured, or if the words "tusk" or "dorsal fin" were worked into the sentence (in both cases without modifications to the original), neither the picture nor the sentence would any longer be a true copy of the original.

So too if either the picture or the sentence lacked an element possessed by the original it would fail to be a true copy.

Furthermore, the elements of the picture just stand for or represent elements of the original. So too the words "bear," "tree," and "is on" stand for elements of the original. This would be even more apparent were we to choose a more terse symbolic idiom to express our sentence, say "b R t" where "b" = *bear*, "t" = *tree*, and "R" = the relation of *being on*.

Moreover, these copies, by arrangement of their elements, *show* the arrangement of elements in the original situation as depicted or described. Thus,

(a)

by its arrangement, *shows* a bear on a tree, as

(b)

by its arrangement, *shows* a tree on a bear.

And "b R t" (i.e., "The bear is on the tree") *shows* by the arrangement of its elements how the real bear and the real tree are getting on, as "t R b" (i.e., "The tree is on the bear") would not. There is no theoretical limit to the complexity a physical situation may manifest while still allowing language to reflect the disposition of its elements. Thus the situation in which an x-ray machine was left in operation overnight in a laboratory causing certain photographic plates to "fog" might be described in the language "There is an object X in the laboratory such that X is an x-ray machine and, because the photographic plates were fogged, X must have been left operating overnight." Or in the terse "mirror-writing" of the logician, $(\exists x) (Xx \cdot (P \supset Ox) \cdot P)$. If this last symbolic assertion could but be held up to the world for comparison, or at least held up to the laboratory for comparison, we would be able to check off each element in the statement $(\exists x, X, P, O)$ against elements in the laboratory such that the arrangement of $\exists x, X, P,$ and O in terms of the symbols "·" (*and*), "⊃" (*if . . . then*) will show the arrangement of the x-ray machines, photographic plates, and selector switches, i.e., the elements of the laboratory situation.

And should one become uneasy about all that is being packed into symbols like X, P, and O in this example, then it is open for him to suggest that these symbols are really logical compounds of very much simpler symbols relating, not to a complex situation such as we have to describe, but to the contents of one's visual sense-field. So that even if one feels uneasy about the possibilities of checking off "$(\exists x) Xx$" with the presence of an actual x-ray tube in the lab—for we have already indicated how this may not be open to everyone, but perhaps just to people who have learned enough to be able to identify x-ray tubes—it could still be argued (as Russell, Wittgenstein, and Wisdom did all at various times argue) that "$(\exists x) Xx$" is an intricate logical compound made up of much simpler elements like "There is an appearance in my visual field of color patches, glassy clear, metallic silver, and coppery brown, composed in circular shape with glassy appendages at either pole, etc." It is the fundamentally simpler observational reports, then, that are direct copies of the contents of the observer's private visual field. Hence the picture account of language goes hand in hand with the sense-datum view of seeing and with the logical construction account of the concepts of science. We have our private visual sensations, we copy these straight out into the basic language of visual sensation, the sentences of which are fed into a logical construction machine, there to be built into our familiar sentences like "A bear is on that tree" and "Someone failed to switch off the x-ray machine last night," etc. A good deal of the philosophy of logical constructions, of sense-data, and of basic languages therefore hangs on this picturing account of the nature of language. The suspected gap between our visual impressions and our knowledge, or between the pictures we have and the language we use, is closed before it ever gets open, by the argument that knowledge and language are just logical constructions out of basic, fundamental statements that are logically related to our sensations as copy to original, as sign to signified. The elements of our lowest-level sentences are in complete correspondence with the elements of our raw visual experiences; the elements of the sentence all represent some element of the observer's mental picture, and the arrangement of the elements in the sentence shows the arrangement of the elements in the observer's private mental picture.

There was another aspect to the copy-original terminology, however. The copy had to be of the same logical type as the original. One can at least try to draw the Mona Lisa's smile, but there is no point in trying to draw the sound of her voice. So the picture idiom must be expanded in the case of language. A picture is a copy of a scene; so too a hi-fi recording is a copy of a sound heard. But language is much more versatile. It can give an account of scenes *and* sounds. It can supply us with logical pictures of scenes *and* of songs, of windmills *and* of whistles, while pictures and recordings can be of either scenes or songs *but not both*.

This is just the beginning of the differences between pictures and language, however.

Reflections, images, pictures, copies are, to use an expression of the 19th Century American philosopher Charles Sanders Peirce, icons. An icon is a sign which has at least one of the properties of that for which it is a sign. It signifies in virtue of this property. This is why the word "copy" was a good one for bringing out how pictures are related to scenes. [And of course the word "scene" itself is most useful in this connection. For scenes may be seen from hilltops, as when from Coit Tower we gaze on the scene San Francisco presents, or we may see the same scene in an art gallery in the form of a painting of San Francisco executed on that same hill. Scenes are somehow common to nature and to the *copies* of nature made by artists.] A drawing of a bear and her cub is an *icon*. For in the drawing, the relative dimensions of the actual bear and the actual cub are shown. Reflections, images, pictures, and maps duplicate most of the spatial properties of what they reflect, image, picture, or map.

Sentences, however, are not (usually) icons. In "The bear is bigger than its cub," "bear" is printed in type of the same size as "cub." We need not write "THE BEAR is bigger than the cub" in order to convey our meaning, whatever the specific problems encountered by the writers of children's books. Nor need we growl "bear" and squeak "cub," however effective this may be on children's radio programs.

Moreover, it is not the case that reflections, images, pictures, and maps are icons to the same degree. A mirror reflection of New York City will not be a copy of New York in the same sense that a picture or a retinal or mental image of New York will be a copy. And all of these would differ from a map of New York. The buildings along Fifth Avenue are very different in color and shape from those to be found on Delancey Street. But this difference would not figure in a map, and only partially in a comparative line drawing. In an obvious way the more iconic a map is, the more like a mirror reflection it is, the less useful is it as a map.

Language is the least iconic of all. Of course, there are a few words like "buzz," "tinkle," and "toot" which, when spoken, sound somewhat like the sounds to which they refer. But these are the exceptions that display the rule. Modern languages are almost wholly conventional in nature, and this is particularly true of scientific languages. They are not primarily iconic: There is nothing in the look of the written word "bear" to remind us of the look of a real bear, nor is there anything in the sound of "bear" to remind us of the scratching and growls of a bear. That "bear" can be used in reference to bears is due to a rule or a convention by which we come to coordinate the word with the thing, rather in the way that a red flag has come to be a signal for danger, though there is nothing inherently dangerous in red flags. (A yellow flag could do the same job given enough publicity in advance of the change.) So too there is little resemblance between actual bishops and the chess tokens of the same name. And we can appraise the rules and conventions governing the moves of a bishop in chess without any reference whatever to such resemblances as there are between actual bishops and chess bishops.

A picture or reflection of a bear may lead us to remark, "Here is a bear." Or we might indicate a mark on a map of Connecticut, saying, "Here is Hamden." Would we ever distinguish a word in a sentence in this way, however? In the sentence "The bear is on the tree" is the word "bear" a bear—a real, live, furry bear? Who would ever say it was?

That icons are like what they signify is signaled by our willingness to say of some icons that they *are* what they represent. At the cinema we say, "Look, there is St. John's Chapel!" and not "Look, that represents, or stands for, St. John's Chapel." But of the *words* "St. John's Chapel" we would not say, "Look, there is St. John's Chapel." Words are not what they signify, with the one exception of the word *"word."*

Drawings may vary. Of a roughly sketched ursoid shape we might say, "That's a bear." Or we might say, "That's supposed to be a bear," or even "That represents a bear." We would almost surely say this of a drawing by an infant, a maniac, or a modern artist. Our tone of voice would, of course, be different in each case.

In general, *icons show* (e.g., a bear climbing a tree). A sentence does not show a bear climbing a tree. A *sentence states,* or can be used in stating (e.g., that a bear is climbing a tree). Showing a bear climbing a tree consists in *representing* a bear and a tree and *arranging* them in a bear-climbing-tree manner. Stating that a bear is climbing a tree consists in *referring* to a bear and *characterizing* it as climbing a tree. Representing and arranging are very different from referring and characterizing.

Needless to say, all these differences are to be found at the logically fundamental level of private pictures (visual sense-data) and basic sentences. Saying that one sees a bear may be rather more than merely registering what occurs in one's visual consciousness. Indeed, we have been trying in ever so many ways to make that very point. For all that I may be visually aware of may be a flat grouping of color expanses, say a brown ursoid patch superimposed on a brown arboid patch. Nevertheless, by the same argument the sentence that would record this awareness, e.g., "I am aware of a brown ursoid patch, etc." would differ from the actual picture registered in the observer's private space in all the ways we have been considering. The picture would be of X . . . ; the sentence would be to the effect that X . . . ; the picture would show X . . . ; the sentence would state that X . . . . The picture would represent elements as arranged in a certain way, the sentence would refer to things characterized in a certain way. All of which is to say that there is a logical gap of immense breadth between pictures and language, a gap which is not closed one millimeter by the focusing of attention on the area of immediate visual experience and its verbal expression. Sense-datum experiences and sense-datum sentences are as far apart as pictures are from statements.

> [In all this I want to avoid the reference to ideographic, hieroglyphic, and "picture languages." It should be clear that I am not dealing with the special and interesting problems raised by ancient Egyptian and Chinese script or the onomatopoetic speech of certain non-literate societies. It is the highly conventionalized vernacular and scientific languages in which *we* read, write, and think that are our concern here. And of these it must be clear that they are logically, even if not genetically, distinct from *icons,* whether they be mirror reflections, drawings, or the pictures registered in the private visual space of observers.]

Some elements of a sentence, e.g., "bear," "tree," can be used to refer. Some elements (e.g., ". . . is on . . .") are used to characterize. Not all the elements of a sentence do the same work, as we have had to learn so painfully in classes on grammar. But *all* the elements of icons just represent, or show. If you cut a picture, e.g.,

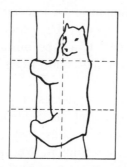,

into six pieces you will end up with six small pictures. But if you cut up the sentence "The bear is on the tree" into six pieces you will not end up with six small sentences. Again, the picture shows something, and each of its elements shows something too. But while the sentence can be used to state something, not all of its elements could, by themselves, state anything. True, "Bear!" shouted to the members of a hunting party will do the work of a statement. Likewise "Tree!" shouted in a tree-felling contest. But it is hard to imagine contexts wherein "the," "is," and "on" would behave like statements. So, while the elements of pictures are tiny pictures, the elements of sentences are not tiny sentences.

This could be developed further, showing how the elements of language are parts of a vocabulary, words with a relatively fixed significance. But this is not true of the elements of pictures, where lines of a given contour could go to form many different types of shape. This would lead us to look upon maps as a half-way house between pictures and modern languages: Unlike pure pictures, maps have the rudiments of a vocabulary in such symbols as ⚒ , ◎ , and ✈ . But though they are partly conventional (notice how we speak of *reading* a map), maps must also be partly iconic if they are to be any help at all. They must show the coastline as it actually is and the relative distances between cities as they actually are. Had philosophers of language paid more attention to the differences between maps and mirror reflections, and the differences among both of these and a conventional modern language, the picture model might have met with less enthusiasm as a clue about the nature of language. For language is not even restricted with respect to logical type, as are pictures. We do not even try to draw the growl of a bear, but we could describe it in language. It is this freedom with respect to language (a freedom largely absent in icons) that makes type mistakes possible: The schoolroom quip "Do not look at me in that tone of voice," is a type mistake. So is "I can find his pituitary gland but not his mind," and "I can find his retina but not his sight."

Hence there are at least two modes of signification. Reflections and pictures are icons, they copy originals. Sentences used as statements are more conventional. They are not icons, they are not copies of originals. Maps are part icon, part conventional. We must learn to read maps as we do books, and as we do not learn to read mirror reflections and pictures (retinal and mental). So there is a logical gulf between pictures and language which *maps* partially bridge. And there is a corresponding gulf between the retinal or mental pictures (i.e., pure visual sensation) and our scientific knowledge which *seeing*, particularly *seeing that,* bridges. As you would expect, therefore, there are thoroughgoing resemblances between *seeing* and *map-reading,* but we will not go on to develop them here.

It must be clear, then, that language is not related to the world as a copy is related to its original. Language and pictures are logically different types of entities. Hence scientific knowledge is a logically different type of entity from mere visual impressions. And yet the two must be intimately strung together; otherwise, scientifically central conceptions like those of *significant observation* and *relevant data* could not even be formed. Knowledge would be blind, our visual sensations a chaotic whirl of blobs, shapes, color patches, indeed, all the things so dear to phenomenalism. For these *logical* reasons, then, *seeing as* and *seeing that* must be active elements in what we call *seeing* in science. Not only do scientists see things as being things of a certain sort, which is just to see that were certain operations performed other actions would follow, but, by the long argument just undertaken, we see that the situation could not be otherwise. It is a matter of logic, not merely a matter of fact, that *seeing as* and *seeing that* are indispensable to what is called, in science, *seeing* or *observing*.

This is the vital point most likely to be obscured by all the philosophical overstatement about knowledge deriving from sense experience, from memory, association, correlation, etc., and the uncritical accounts of observation as a simple opening of one's eyes and looking. Memorizing, associating, correlating, and comparing our mental pictures of the world may be undertaken *ad infinitum* without one step having been made towards scientific knowledge, i.e., some proposition or set of propositions known to be true. It would be like shuffling a lot of photographs of whales waiting for the statement "Whales are mammals" to come forth. So long as the linguistically formative element is left out of our examinations of visual experience and observation, we will endure accounts of science that overstress by far the role of sensation and low-grade experiment in the actual accretion of our knowledge of nature. Science will be forever represented as a repetitious, monotonous concatenation of sensations or of school laboratory experiments. But of course very much more is involved. And appreciating the complexities in the concept of *seeing* is a first step in seeing how much more is involved.

I should like to conclude this chapter with an aside that is calculated to support (not by argument but by persuasion) the contention that seeing is a *bridge* concept spanning the logical gulf between our visual pictures and our knowledge, just as maps span the gulf between landscapes and descriptions.

While we can *read* maps, and interpret them, we can also see the countryside through them, *observing* the location of cities and rivers. The language of knowing is much richer than this, however, in idioms drawn from seeing, perceiving, detecting, illumination, and light. Indeed, when it is seen how much our ways of talking about our knowledge are drawn from the language of vision, it may be more readily agreed that into the concept of *seeing* itself flow many more factors than merely what meets the eye. Specifically, seeing and knowing are interdependent concepts, for consider how their idioms intermingle.

People are said to *see that* a theorem is incorrect, or they *notice* a contradiction, *perceive* that an argument is unsound, or they *observe* that the diagonal of a square is incommensurable with the side, they take a *dim view* of things, throw *light* on matters, are sometimes *blind* to the truth, or do not see things clearly. They have *outlooks,* and *insights,* make *searching* criticisms after having *surveyed* the scene. They are sometimes *shortsighted, clouded* in unreason, and *will not* see the *light;* they get lost in *fogs* and speak of the *dark ages*—and the En*light*enment, Dominus Illuminateo Mea! But *certain bright* people, certain *visionaries,* get *brilliant* ideas, *flashes* of comprehension, and by their *illuminating* pronouncements *fire* our imaginations and give us *glimmers* of understanding. We are made to *reflect, introspect,* but left to ourselves we begin to chase *shadows,* we give distorted *pictures* of things, make *obscure* remarks, draft *illusory* programs, and lo, a *chink* in the wall of our ignorance, a *spark* of genius is fanned, we *see* through our problems, everything is pellucidly *clear,* and we *glow* with anticipation of the future. Perhaps we even *burn* with a hard, gemlike flame.

Need anything further be said about the intimate connection of seeing and knowing? Oh, and incidentally, I know several people who, after being fitted for reading glasses for the first time, said that they could not *think* with their spectacles on.

9 | Spectacles behind the Eyes

IT HAS been said of Sir Arthur Eddington, whose philosophy of science included many Kantian *a priori* considerations, that he had been very nearsighted all his life. At a late age he was fitted for spectacles and for the first time began really to take in the visual data. The implications of this uncharitable gossip are of course that the *a priori* elements of Eddington's philosophy were primarily the result of myopia and only secondarily the result of methodological and scientific considerations.

This observation has no real bite because it is mostly unjust, but there are a few teeth in it. Eddington did rather underemphasize the importance of experiment, the laboratory, and observation in his account of the total conceptual framework of physical science.

The substance of our discussion so far is in similar danger. I have been stressing how very much more is involved in seeing, in observation, than the mere having of visual sensations. The sort of *seeing* that dominates laboratory experience is a seeing of objects or events *as* objects or events of a certain variety. It is *seeing that,* were a certain sort of operation performed with respect to that object, a certain sort of reaction would follow. It is a seeing which gets its cast and its hue from our already established knowledge and from the shape of the language in which that knowledge is expressed. In short we usually *see* through spectacles made of our past experience, our knowledge, and tinted and mottled by the logical forms of our special languages and notations. Seeing is what I shall call a "theory-laden" operation, about which I shall have increasingly more to say.

And in all this I have been contrasting such seeing with the pure-visual-experience kind of seeing we encounter in the oculist's office or at the limits of consciousness. We encounter it in research science, too, at the limits of optical resolution or the limits of our conceptual framework. Here the data before us are often wonderfully confused. The researcher is usually clueless as to the nature of the phenomenon before him. He may not know *what* he is looking at, and so he can neither see it *as* a certain kind of phenomenon nor see *that,* were he to perform in certain ways with respect to it, it would in turn perform in certain ways. This is precisely what must be found out, by careful observation and well-planned experiment. If these latter are successful in revealing something of the nature of the phenomenon, then we will have come to see it in our more usual and more complicated way.

So just as Eddington's underemphasis on experiment and systematic observation led him to a position that invited attack from many quarters, so my account of seeing and observation (given in the heat of trying to disclose the many facets of these notions) may have made it appear that I stress too little the seeing and observation of research science, the seeing that is done best when it becomes a lustreless, rigid description of what meets the eye. And this may have inclined you to think of me as severely myopic and as speaking only of the observation in science that depends least on the eyes, and of the observational situations that are least relevant to pure research.

This must be put right now. For I had not the slightest intention of underrating "phenomenal" observation. I wished only to show how very different it was from less spectacular, more ordinary sorts of observations; it is in fact a sophisticated, painstaking discipline, something that must be learned and practiced. It is something we must develop from our ordinary sorts of seeing, and not that from which our ordinary sort of seeing is developed. It is posterior, temporally and logically, to ordinary seeing, and only accessible when training and care have conspired to render a man capable of behaving nearly like a camera. Let us discuss now how very difficult it sometimes is to acquire this skill, and how the theory-loaded activity called *seeing* is forever intruding and biasing what we hope will be a pure and simple optical registration of some physical phenomenon. And after exploring these difficulties in the way of our becoming good laboratory cameras, let us inquire finally why one should wish so to observe the world at all. What is gained, what is it all in aid of? What is the goal of seeing the data phenomenonally? The answer is, I think, that by viewing the world temporarily as if we were cameras, we may arrive at clues about the nature of the phenomenon before us that will at last permit us to see it as a phenomenon of some particular sort; we will come to see that, when this is done to it, that will follow. In short, the goal of rigorous phenomenal observation in research science is that we shall learn enough to be able to see new phenomena in the more familiar, theory-laden sense of "see." We hope ultimately to view the phenomenon as being within an intelligible framework of what we already know, say, and see of the world. By thinking of ourselves as research cameras for a period we may acquire what the camera lacks, *vision into the workings of nature*. And for this sort of vision to be useful it must be fallible. An infallible observation is a sterile observation; it holds in itself no promise for our future understanding, in much the same way that a picture, in itself, has no implications for our scientific knowledge.

First, then, on the difficulties people encounter in trying to see in a fashion other than that in which we ordinarily see, in science and in everyday life. Coming to see as we ordinarily see is, of course, an extraordinarily complex process. The complexities are revealed to us, not only in the conceptual and logical arguments that have been our concern in the past few chapters, but in the consideration of matters of fact as well. For example, studies of adult people congenitally blind because of cataract who were able to see after operation have shown that learning to see, in *any* sense of the word, is a thoroughly exacting and laborious process. It took patients at least a month to see even a few objects as others of us would see them. After 13 days a patient could not distinguish between a square and a triangle without counting the angles. Although a cube of sugar could be identified when resting on a table, it was not recognized when suspended by a thread against a different background. (How reminiscent of Locke's conjecture about the cured blind man's problems of distinguishing spheres from cubes!)[1] Of course, these people can see in the sense-datum sense of "see," but can they see anything?

And of course we all encountered a similar ordeal in our infancy. So that coming to see objects in the effortless, painless way that we now see each other is a complicated business which, once mastered, requires

---

[1] Locke, *Essays Concerning Human Understanding, op. cit.*, pp. 75–6.

nothing less than whisky, opium, ether, or a punch on the jaw to alter—or, alternatively, nothing less than a long training in natural science.

The complications are increased, evidently, by the remarkable ways in which we select and organize the elements of a mere stimulus pattern, like the duck-rabbit or the wife-mother-in-law. For not only is it possible that two normal observers may see different things in these figures, it is more likely that neither observer will see in his own visual field such things as his own nose or his own cheek. We ignore those parts of a stimulus pattern that are not particularly useful or pertinent at a given time, those things that do not mesh with our expectations, or our knowledge, or our current interests. The reasons how and why this is so are all obviously within reach of our earlier examination of *seeing as* and *seeing that*.

It is for reasons somewhat like these that children are thought to be specially observant. They notice everything. Having, in a way, rather less knowledge and experience than we have had, they comment on things much more at random. Their visual interests are splayed far and wide instead of being organized as ours are (for better or for worse) around a rather limited number of visual and intellectual focal points. Through the pains of an education we have come to save ourselves trouble by just learning which are those parts of a familiar visual stimulus pattern that we can afford to ignore. Most of us ignore our noses and our cheeks; our own noses and cheeks are seldom worthy of attention. We have *all* learned that much. We usually ignore the ticking of clocks, and street noises. Some of us will ignore the bump on the back of the duck's head, others of us will dwell on it as the mouth of a rabbit. Galileo would have learned enough physics to have come to ignore minor differences in the color of his rolling and free-falling spheres. And, indeed, the identity badge of every modern scientist consists of those things he ignores among his visual data. For the physicist, most animals are just material objects in motion, or assemblages of levers, or systems of wires, pulleys, and elastic layers of insulating material. For the chemist, they are energy-factories, beautifully efficient laboratories. For the biologist, animals are very much more than this.

As Dr. M. L. Johnson says, "This way of behaving makes [quick and easy] the seeing of, noticing of, reacting to, familiar things which can fit into a pre-existent perceptual scheme . . ."[2] And it must be stressed

---

[2] M. L. Johnson, "Seeing's Believing," *op. cit.,* p. 62.

that a very great deal of the seeing, noticing, and observing in science is the seeing, noticing, and observing of *familiar things,* however hard certain philosophers try to make it sound as if the eyes of research scientists were continually being visited with strange and unconnected near-apparitions and appearances.

But of course this visual facility has its price. We cannot regard this more usual form of seeing as a wholly unmixed blessing, however much of a chaotic hell our visual life would be without it. For it does incline us sometimes to overlook certain discrepancies between what is there to be seen and what we ourselves see. By its use, as by the use of any efficient adaptation to our environment, we can occasionally go wrong.

Many experiments have shown how, e.g., the shape, size, color, and position of objects are, as it were, "projected" onto them by the observer. The perception of color and shape depends not alone on the thing looked at but partly on past experiences of the color and shape of similar and dissimilar things. This is more than just the consideration that in learning "color" words and "shape" words we must countenance the word as correlated not only with things of the given color, say *red,* but also with green things, black things, etc., at which time it is pronounced "Not red." For only in this way could one recognize a misapplication of the word; and this is essential to learning the word's use or, what is the same, its meaning. It is more than just this, for we are concerned here also with the phenomena encountered when subjects are momentarily shown such things as playing cards whose colors have been reversed. A red six of clubs is very often reported as being a normal card, either a black six of clubs or a red six of diamonds or hearts, while others see the cards as purple or brown. Rather than see such a monstrosity as a red club, people will change the color to what it "ought" to be—black—or they will compromise with brown or purple, or even change the suit to fit the color, calling it a *diamond* or a *heart.*

Playing cards of different sizes when shown to most observers are not seen as such but as being at distances varying inversely with their size. A normal card, one half as big, and another twice as big, all at the same distance from the observer, are seen as three normal cards at different distances from the observer. He "alters" the distance at which the cards appear to be placed in order to make all playing cards conform to the standard-sized card with which he is familiar.

154 | *Chapter 9*

This propensity to "alter" the details of one's visual field in order to get things sorted out intellectually is particularly intriguing in the demonstrations of Dr. A. Ames, Jr., of the Hanover Institute.³ An observer is brought to a peep hole through which he sees the interior of a room apparently shaped like this:

---
³ Ames' distorted room experiments are described in S. Howard Bartley, *Principles of Perception* (New York: Harper and Row, 1958), pp. 216ff.

The observer imagines the floor plan to be roughly square. In fact the plan of the room is this shape:

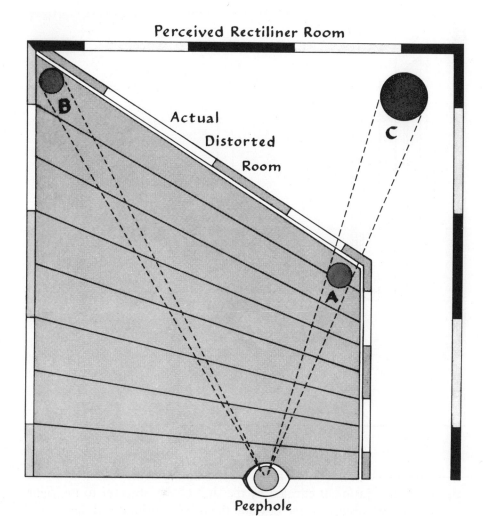

156 | *Chapter 9*

The real shape of the room is this:

But of course it appears as in the initial figure because of the perspectival adjustment effected by the shift of the observer to the right.

With such objects as the boy and the dog involved it is quite natural to say that they are different sized objects. But when, on exchanging places, they appear in this way:

we ought to get suspicious. And almost everyone who is subjected to this view does get suspicious, but not in the right way. They imagine that mirrors or lenses are distorting things somewhere, but they never see the room for what it is, itself distorted in every dimension. Dr. Ames' interpretation of this phenomenon is that we wish to keep our environment relatively stable if possible, so that we can move about in it with confidence and surety. And when we are forced to admit an incongruity we prefer to restrict the incongruity to moving objects like people (or their appearances) rather than static ones like rooms. Remember how the stuff of nightmares, surrealist art, and Hollywood psychological-thrillers consists in having the earth and sky go all syrupy with buildings and trees shrinking or expanding in as eccentric a way as possible. We can adjust to queer objects, but we are genuinely unsettled by an unpredictable environment.

Dr. Ames and other psychologists take this reaction as an indication that we are always construing the visual stimulus pattern in terms of past experience and knowledge as well as present and future needs.[4] These bizarre cases with their mildly shocking visual effects, like the conjurer's tricks we mentioned earlier, tend to bring home to us what is at work in the non-bizarre, ordinary cases. Dr. M. L. Johnson puts it as follows: "As we go through life we are continually . . . testing our assumptions by acting on them, and . . . we modify them according to our experience. Consequently, if we see things often enough we learn to see them reliably. The difficulty is with seeing new things . . ."[5]

But before we go on to consider this difficulty of seeing new things let us consider further some of the effects of past experience and knowledge on what we see. For until now our arguments have been meant to be mainly logical. We have urged that experience, knowledge, language MUST play a role in visual perception. Now we are beginning to consider the factual issue of *how* they play such a role. We can do no better here than to review some of the findings of experimental psychologists; they have been working on these problems for a long time. And though these are matters of fact—professional philosophers might say "mere matters of fact"—they are nonetheless of interest in that they reinforce and delimit the strictly logical inquiries to which we have been thus far addressed and to which we shall return shortly.

Psychologists rush in where philosophers fear to tread. For several chapters I have been dancing all around the thesis that an important factor in how we see the world is the knowledge and experience we bring with us to observational situations. Our knowledge, our experience, and (as I shall argue with gusto later) our *language* are, as it were, spectacles that we wear *behind* our eyes. The visible world is an entity that has filtered through to us by means not always obvious but always marked. This is the position that the logic of many of our observation concepts has brought me to, sometimes quite reluctantly and never without some qualms.

Consider now the verve with which the psychologist M. D. Vernon

---

[4] W. H. Dember, *The Psychology of Perception* (New York: Holt, Rinehart and Winston, 1960), p. 267: "The Ames demonstrations . . . seem most plausibly accounted for by reference to the observer's past experience."

[5] M. L. Johnson, *op. cit.*, p. 65.

expressed a parallel thesis in her book *A Further Study of Visual Perception*: ". . . the phenomenal qualities of the external world . . . are to a great extent determined by certain schemata common to the minds of most observers, yet it is equally clear that they are in fact attributed by the observer to the external world and the objects within it . . ."[6] The experimental reasons for this surety are many:

Consider the concept of *attention*. The situation in which our attention is drawn one way rather than another, or in which it fades altogether, is well known. Spectacular phenomena, like the sunrise or a sudden movement of an object before us, can claim our visual attention almost exclusively. A boring writer can have a well-known opposite effect, details of which I dare not labor here.

And, of course, several of the figures we have already considered elicited differences with respect to what you were able to attend to. The black and white crosses and the goblet and the faces were such that you might at one time see one patch as the *figure* and the other as the *ground*, while later these would reverse. With most of us the dark patches tend to dominate our initial confrontation with a figure, inclining us initially to see a black cross on a white ground. The psychologist C. H. Graham argues that the observer's previous experience with printed material and black and white drawings creates a "set" or a visual disposition for seeing black as figural.[7] Symmetry can also be an attention claimer. P. Bahnsen showed to his subjects two black and white figures in one of which the white pattern was symmetrical, in the other the black.[8] Eighty-nine percent of all observers saw the symmetrical pattern as the figure.

Recall our shifting perspective figures, the cube, the tea tray, the staircase, and the tunnel. Our attention is drawn sometimes one way, sometimes the other way. J. C. Flugel showed experimentally that this alternation of attention was due neither to eye movements (scanning) nor to local retinal fatigue.[9] Vernon remarks that our more usual everyday shifts in attention from this object to that one are not of the same nature as are the spontaneous fluctuations of awareness so far considered.[10] She holds, rightly I think, that outside the laboratory shifting

---

[6] M. D. Vernon, *A Further Study of Visual Perception* (Cambridge: Cambridge University Press, 1952), p. 200.

[7] C. H. Graham, *Journal of General Psychology*, Vol. 2 (1929), p. 470.

[8] P. Bahnsen, *Zeitschrift für Psychologie*, Vol. 108 (1928), p. 129.

[9] J. C. Flugel, *British Journal of Psychology*, Vol. 5 (1904), p. 357, Vol. 6 (1904), p. 60.

[10] M. D. Vernon, *op. cit.*, p. 208.

perspective and shifting aspect situations are rare. But I have, of course, been at pains to stress similarities between these artificial cases and other situations very familiar to observational science. Thus, the sun may be seen as fixed or as revolutionary, protozoa are seen as one-celled or non-celled, man may be seen as a degenerate angel or an exalted ape (and by some very astute observers as man), and the universe can be seen as winding up or running down, as a stage in a great explosion or as the state of an intricately well-balanced system. These are not out of the *same* basket as the attention shifts we have so far been noting. But neither are they so wholly dissimilar that any talk of resemblance between them is absurd.

"In general," says Vernon "awareness is directed to those parts of the [stimulus] field which are of the greatest interest to the observer. . . . This canalization of awareness . . . can be achieved only by learning and maturation. In the infant it is conspicuously absent. . . . As interests develop so does the power of attention. . . ."[11]

It is clear to everyone who has played children's games or taken university examinations that the voluntary direction of one's awareness or attention is improved, not only by experience and motivation, but also by "knowing what to look for." One's efficiency in reacting to a stimulus is markedly improved when he knows when and where and how his attention should be focused. The person who is just generally alert is slower and less efficient in his every reaction to external stimuli than the man who is *set* for them. This, apparently, is why intuition of one sort or another is usually necessary for getting the beginner to direct his awareness towards certain features of a situation rather than to maintain a diffused alertness to the situation as a whole. He may have to look for certain visual stimuli which are so small or weak or lacking in organization or figure that they are difficult to perceive at any time, much less when he is not specifically ready for them. You've got to be looking just where Dr. Pantin is looking before you can spot the worm he has probably already identified and classified. You may indeed have to select and isolate certain features from a very complicated visual pattern, observing these alone and ignoring the remainder. What occurs in such cases of *preparing to see* comes out in certain psychological experiments on *set*, a few of which we will now consider:

---

[11] *Ibid.*

O. Külpe in 1904 urged that what is perceived in any given field is controlled by a "determining tendency." This he called *"Aufgabe."* [12] It arises in large measure from the character of the given perceptual task we assign ourselves. In a way, the questions we ask about the visual field before us determine what we see in just the way that our hypotheses about nature determine what are to be relevant and what are to be irrelevant observations. Külpe exposed to his observers groups of four nonsense syllables printed in four different colors and arranged in different groupings; he asked them to observe and reproduce as many as possible, and on certain occasions gave them one of four special tasks:

1) To discover the number of letters visible;
2) To note the colors and position of the colors;
3) To note the form and arrangement of the syllables;
4) To discover as many letters as possible, noting their spatial position.

He found that the number and nature of the elements reproduced was largely a function of the assigned task; the elements not accentuated by the special question were frequently forgotten, and the observer sometimes thought that they had never even appeared. J. J. Gibson puts it that a "set" to perceive a particular stimulus may give it a "prior entry" over other visual stimuli of the same type.[13]

This is not to say that we have here a completely all-or-nothing reaction to the elements of the visual world. E. G. Boring, in 1924, found that observers when confronted momentarily with two bright rectangles were able to give some sort of secondary judgment as to the attribute which they had not been expected to notice, though naturally this judgment was not as accurate as the primary judgment.[14] H. J. Kingsley discovered that if observers were instructed to search for a given feature in a complex picture they often developed a perceptual *set* to see that object in a particular way.[15] If the object then appeared in a different way it might be overlooked. Thus, when told to look for a *turkey*, observers looked for a whole turkey and overlooked a turkey's head. And even when shown a picture of the object beforehand, they might still

---

[12] O. Külpe, quoted in Vernon, *op. cit.*, p. 215.
[13] J. J. Gibson, *Psychological Bulletin*, Vol. 38 (1941), p. 781.
[14] E. G. Boring, *American Journal of Psychology*, Vol. 35 (1924), p. 301.
[15] H. J. Kingsley, *American Journal of Psychology*, Vol. 44 (1932), p. 314.

overlook it if it appeared subsequently in a general setting different from that expected.

It has been shown experimentally that the greater the number of features an observer is instructed to note, the less efficient the perception. It has also been established that the more narrowly concentrated the awareness, the greater the accuracy of perception within it and the less the accuracy of perception outside it.

M. J. Zigler presented to his observers geometrical figures which were gradually or instantaneously replaced by other figures.[16] He found that the figure actually perceived after the change differed according to whether the observer assumed a *set* towards *form* or towards *object*. Again, the observer may be *set* to see the stimulus as the "real object," or to "attend predominantly to the retinal impression" (Vernon) or the visual sensation. These *sets* were found to depend partly on the attitude spontaneously adopted by the observer, which was to some extent a function of his intelligence and his training, as well as on the method of experiment and the experimental instructions.

In 1934 I. M. Bentley argued that a change of *set* is likely to occur whenever the process of simple perception and bare apprehension (i.e., camera reaction) passes over into that of "inspection" or scrutiny.[17]

> Here self-instructions appear, accompanied by general feelings of strain, to see details, to *identify* the stimulus object, to compare it with past experiences, and so on. The exact nature of the self-instruction is determined by the experimental instructions, the nature of the material, and by the observer's underlying attitude towards the experimental situation. . . . But with prolonged repetition of the experiments the observer ceased to reinforce the experimenter's instructions with self-instruction to inspect the stimuli closely. Strain was decreased and the performance proceeded with greater ease, precision and accuracy . . . both experimenter and observer should . . . aim at obtaining immediate perception and at discouraging the inspection process. But it is doubtful if this can occur without prolonged practice on the part of the observer, however favourable the experimental instructions. In perceptual experiments in which the observer is required to report minute details, or to obtain the full meaning and significance of the percept, inspection must be a *sine qua non*. Here the observer's self-instructions will operate

---

[16] M. J. Zigler, *American Journal of Psychology*, Vol. 31 (1920), p. 273.

[17] I. M. Bentley, *The New Field of Psychology* (New York: D. Appleton-Century, 1934), *passim*.

most powerfully and the nature of the percept will be much influenced by their effectiveness and cogency.[18]

This sounds familiar. It sounds very like the situation that obtains within research science where the data before one can sometimes be marvelously conglomerate. If the phenomenalist claims this as his province, I have expressed my willingness to let him have it that here is a kind of *seeing* typically unlike the ordinary seeing we have been discussing. I have wished only to urge that this "phenomenal" seeing is vicarious in our logically and epistemologically more fundamental and more usual variety of seeing. But now these experimental psychologists have raised the question of whether even this is right. I have felt that there may be a case for the camera-model of seeing at the frontiers of research science. The efficacy of this model is seriously questioned even here in the finding that when self-instruction, inspection, and the straining to identify mark the observational situation, its whole character changes, its *set* is altered from that in which a merely passive, supremely sensitive camera is registering what lies before it.

Set operates with respect to visual perception in science much as it does in familiar everyday experiences. The difference between reacting to visual stimuli in terms of a visual set and reacting to them like an electric eye is like the difference between catching the baseball when you've been waiting for it to come your way and having it strike the palm of your hand while you are strolling absent-mindedly across the outfield. Scientific observation, I contend, is distinctly more like the former than the latter.

In an aside, Vernon makes the following remarks about experimental psychology; their applications for experimental science in general should be clear.

> . . . it is also important to remember the effects upon results obtained which can occur in response to manipulation of instructions and to experimental setting; since these effects may assume some prominence if experimenter or observers have a strong theoretical bias. Frequently such a bias may have been created in the experimenter's mind by the theories of a particular 'school' . . . Such theories undoubtedly shape the training of students; hence the outlook of the latter and the techniques they adopt in carrying out experimental work. Thus they may almost

---

[18] Vernon, *op. cit.*, p. 220.

unconsciously design their methods and instructions in such a way as to influence their experimental results . . .[19]

Sir Ronald Fisher has shown us how Mendel's statistics are too good to be true! [20]

We will consider in a moment some examples of this from natural science.

Vernon continues:

> But though any such general influences on the observer's mind may be avoided, individual differences of previous knowledge, experience and familiarity are unavoidable, and may be expected to operate in any perceptual experiment . . . perception as we know it cannot take place at all without some familiarity with previously encountered relevant situations . . . in so far as these experiences are similar for different observers, as they are in great measure for individuals of the same type of upbringing and social background, then the perceptual experiences of these individuals will be similar. But the effect of familiarity, especially of prolonged familiarity, with experiences and situations peculiar to certain observers will be to cause their perceptual experiences to diverge from those of observers who have not become familiar with such situations . . . The city-bred child does not recognize the animals and flowers which are readily perceived by the country child. The European does not perceive the tracks of wild animals which are obvious to the African native . . .[21]

From this Vernon goes on to one of the most important statements in her book:

> But familiarity is something more than a mere acquaintance with a particular situation or with the material of a laboratory experiment. It denotes the existence of some cognitive scheme of experienced phenomena in which the present percept fits. It is the nature of the relationships to such a scheme which supplies the meaning of the percept . . .[22]

And of course this should not surprise us, for we have seen that the logic of the concepts *see* and *observe* is such that the situation

---

[19] *Ibid.*, p. 223.
[20] Sir R. A. Fisher, "Has Mendel's Work Been Re-discovered?", *Annals of Science*, Vol. 1 (1936), pp. 115–37.
[21] Vernon, *op. cit.*, p. 224.
[22] *Ibid.*, p. 224.

described by psychologists could not really have been otherwise, not at least so long as "see" and "observe" are taken to mean to us in ordinary contexts what they do mean.

Feingold and Kingsley found that even familiar objects may go unrecognized and unnoticed if the setting in which they appear is strange.[23] It is necessary for the whole setting to be familiar if perception and recognition are to be easy and effortless, if indeed we are to *see* the perceptual object and not merely *have* it. Only after what Sir F. C. S. Bartlett calls "the effort after meaning" has been resolved does a visuo-perceptual situation constitute a case of observing, seeing, witnessing, and noticing in anything like the usual meanings of these words.[24]

Again, if an observer, prior to a presentation of the wife-mother-in-law, is shown one phase of the picture in isolation, then that same phase will predominate when the whole figure is seen, to the exclusion of the other phase. This was established by Leeper in 1935.[25]

And if an observer is "set" to perceive the names of animals, he perceives the nonsense syllables *sael* and *wharl* as *seal* and *whale*. If, however, he is set to perceive something to do with boats, he perceives them as *sail* and *wharf*.[26]

The effects of familiarity are greater, however, if some comprehensive scheme of knowledge has been established into which the present visual experience can be fitted. Thus C. Fox carried out this experiment: He showed students pictures of medieval armor and required them to report afterwards as much as they could remember of the pictures.[27] But he had first given instruction to some of the students about details of the construction, ornamentation, and uses of armor. The students who had had instruction were much more active in their subsequent perception of the pictures and reported much more about them than did those who had received no instruction. This only occurred, however, when the students had thoroughly understood and digested the previous instruction; otherwise they were muddled by it and in no way assisted in their perceptions. As the psychologist M. Wertheimer said, "The role of past experience is of high importance, but

---

[23] *Ibid.*, p. 226.
[24] Sir F. C. S. Bartlett, *British Journal of Psychology*, Vol. 8 (1916), p. 222.
[25] R. Leeper, *Journal of Genetic Psychology*, Vol. 46 (1935), p. 41.
[26] E. Siipola, *Psychological Monographs*, Vol. 46, No. 210 (1935), p. 27.
[27] C. Fox, *British Journal of Psychology*, Vol. 15 (1924), p. 1.

what matters is *what* one has gained from experience—blind un-understood connections, or insight into structural inner relatedness." [28]

All this work of experimental psychology runs parallel to our own earlier conceptual investigations. We found that the unqualified *pictures* of either retinal or sense-datum seeing were at a logical remove from both what we ordinarily say we see and what we ordinarily say we know. We argued that seeing and knowing *must* be this way. That's the sort of *concepts* they are. The psychologists show that seeing and knowing *are* this way. But they do more. They give us some insight into the common types of error we encounter in the laboratory every day. For it is in a university scientific education that certain new conceptual and perceptual frames of reference are expected to be instilled in the undergraduate. Failures to observe errors in observation, misinterpretations, clashes of opinion are just what we might expect. And that's just what we get. As G. H. Lewes put it in 1879, ". . . the new object presented to Sense, or the new idea presented to Thought, must also be *soluble in old experiences,* be recognized as like them, otherwise it will be unperceived, uncomprehended." [29] Part of the purpose of elementary science is to give a man a set of experiences into which much of his new experience later on in research will be soluble. But this is not, of course, an easy task for anyone concerned.

Thus the sense of floundering bewilderment the freshman feels on the first morning, e.g., on confronting a prepared section of kidney—a sense not completely unlike what the recovered blind man must experience when first opening his eyes to a world of shapes, sizes, and colors, all devoid of meaning to him. In such an unfamiliar, complicated, featureless muddle, it seems almost impossible to distinguish what is significant from what is irrelevant. And this poses a critical difficulty for the demonstrator. He must help the student to learn what a kidney section looks like. He must give his pupils a perceptual framework, or *set,* in which such things as sections of kidney will stand out as recognizable. And yet he would like them to make thoroughly unbiased observations, to see what is there to be seen, and not just what they are supposed to

---

[28] M. Wertheimer, *Productive Thinking* (New York: Harper, 1945), p. 62; quoted in Vernon, *op. cit.,* p. 231.

[29] G. H. Lewes, *Problems of Life and Mind* (Boston: Houghton, Osgood and Co., 1879–80), Vol. II, p. 108.

Spectacles behind the Eyes | 167

see. For all of us tend to see what we expect to see. "I can't draw," said the student botanist to Professor Briggs. "You can't see," he replied.

Freshmen were asked to identify and write illustrated notes on specimen X (a transverse section of hydra). Those who identified X correctly drew what looked much more like textbook drawings of hydra than the specimen before them. Those who failed to identify X produced drawings superior in every respect. Apparently, once having identified X as hydra, and having no reason to doubt this, the student begins to see the specimen through those elusive spectacles he wears behind his eyes, spectacles ground, shaped, and tinted by textbooks and illustrated lectures in invertebrate zoology.

Indeed, a textbook diagram, in biology, in chemistry, or in physics, *is* a kind of perceptual scheme. It provides a *set* that influences one's visual experience markedly. In another context Dr. Stephen Toulmin makes this point in a striking way as concerns the ray-diagrams of geometrical optics.[30] A diagram, like the perceptual scheme of organization of a visual experience, is not a discrete special picture of a particular phenomenon. It is generalized and schematized, a kind of average shadow-casting situation or an average hydra section. And it is just as difficult to see the unique and individuating marks of a laboratory phenomenon perceived in terms of an average or diagrammatic conception of what the phenomenon is *really* like as it is for a tourist in Britain to detect differences in Englishmen when he sees them all through the "average-Englishman" spectacles that were ground for him in his homeland, or in Hollywood films.

Certain details are quickly—perhaps too quickly—seen as irrelevant details. And we all know what a very short step it is from that position to the making of a serious error in observation, identification, or interpretation.

Hydra is usually presented as a *type* animal, the paradigm case of a two-layered body wall. Our diagrams, our perceptual schemes, of *hydra* help us to see it as the contemporary textbook writer sees it: an early stage in an early chapter of an early book. The difficulty is, however, that this may prevent us from seeing anything else, such as something unprecedented and peculiar to this particular hydra before us. Thus

---

[30] Stephen Toulmin, *The Philosophy of Science* (London: Hutchinson & Co., Ltd., 1953), pp. 25–8.

168 | *Chapter 9*

preconceptions as to what a natural phenomenon *ought* to look like, and instructions as to what to look for, have this double effect: They assist and inhibit observation. This is what Dr. M. L. Johnson calls "the dilemma of learning, and teaching, to be observant." [31] For all the logical reasons we have been considering in past chapters we cannot be mere cameras in the laboratory. We cannot, both because we are psychologically incapable of doing so and because *doing so* is itself a conceptual impossibility. But in so far as we do depart from being mere cameras, we run the increasing risk of warping the visual data to accommodate to our intellectual framework, or our perceptual set. This is a dilemma, and I feel an irresolvable one. It is impossible that we should be scientific cameras and hence impossible that observation should be foolproof and without risk. But vigilance is a small price to pay for the ease with which we can now see the visual world as intelligible and as the province of future scientific research. Practised observers as well as freshmen have been notorious for *deviations*. Thus thirteen centuries of expert observation failed to disclose the error in Galen's contention that the septum between the ventricles of the heart is perforated. And how many anatomists before Harvey saw, but did not *see,* those little one-way valves in the veins of the head and neck? Only people who have never found themselves face to face with a really new fact can laugh at the inability of medieval observers to see what was before their eyes.

And so I should like, in a way, to agree with those philosophers and psychologists who by rigorous deductive[32] argument show that the observation often required in research science is different from our ordinary, effortless observation of bicycles, bears, and boxes. I eagerly agree that these are different cases of seeing. But I dissent from every other reflection made on this difference in the name of sense-datum phenomenalism or mechano-set neurophysiology. For the seeing required in research is not primitive—logically, psychologically, sociologically, or historically—to our ordinary case of seeing, but just the other way round. Nor is seeing the effortless registration of the visual field that some commentators have supposed, researchers being little more than

---

[31] M. L. Johnson, *op. cit.,* p. 68.
[32] By the time the manuscript reached its typed form, this word appeared as "seductive." Indeed, it may be that "seductive argument" was what Hanson intended here!—Ed.

bits of optical litmus paper. The "receptacle" theories would have it that *retinal seeing* or sense-datum seeing is an automatic, effortless, unthinking affair, while our ordinary seeing is riddled with *ex post visuo* mental processes: thinking, association, interpretation. Here I have been adducing considerations calculated to support first the opposite thesis. Seeing bicycles, bears, ducks, and x-ray tubes is effortless, and effortless because what is called *interpretation,* i.e., seeing as and seeing that, is *built into* the very concept of seeing. Where this is not the case, as in research, the interpretation is something we have to bring to the visual situation. This takes effort; it is a strain; we have to think, sometimes hard and long. And this is all something we have to learn to do. We have to learn to do it because, in a way, it goes against the grain of our ordinary notions of seeing; it goes against the grain logically and psychologically. Again we may note that where Aristotle had appeared to be so obviously wrong there is some wisdom in his observation nonetheless. For in urging that vision is affected by our sending out emanations from our eyes he at least recognized our very important and very active contribution to the whole business of seeing. This is an old but not very simple truth that has escaped many of our ultramodern philosophers and scientists.

Most important, the *goal* of trying to see as the researcher must force himself to see—like a controlled camera (his brain *using* his eyes) —is that we will one day see as-yet-unknown phenomena in the familiar way that schoolchildren now see the cardio-vascular system, thanks to the unsettling observational labors of Vesalius and Harvey. In scientific observation the visual stimulus pattern is interpreted, not merely received; we interact with the visible world and do not merely inspect it. And for the end in view that our knowledge might one day reach the stage when the painful looking and interpreting of today's research science becomes the effortless seeing of tomorrow's schoolchildren.

And with that inspiring sentiment we leave the swamp called *perception;* we got in rather more deeply at times than had been anticipated. Doubtless the effects of this will be apparent in our future discussions of facts, hypotheses, and causality. But there is one point that I do very much hope will stick with you. The observational situations of research science today, in relativity physics, quantum mechanics, and microbiology, are not the exotic anomalies that so many commentators have described them as. Yet it is true that in these reaches of science we cannot think of the observer as an impartial, objective spectator on the physical or biological scene. The observer *interacts* with his data, sometimes in a quite unpredictable way: Observation is not carried on from the stalls, it is now itself a factor on the scientific stage. In relativistic astronomy the observer's physical state is of fundamental importance to the adequate estimation of the behavior of other masses in space. In quantum physics the very instruments of observation perturb and interfere with the specimen particle to a degree calculable only in terms of a most subtle and complicated probability theory. And in the microbiology that proceeds within dimension ranges very near to the limits of optical resolution, and when the observer may very well find himself confronted with a visual complex whose elements he can only barely sort out as *ground* and *figure,* here too the contribution by the scientist to the identification and characterization of his data is considerable. Observing in these contexts is rather like inspecting a wrist-watch factory while pushing a large electromagnet.

Too often these situations are remarked with an air of drama and novelty, making it seem that they are situations *in principle* different from classical conceptions of our world. And in many ways this is true. But as regards the concept of *observations* these frontier situations are paradigms of what obtains whenever we observe, see, witness, or notice anything, even humble things like boxes, bears, and bicycles. So classical conceptions of observation, evidence, and data do not apply in fundamental research science because *in principle* they do not apply anywhere. There is just as much involved, and the same kind of thing involved, when we see boxes and bicycles as when we see Koala bears and *Rhyncodemus bilineatus,* or the shift to the red of Andromeda and the track of a meson. In these latter cases observation does not proceed at all according to the received conceptions of it. People are brought up short by the realization of this fact only because they have not seen that observation does not proceed *anywhere* according to those received impressions. To get you to see this has been the purpose thus far. So the next time you experience bewilderment in the laboratory, ask yourself again whether a 13th Century and a 20th Century astronomer see the same thing in the east at dawn.

10 | Can We See Facts?

WE HAVE discussed at length the concepts of *seeing* and *observing*. The intellectual and linguistic character of seeing was remarked in such a way that we could at least detect some justice in the assertion that our two astronomers, the 13th Century man and the 20th Century man, do not see the same thing in the east at dawn. In just this way two doctors may not see the same thing when looking at an x-ray photograph. Nor will two microbiologists necessarily see the same thing when looking at a protozoon, particularly if one calls it a one-celled organism and the other calls it a non-celled organism.

How will these investigations affect our appreciation of what we call *the facts?*

We encountered in previous chapters the tough-minded individual who, when asked what *observation* was, barked back that it was *just opening one's eyes and looking.* It was obvious to him that a 13th Century astronomer and a 20th Century astronomer see precisely the same thing in the east at dawn—and let's have no nonsense about it.

This same chap would probably have *facts* taped pretty well, too. Facts? Why they are just the things that happen: the hard, cold, stubborn facts, the sheer, physical, plain, and unvarnished facts, the observable facts out there for all of us to see, come up against, trip over. You know, we face the facts, collect them: the little, detached, lawless, particular, and individual facts. Facts, in short, are just chunks of the material world; sticks, stones, boxes, and bears.

And of course many tough-minded "empiricist" philosophers, e.g.,

Hume, John Stuart Mill, Bertrand Russell, as well as the Wittgenstein and Wisdom of 40 and 30 years ago, have thought of facts in terms not too dissimilar to this point of view. Consider how Mill speaks of seeing facts, of separating the facts in nature, how Russell speaks of the world containing facts like the condition of the weather and the death of men, and how perception goes straight to the facts, how Professor Austin of Oxford speaks of facts as three-dimensional states of affairs, and how the biological scientist J. A. Thomson tells us to begin our scientific research by collecting the data, which ought to be facts.

Please understand, I am not going to complain about the way these gentlemen, and many others, use language. We may all speak as we please, provided we are understood. There is a conceptual muddle underlying this talk, however, and it is a muddle that ties in intimately with the views of *seeing* and *observation* already examined. The form of language here is singled out because it is symptomatic of a serious conceptual ailment, an ailment that consists in regarding our knowledge of the world as an uneasy conspiracy between our sense experience and our thinking, between our visual impressions and our brains, between the world of objects, events, situations and what we are able to make of them, between reality and language. But there is no such conspiracy because these aspects of our experience have always been well and truly bridged together. What we call *seeing* is a composite of visual sense experience and what we know. Hence *seeing as* and *seeing that*.

Now what about facts? What sorts of logical span are facts, if indeed they can be spoken of in this way at all?

One obvious point straightaway. It is perhaps not serious in itself, but as a logical indicator it is indispensable. Of those of us who so readily speak of observing the facts, looking at them, collecting them, etc.—and most of us *do* so speak—has any ever asked what observation of a fact would be like? What do facts look like? In what receptacle might we collect them? I can photograph *objects* like x-ray tubes, or *events* like fluorescence, or *situations* like the set-up of an x-ray diffraction experiment. But what sort of photograph would a photograph of a fact be? Asking the question in this way is like biting the forbidden apple. Facts can never again be regarded in the fat, dumb, and happy way we looked at them before. It is like discovering of a close friend that he was educated at Oxford or Princeton.

Of course the *fact* concept is much more complicated than we would expect, perhaps even more complicated than we found *seeing* to be.

For we have come to appreciate how seeing the sun is more than the mere experiencing of a brilliant disc in one's visual field. Seeing the sun is appreciating our visual experience as an experience *of* the sun, it is *seeing that* the object in view will behave in all those characteristic ways in which experience, intuition, and reflection have led us to expect the sun to behave. But can it even be in this comprehensive sense of *seeing* that we might see *the fact* that the sun is on the horizon?

We may very well stare eastward at dawn, and we may see the sun. We can describe what we see and say what the star we call the sun looks like. But how can we describe the fact that the sun is on the horizon? How shall we say it looks? The question is an impossible one.

Of course, if it were midnight, if no one did see, or could see, the sun on the horizon, then (obviously) it would not be a fact that the sun was on the horizon. So, naturally, what we see in the east at dawn is highly relevant to the question of whether or not it is a fact that the sun is on the horizon. But what we see does not constitute this fact. So while there may be a point to the suggestion that a 13th Century and a 20th Century astronomer do not see the same thing in the east at dawn, it can be said pretty firmly, I think, that *neither* of them sees the *fact* that the sun is on the horizon. We do not see facts; they are not there to be seen.

"But hold on," it may be countered, "It is just what we *mean* by referring to something as a *fact,* that it is pre-eminently objective, hard, physical, *out-there-to-be-seen*. What makes it a fact that the sun is on the horizon is the sun's being on the horizon, *pure and simple*. And that the sun is on the horizon is something about the physical world that is not to be put off by any amount of logical conjuring." And so forth.

I think we must inquire further into this. Our attention will be directed to what I think is a remarkable and curious compatibility between *facts* and the language at hand. For consider: What would an inexpressible fact be like? And by this I mean not merely a fact the details of which cannot be understood, as the 13th Century astronomer may not have understood the joggings and hesitations of Mercury. I am speaking of a fact which is *in principle* inexpressible, a fact which is somehow constitutionally impervious to being described. What would reference to such a fact consist in? What difference would it make? Is there any situation inside or outside science wherein we are directed to certain facts, but with the apology that these facts elude and always will elude linguistic expression? As-yet-unknown facts, what some philosophers have called *future facts,* elude linguistic expression, naturally enough. But this is to say no more than that we cannot now give the scores of the coming summer's baseball games. In this sense our 13th Century astronomer could not set out the facts of a 20th Century astronomer. But could a fact, once discovered and spoken of *as known,* elude linguistic expression? Could we ever be said to know what are the facts when no way is available of saying what they are?

This strikes me as an impressive feature of facts, of knowledge, of language, of the world. Let us pursue it.

Just for the purpose of a certain lecture at Cambridge University I arose early one morning and witnessed the dawn, which took place at the indecent time of 7:37 a.m. Let us imagine that it is 7:37 now, and imagine too that you are with me standing on Madingly Hill (we must travel that far, I fear; the sun couldn't be seen in the section of Cambridge where I lived until about 10:30).

Then I might say to you, or you to me, or me to myself, "Whether or not there is a language in which to express it, it is a fact that the sun is on the horizon." And it is a fact that the grass at my feet is green. Whether everyone, or no one, knows these things it seems a stubborn, brilliant, brutish fact that the sun is up, and that the grass is green. Furthermore, the facts are what they are, independent of any further considerations, linguistic, epistemological, or psychological. The sun simply is on the horizon. The grass just grows green in England. That's the way the world is: Those are the facts, objective, extra-linguistic, independent of our theories or modes of expression. We have only to open our eyes and look.

Does not the sentence "The sun is round" state a fact which is largely invariant with respect to the possible contexts in which it might come up for consideration? How could this be otherwise? If we see the bright, round sun before us, then the fact is that the sun is round, no matter how we express this fact. Or so it would appear.

And what about the sentence "The sun is bright yellow"? Is this any different? One difference is immediately apparent. Naturally, if a man were blind to the color yellow, his stating either "The sun is bright yellow" or "The sun is not bright yellow" would not in the least affect or alter the fact that the sun is bright yellow. Suppose, however, that an over-ambitious American physicist were to explode an A-bomb, an

H-bomb, or an X-bomb without having taken the proper precautions as regards the diffusion of radioactive material into the terrestrial atmosphere. Suppose further that one of the radioactive by-products of the explosion was totally destructive of the human retina's sensitivity to yellow and green light; that is, the light-sensitive cells of all men's eyes were rendered electrochemically inert to light radiation of the order of Å 6000–Å 4500.

Under these conditions what would be the significance of expressions like "This is yellow" or "That is green"? Quite clearly they would have the force that certain of our color statements now have, e.g., "This is ultraviolet" or "That is infrared." In other words, the locus of the significance of talk about yellow and green would have been shifted by the catastrophe from one involving the awareness of specific retinal reactions—color sensations—to one wholly dependent on pointer readings, photocell reactions, and chemical changes. Carried further, we might reflect upon how quiet the world would be if men were deaf, and the consequent status of facts such as the fact that a body's acceleration through air beyond the speed designated by "Mach 1" is accompanied by a loud *bang*. This is a variant on the familiar adolescent question about how things must look to insects with their compound eyes, or to dogs with their insensitivity to color, or to chameleons with their individually articulated eyes and their two distinct and discrete fields of vision. Wittgenstein puts it in his provocative and oracular way: "The ether was filled with vibrations, but the world was dark. Then man opened his seeing eye, and there was light." [1]

We say, of course, "The sun *is* yellow," "The grass *is* green," "Sugar *is* sweet," "Bears *are* furry." In this adjectival idiom, yellowness, greenness, and sweetness seem to be properties which inhere (in an almost passive way) in suns, or sunlight, in blades of grass, lumps of sugar, and bears. Were it not sweet it would not be sugar; not furry, not a bear; not green, not grass. These properties appear solidly to be built into the object of which we are speaking, and into the logic of how we speak.

But just for fun let us tinker somewhat with the form of speech: Let us convey such information, not adjectivally as is usual in our English, but verbally instead (as is the practice in the Arabic and Russian tongues and occasionally in Italian). Let us say then "The sun yellows," "The grass greens," "Sugar sweetens," "Bears fur," much as

---
[1] Wittgenstein, *Investigations*, p. 184. (Hanson's translation—Ed.)

we now would say "The star twinkles," "The droplet glistens," "The diamond sparkles." And let us suppose further that this *verbal* idiom were the only way of conveying this variety of information, as is the case in Arabic and Russian. In any case we can go along with Wittgenstein when he supposes that "it would be possible to imagine people who as it were thought much more definitely than we, and used different words . . ."[2]

Now in language so modified would it be stated as a fact that the sun *is* yellow? Could this be stated as a fact? Could we assert as facts that the grass *is* green and that sugar *is* sweet? That it is bright yellow seems a decidedly passive thing to say about the sun. It is to say that its color is yellow much in the way that its shape is round and its distance from us great. It suggests that yellow inheres in the sun as it does in a buttercup. "Yellow" is the name, then, of a constituent, a component characteristic of the sun.

"The sun yellows," however, seems to give us an account of what the sun does. Just as its surface burns and its spots appear and disappear, so it yellows. And in this idiom the grass would no longer merely *be* green, it would *green*. Sugar would cease just being sweet; it would sweeten. College buildings would grey. Academicians would no longer traverse merely flat green lawns. Now the grass would send forth its greenness; it would radiate greenness, and crossing a lawn would be like wading through a pool of green light. Students of colleges would no longer walk through cold, lifeless grey stone and masonry, for the buildings would now be characterized as if they pulsed with greyness, emitting it, twinkling with it, dispersing it into the courts and staircases. Instead of just tasting the sweetness of sugar cubes (in the manner in which we might just count the sides of such cubes) we might, with speech, be inclined to experience the sweetness of sugar as something acting upon us, something that sugar does to us. Sweetness would be thought to fill our mouths as candlelight fills a hall. And so the sun at dawn would not merely show itself as a passive yellow sphere. Saying "The sun yellows"—having it in the same logical and grammatical basket as "The bird flies" and "The bear climbs"—would encourage us to appreciate the dawn as a kind of yellow surge over the horizon, a flood of color that envelopes us and the earth around us.

Mind you, this is not merely the result of linguistic legislation, our

---

[2] *Ibid.*, p. 188.

conceptions being much the same as before but we being just unable to express them in the altered language. The suggestion I am making is much more radical. The great German philosopher Kant once remarked about the Latin tongue that Latins have only two words in a certain connection where Germans have three, hence Latins lack a concept Germans possess. Thus our "new" language in dispensing with the passive idiom of our speech dispenses also with a concept. Here is the converse situation to the one Wittgenstein asked us to imagine; here are people who think less definitely than we, and are limited to a single word where we have a choice of different ones. It is not a case therefore of our thinking exactly as we did before but speaking differently. For, as we saw, the logic of speech and that of thought are extremely intimate, and having to speak with all our color words as if they were verbs just is to come to think of colors as activities and of colored things as agents. It is in this connection that the complaint of grammarians that our language is losing words (and failing to make what were clear verbal distinctions 100 years ago) has a philosophical moral. For if this is true of the course of our language then we are gradually losing *concepts*, too, just as surely as in this example I have invented.

What if information about colors were conveyed adverbially instead of adjectivally as it usually is, or verbally as we have supposed? Suppose that then we would say, "The sun glows yellowly," or "The grass glitters greenly," or "John's Chapel twinkles greyly." We would have to speak of tastes and smells in this idiom too: "Sugar tastes sweetly," "Roses smell sweetly."

If there were no other way of speaking, if everyone spoke in this manner, would there be the slightest temptation to insist on its being a fact that the morning sun *is* yellow, or that the grass *is* green, or that John's Chapel *is* grey, or that sugar *is* sweet? And how could such facts be articulated at all? Would they be the inexpressible facts about whose existence we had such doubts earlier? Is there at the present time and in our present language any temptation whatever to insist on its being a fact that John's Chapel's bells *are* loudly, or that John's Chapel's fires *are* brightly? And what facts would these expressions locate?

"But surely," comes the overdue reply, "No matter how we choose to speak we could always see the difference between the activities of a bear while climbing and the activities of grass while it is greening (to use our modified idiom). Language could never blind us to seeing the differences between the manner in which grass brings its color to us and the manner in which a dog brings his ball to us."

Well, I am not sure about this. The objection has teeth in it, I will readily grant. And the whole business of trying to gesture in *our* language at what cannot be said, or thought, in our language is a very difficult undertaking: It reminds one of Wittgenstein's question of whether the eye can see the limits of vision.[3] But with your indulgence I will carry on in this enterprise for I believe that some light may yet come to fall on the shadowy concept we have entitled "facts."

With respect, then, to these so-called "secondary qualities" of things —colors, odors, flavors, sounds—it would appear as not altogether implausible to suggest that the ways in which we perceive things are not unrelated to the manner in which our language is constructed and expressed. Poets and writers can exploit this quite skillfully, e.g., ". . . a bright, green leave ago . . . ," ". . . his stare broke the silence . . ." "The morning sun was snagged in the branches of the great oak tree . . . ," etc. And at least part of the effect of such language is to make things seem just a little different from what they ordinarily would seem. Indeed, if we are conditioned by expressions like these, a leaf, a man's countenance, the sun may actually look different to us. Thus, even more radically, while it may be a fact for us that the grass *is* green, for a Russian, or an Arab, the fact may be that the grass greens, or that the sun yellows, or that sugar sweetens. For English-speaking peoples the fact that St. John's Chapel is grey may be an extremely passive feature of the chapel and its constituents. Expressions like "bleak, grey stone," "cold, lifeless masonry," etc., are often close to hand when one speaks of the greyness of a college building. To those whose form of expression is different, however, the fact that John's Chapel greys may be a phase of what the chapel does, a stage in its color behavior.

These *facts,* moreover, may be mutually exclusive. To accept one rendition of a "fact" may be to make acceptance of an alternative rendition impossible. This calls to mind those cases of visual perception

---

[3] Ludwig Wittgenstein, *Tractatus Logico-Philosophicus,* tr. by D. F. Pears and B. F. McGuinness (London: Routledge and Kegan Paul, 1961), p. 117 (Props. 5.632–5.633): "The subject does not belong to the world: rather, it is a limit of the world. Where *in* the world is a metaphysical subject to be found? You will say that this is exactly like the case of the eye and the visual field. But really you do *not* see the eye. And nothing *in the visual field* allows you to infer that it is seen by an eye."

wherein a man once having seen a figure as a duck can no longer see it as anything else. Or, again, once we have located the face in the drawing of the tree we have the greatest difficulty in seeing the tree without the face thereafter.

Must it be otherwise with what are sometimes called the primary qualities of objects, i.e., shape, dimensions, weight, etc.? We remarked earlier that the sentence "The sun is round" stated a fact invariant to the context in which it is presented for consideration. And we say that St. John's Chapel *is* rectangular, that John's lawns *are* square, that sugar lumps *are* cubes. But let us tinker a little more. Try treating roundness, rectangularity, squareness, and cubeness verbally. Let us say now that the sun *rounds*, that John's Chapel *rectangulates*, that the lawns *square,* and that sugar *cubes*. Would such statements as these state the facts that the sun *is* round, or that sugar lumps *are* cubic?

To say that the sun is round, or that John's Chapel is rectangular, is not, I suggest, to say the same thing as that the sun rounds or that John's Chapel rectangulates. There is a distinct suggestion of activity in the latter which is, to our ears, absent in the former.

Here looms an unfortunately woolly idea which I will do no more than indicate very roughly. Could it be that the idiom in which one is prone to speak about the sun, or a chapel, is also the idiom in which he thinks about the sun or the Chapel? There are strong reasons for saying this, particularly as we are concerned with the thinking of natural scientists, though perhaps one ought not to say it *too strongly*. Could it also be that, since a man usually thinks in the idiom in which he speaks, perhaps he also *sees* in the same idiom? This is thrown out as a metaphysical suggestion (which does not mean that it is only fit for being thrown out, i.e., being disposed of). You will remember how much importance we came to attach to the concept of *seeing that* in our earlier chapters, and perhaps you will remember too how influential our language-forms were to the ways in which we express what we know when we use forms of speech like "He saw that it could float," "He saw that a contradiction was immanent," etc.

Now the user of an "active" language like Russian or Arabic, i.e., a language in which the properties of things are set out verbally instead of adjectivally, would speak of the sun as rounding and of the chapel as rectangulating. And perhaps such a man will actually see the sun as *rounding;* facing east at dawn he will describe what he sees in the statement "The sun rounds," and this language will aptly express what he sees: Our local star sending forth roundness, actively and incessantly arranging itself into a sphere. Or are we to say of such a person that he knows some fact he cannot express in language? [I invite you to reflect again on what we should think of someone who broadcast that he knew a certain inexpressible fact, a fact beyond the power of his language to express. We should very likely question in what sense it was fact, and in what manner he came to know it. We might even question whether *"fact"* was really the word he wanted; perhaps *"feeling,"* or *"hunch,"* or *"suspicion"* would be more suitable in these cases where communication boggles, but not *"fact,"* surely.]

And for such a person could it be a fact that the sun *is* round, in the same lustreless manner that New Haven *is* east of New York? Could it be a fact for him that John's Chapel *is* rectangular, a thing he could not say or even think, and *perhaps* not even see. Perhaps he sees it as *rectangulating*, continually pressing itself into an invisible rectangular mold.

"The chapel tower stands squarely." Does this state the fact that the tower *is* square? The idiom suggests that the tower stands like a sentinel over St. John's College. It stands foursquare against the ravages of time, tide, thunder, lightning, and 30,000 visitors a year. "The tower is square" seems flaccid and passive by comparison.

If *everyone* said of sugar that it lumped *cubicly*, what exactly would be the force of insisting that, irrespective of the ways in which everyone *does* use language (irrespective indeed, of the very existence or non-existence of language itself), it is still a fact that sugar lumps *are* cubic? Even if such a thing could be said, it could not be understood, it would be like something out of Lewis Carroll or Edward Lear. And were all men habitually to say that John's Chapel *rectangulates* what would be the point of asserting, if one could assert it, that irrespective of current linguistic habits it is still a fact that St. John's Chapel *is* rectangular? What could be the force of such an assertion?

Now it is not just a question of "Say what you please, the data are always the same." For in a language in which *all* characterization of objects was assimilated to the verbal paradigm of "The bear runs," "The cat sleeps," "The sun shines," etc., the *properties* of objects would just be the things *done* by objects. The squareness of the box would just be something the box does, the greenness of the grass an activity of the grass. If there were a *conceptual* distinction to be made between the sense in which the bear is doing something when climbing and the sense in which it is doing something when furring, then the machinery for making the distinction ought to show itself in language. To paraphrase Kant: If a distinction cannot be made in language it cannot be made conceptually.

"But surely," comes the inevitable objection, "everyone can see the difference between the bear's tree-climbing activities, and his fur-having activities (if indeed this talk of having fur being an activity can even be tolerated)."

This need not be so. Malinowski (the anthropologist) remarked on how far the Trobriand Islanders construed a young potato, a mature potato, and an old potato as three different *types of entity*.[4] We might have said "But everyone can see that these are just three stages in the development of a single potato, or a single type of potato, whether or not different names are assigned to each stage." And yet the Trobrianders' use of different words for the three stages of the potato signaled their conception as being of three different types of entity (of course, in English different words do not necessarily imply distinct entities; for example, a colt and a mare are not thought to be different kinds of things without any connection between them). This is one of a large class of hard lessons that every student of anthropology must master. And it puts strength into Wittgenstein's conjecture that there might be people who thought much more definitely than we, and used a variety of words where we use only one.

Conversely, there might be a society whose members counted in the following manner: "One, two, three, few, many." Would we insist to such people that whether or not *they* can say so, think so, perceive so, it is a fact that John's Chapel *has four spires* and that every gallon *has eight pints*? How could we insist on this when the expressions "four spires" and "eight pints" elude their language?

Or shall it be the case that, for them, the facts that John's Chapel has *four* spires and that gallons contain *eight* pints are inexpressible? But would this be any different from saying that, for them, these are not facts at all?

To this our inevitable backbencher will say: "These considerations

---

[4] Bronislaw Malinowski, *Coral Gardens and Their Magic*, Vol. II (Bloomington: Indiana University Press, 1965), Part V, Div. III, pp. 98–107.

would do nothing more than reveal the paucity of such a primitive language. What makes it a fact that John's Chapel has four spires, or that gallons contain eight pints, even in a world of men who speak, think, and perhaps even perceive along the lines of "One, two, three, few, many," is that *if there were a language* (like English, for example) in which one could state what is the case with Cambridge college chapels and gallons of milk, such a language would include the statements of fact: "St. John's Chapel tower has four spires" and "Gallons of milk contain eight pints."

But if this answer is right it must cut into every language we are familiar with today. There might be millions of things about Cambridge chapels and gallons of milk that cannot be expressed in English either (or has absolute universality of expression been achieved in our language, or any other?). Perhaps quite striking features of the subjects of our scientific study are likewise unnoticed, being out of the effective focus of *our* language and notation. Perhaps our ways of speaking and thinking in microbiology have just blinded us to aspects of life on a small scale rather in the way that we may be blind to the rabbit in when seeing it as a duck. Perhaps our taxonomic policies have put up fences in our minds, rendering us incapable of detecting connections between species now regarded as quite thoroughly independent. Perhaps our types of notation in physics may occasionally render us insensitive to features of the material world beneath our very noses. You will remember the inability of Galileo's predecessors to translate their thoughts into mathematical expressions. Could these predecessors have seen what Galileo saw, discovered what he discovered, detected the facts he detected? And before the Newton-Leibniz calculus certain events certainly could not be seen as they are seen now; our knowledge of the facts of nature was altered considerably by this powerful formal instrument. And due to the strictures of another recent language, the language of quantum theory, the facts of the microphysical world are now understood in a manner profoundly different from what we should have predicted in 1920.

I am not saying that there *are* aspects of Cambridge chapels, protozoa, and subatomic entities that elude description in the languages available to us. My point is only that it is not logically impossible that there might be. And if this point is sound we can see that it is not logically impossible that we *might* have come to think about the physical world very differently from the way we actually think. We might have come to see different aspects of it than we see, to know different facts about it than we know. And this is not identical with the conjecture that the *physical world* might have been different from what it actually is. Given the same physical world we might have (*logically* might have) come to speak of it differently, might have come to think of it differently, might have come to *see* it differently than we do now. What, then, are we to say of the "blotting paper" view of human knowledge which regards *facts* as the super-objective, ultra-impersonal realities: three-dimensional states of affairs merely copied in our language and in our eyes? It might be that our language, in the form of what we *know*, puts an indelible stamp on what we see, and on what we appreciate as the facts of nature.

So it might at least be argued that the character of what we call "the facts" is affected fundamentally by the logical-grammatical peculiarities of the language in which those facts are expressed. Those peculiarities provide a "set," as it were, a context in which the world looks one way as opposed to another way, or in terms of which the facts are construed in one way rather than in another. In other words, the logical and grammatical traits of our several scientific languages, notations, and symbol-clusters may affect how we see the world, or what we understand to be the facts about the world. They may do this in just the way that an assemblage of ducks may incline us to see the duck-rabbit one way, while if the figure is mounted amidst rabbits we may see it in another way. It is in appreciation of something like this point that classicists will forever lament the essential untranslatability of passages in Plato and Aristotle. Our intellectual and linguistic context is too different from that in which these great Greeks lived and spoke and thought. Trying to see the point of what Aristotle says on *causation,* for instance, is something like trying to see the duck-rabbit as a rabbit when it is surrounded by ducks. Similarly with some of the phenomena of research science: Trying to see a paramecium as a non-celled organism, or trying (as our scientific ancestors had to try) to see man as a superior ape rather than a degenerate God, or trying to see light as corpuscular, or the planetary orbits as grooves in space often requires a training and an effort that not even every science graduate is up to.

But now this is like trying to see the limits of vision, and to proceed further would be to go into the dark altogether.

I would like to conclude, however, with some slightly oracular pronouncements about what will be the content of the next chapter. Until now I have been dancing all around the thesis that there is a very intimate relationship between the ways in which we speak and write and the ways in which we think, and between both of these and the ways in which we see the world.

The language in which we speak and think (this includes mathematical and technical languages as well as our everyday prose) and the circumstances in which we find ourselves speaking and thinking in that language contribute to the formation of, and participate in the constitution of, what and how we think and hence what and how we actually perceive. This is not to say that our language *produces* what we think about, or produces what we perceive, any more than the plaster mold produces the bronze statue, or the recipe produces the cake. It is rather just to suggest that perhaps the (logico-grammatical) form of language exercises some formative control over our thinking and over our perceiving, and over what we are inclined to state as the facts (and indeed *how* we state those facts). In particular I wish to stress in the next chapter how what we call "facts" are almost always stated in *that-clauses*, that same linguistic element we encountered in *seeing* when we considered SEEING THAT. We employ therefore a linguistic idiom for stating the facts which is logically and grammatically identical in form with the idiom we employ when saying we see that a certain visual object is the kind of object it is. This is, of course, a fundamental factor when one directs attention to how remarkably language and facts fit together, how the world and what we say of the world are made to mesh.

My suggestion will be the shamelessly metaphysical one that the languages we use somehow serve to screen the world into two conceptual hemispheres, the noticed and the unnoticed. Between these is a vaguely defined "strip," wherein discovery can take place. It is here that the great efforts of scientists to see things differently is exerted; it is here that they wrench themselves from seeing paramecia as one-celled organisms, men as degenerate gods, light radiation as mere wave propagation, matter as merely hard and locatable, and planetary motion as if it consisted of the paths of whirling grapefruit held "in" by invisible chains. Within this strip, language is fitted to what comes to our attention in discovery. Discoveries change language, our ways of thinking and seeing. But at any given time it is our language and our ways of thinking and seeing that control what kinds of things we would be capable of appreciating as discoveries, the possible range of things discovered. Had they found a scroll containing a complete account of Newton's theory of fluxions, ancient Greek mathematicians could have made little of it, nor could 17th Century astronomers have been expected to notice the apparent shift of the fixed stars during a solar eclipse, a phenomenon as available to their retinas as it is to ours. On the other hand a discovery of such magnitude that it scrapped every shred of the knowledge current at the time is an impossibility, I think a logical impossibility.

I should like rudely to paraphrase Goethe who said, you will remember, that we can see only what we know: I should like to say, "We can see only what we can say or can say to some extent." We can perceive only what we can express, or can to some extent express. What would seeing the utterly unsayable be like? What would perception of the inexpressible be like? Who will indicate for us the facts that in principle elude statement?

Long ago Aristotle remarked the similarities in the forms of language and the forms of the world, the categories of his logic and the categories of being. This similarity he signaled with a single word applied to both language and the world; a word that might be translated "intelligible principle." [5] The intelligible world, the ideal of science, is the brute world as it comes through the sieve of language: it is (in part) the visible world as we *see* it, it is the world of objects, events, and situations as we can state facts about it. So perhaps after all there are really no insuperable philosophical or logical problems about how language and the world ever got together. Nor is it any wonder that our statements so often fit the facts. They were made for each other.

---

[5] The concept of "logos," or "intelligible principle," referred to here figures largely in Aristotle's metaphysics and theory of knowledge. See, for instance, *Metaphysics*, VII, 3, 1029b3–12, especially in the translation of Hugh T. Tredennick, Loeb Classical Library edition (Cambridge, Mass.: Harvard University Press, 1961), Vol. 1, pp. 318–9.

11 | Facts and Seeing That

I TRIED in the last chapter slightly to unfix your views about what we ordinarily call *the facts*. By tinkering with the ordinary language in which we so often make statements of fact, I hoped to suggest that the whole character of what we refer to as *facts* might alter significantly with each change in the fact-stating language. And, of course, the obvious moral of this applies with equal force to the technical and mathematical languages of systematic science; the logical character of the notation and syntax we use to express ourselves is not an inconsiderable factor in the formation of our conception of the physical world. This point has been sharply made in modern physics—especially in elementary particle theory, where, *perhaps*, all of the limitations placed on our conceptions of what the microphysical world is like are really limitations arising out of the linguistic features of the formal languages available. This is certainly the case with the infamous and much misunderstood uncertainty relations, for example: Though this feature of quantum theory is very often announced as constituting a limit to the possibility of *observation* within microphysics, the statement is true in a rather different sense than is often supposed. There have never been any experiments or any observations pertinent to the establishment of the uncertainty relations. Nor could there be such. These relations are a logical consequence of the language of quantum physics, a language the utility of which is now very well established by experiment. Or, to put it another way, the uncertainty relations would result from *any* attempt to synchronize the symbolic description of a wave process with the symbolic description of a particle's state. All of the several quantum languages agree in this, though they put it in different ways. Some will treat it as a logical effect of the wave-packet conception of an elementary particle; others will talk of non-commutativity of operators. It all comes to the same thing. There is a *logical-linguistic* obstacle in the way of our describing with precision the total state of an elementary particle. So, of course, if this is a logical limit to description, it is *ipso facto* a limit on observation. Observation of the impossible is impossible.

So too in biology, where certain perhaps quite arbitrary taxonomic distinctions might very well be shaping and determining what the next generation will construe as the facts of life.

That is the theme. Here I will try a few variations on it, but always in the same key: The shape of the facts is determined largely by the language in which we state those facts. But please remember that I am not saying that language *produces* facts or that facts are just linguistic entities. That would be wrong, I think. The recipe does not produce the cake; nor is a cake just a kind of recipe. But the traces of the recipe are everywhere to be found in the cake. So too the traces of the language in which we state the facts are never absent from our conceptions as to what the character of the facts actually is. That will be my argument, one which I predict some readers will find unsatisfactory. This chapter, incidentally, will be our deepest excursion into philosophical analysis so far. In succeeding chapters we will begin working our way back to the surface, back to common sense and science. But I hope that the things you are familiar with in science will look a little different after this long journey, just as the cave never looked quite the same again to the man who (in Plato's *Republic*) had been able to catch a glimpse of the real sun. [You will notice the change in direction: Where the Platonic character ascended, we have (more modestly) descended. But about that I will say no more.]

We saw how facts were not *objects* in the world, like microscopes or galvanometers. Many philosophers (and still more scientists) talk of facts as if they were all around us, cluttering up the world, there to be seen, photographed, perhaps even kicked. We remarked on the crudity of this view, however, asking what a fact could possibly look like. We may see the sun, or bears, or the grass, or college chapels, and we may state facts about these things. But when do we ever *see* facts about the sun, or bears, or the grass, or college chapels? It is clear that we would rarely, if ever, speak of seeing facts, or of locating them, weighing them, or measuring them, all statements we might very well make concerning ordinary physical objects. So, despite some careless remarks of John Stuart Mill, Lord Bertrand Russell, and quite a few scientists, we will score off any obvious resemblance between facts and physical objects. These are entities existing on different logical levels.

Again, some philosophers and scientists (Alfred North Whitehead being an admirable combination of both) have seen the trap in merely equating facts and objects, so they say that facts are *events*. The sun is not a fact, it is the sunrise which is the fact. The x-ray tube is not a fact, its fluorescence is. The electron is not a fact, its spin is the fact.

This rendering is also objectionable, however. We can certainly locate events in space and time. The sunrise occurs at a certain time and at a certain place on the horizon (wherever *that* is!). The x-ray fluorescence too occurs at a time when a certain circuit is closed and in the area around the instrument's anode. And, though there are certain acute difficulties concerning the location of an electron, we can quite easily say with accuracy where and when it passed through an electromagnetic field (even though we must alter its total energy in order to do so, and *"must"* here is a logical *"must"*).

But are facts locatable within such terms of reference? Is the *fact* that the sun rose today at 7:26 a.m. something that came into existence at 7:26 a.m.? Do facts about x-ray fluorescence spring to life when circuits are closed, and do they reside near the anode? And do facts about electrons occur in magnetic fields? Are they created and annihilated in discontinuous fashion as is the electron? Do they spin, deflect, collapse, etc.? Enough! The things we say of events—that they occurred here and now or there and then—sound grotesque when said of facts. And we may decline therefore to say of facts that they are just events.

Another approach is to lay it down that facts are merely situations or states of affairs or sets of circumstances. This is an extremely attractive position, and one adopted by some very broad-browed philosophers at one time or another, such men for example as Professor Wittgenstein, and Professor Wisdom of Cambridge and Professor Austin of Oxford. Nonetheless the identification breaks down here too, I feel, and for reasons like those advanced earlier regarding the equation of facts with objects on the one hand and with events on the other. For while we may speak of the situation in the Far East or the state of affairs at Brookhaven, that is, we can locate the situation *as in* the Far East and the state of affairs *as in* Brookhaven, I doubt that facts can plausibly be located in this way. In which of the many corners and closets of the Brookhaven laboratories are we to hunt for the situations called *facts?* And though the situation in the Far East may be recorded on cinematograph film I wonder whether the facts about the Far East will at the same time be recorded. If facts are recorded, I think that it must be in a different way from the way in which situations are recorded. Situations are tense, dramatic, exciting, and they are sometimes said to deteriorate. Facts, however, are not tense, dramatic, or exciting in anything like the same sense, and it seems strange to speak of facts deteriorating. Nonetheless, though I would divorce the *fact concept* from the situation, state-of-affairs, and set-of-circumstances concepts, they are sufficiently alike (logically) to remain on close terms with each other. And doubtless there are occasions when our language does not trouble to distinguish them at all.

Another quite familiar rendition of facts is as *true statements*. On this view, the facts about the sun, the x-ray tube, and the electron would be construed only as the statements that could be made truly of the sun, the x-ray tube, and the electron. This certainly avoids the logical snares of regarding facts as if they were just objects like the sun, the x-ray tube, and the electron; and it sidesteps the view that facts are just events like the sunrise, fluorescence, and electron spin. It is further commendable in that it does not consider as facts the situations, states of affairs, and circumstances in which objects like the sun, x-ray tubes, and electrons may be involved. Rather we have it laid down for us that facts are true statements, as for instance in a work called *Science: Its Method and Philosophy*, by G. Burniston Brown (Reader in Physics at the University of London)[1], and also in a book on epistemology by Mr. A. D. Woozley of Oxford.[2]

But surely there are many contexts in which we would refuse to substitute these two types of expression one for the other. There are many scientific contexts, for example, in which when we refer a man to the facts we are not by any stretch of the imagination referring him directly and exclusively to any true statements. When a student of physics

---

[1] G. Burniston Brown, *Science: Its Method and Philosophy* (London: Allen & Unwin, 1950), p. 38.
[2] A. D. Woozley, *Theory of Knowledge, An Introduction* (New York: Hutchinson's University Library, 1949), p. 172.

is referred to the facts that an electron beam deflects in a transverse magnetic field and that the beam will diffract through an aluminum foil target, his attention is not being directed primarily to any true statements. He may not have been asked to consider any statements at all, only these facts. And when perusing a written report of these spectacular experiments he may not doubt at all that the statements he reads are true, and yet he may long to face the facts themselves instead of just the account given of them in the language of the report. Stating the facts, of course, *may* be very like, even identical with, making true statements. So too with learning and discovering facts. We can do this sometimes in a laboratory, and sometimes in a textbook. But facing the facts is not at all like facing the true statements (or, as Mr. Alexander might put it in his refreshing way, laboratory research is not a kind of creative writing, science is not journalism, though—as we saw in the last chapter—the facts we do face may draw much of their character from the language in which they are stated).

We should be rather wary of any too-hasty assimilation of the *fact* concept to familiar concepts like *object*, *event*, *situation* or *state of affairs*, or *true statement*. To paraphrase the later Wittgenstein: We will go wrong if we try to reduce *facts* to any simple relation of things in the world.

But what are *facts*, then? If they are neither objects, nor events, nor situations, nor statements, what in blazes are they? They are already beginning to fade away, like an old retired general. They are moving rapidly towards that corner of the universe of discourse where only the ineffable, undiscoverable, invisible, intangible inventions of philosophers reside.

Well, I think that so long as we ask questions about *facts* as if we were asking questions about the furniture of the world, we are likely to end up with just this will-o'-the-wisp feeling about facts. So long as we ask, "What are facts?" in the same tone of voice as we might ask, "What are plesiosaurs?", "What is segmentation?", and "What is symbiosis?", we will be inclined to think that there must be an answer forthcoming telling us what kind of *thing* a fact is.

But we all know quite well how an ill-put question can twist an answer beyond recognition. The prosecutor knows quite well what he is doing when he asks the defendant "When, sir, did you leave off beating your wife?" If the defendant sputters some inarticulate denial of *ever* having beaten his spouse, the attorney interrupts with "Come, sir, we are only after *facts* here, on what *date* did you leave off beating your wife?" What can poor old Snooks possibly say that will not condemn him out of hand? So too, scientific debates are known to have been conducted along similar lines: "Come, sir, a physical entity must be either a wave or a particle. Which?" "Come, sir, either we are distinguished from the brutes by our reason and our souls or we are no better than chimpanzees. Which is it?" "Come, sir, it's either alive or dead, don't mumble to me about reproduction and crystallization." "Come, sir, it's either a black-backed gull or a herring gull. Now choose." "Come, sir, a physical entity is either penetrable or impenetrable, either a material *object* or a liquid, vapor, or gas. Now which is it to be?" We can easily imagine the Dickensian barrister crowding his opponent with such queries. (And how often have contemporary philosophers been crowded by the question, "Come, sir, are you a scientist or a linguist?")

In just such a way the question, "Come, sir, what is a fact?" crowds any honest attempt to answer. For if you give the simple sort of answer for which the question begs you are sure to go wrong. But if you try to bandy the question with remarks like, "Well, it's all rather more complicated than the question suggests," you will be charged with obfuscation, deviation, and other sorts of conversational misconduct.

Well, I must leave myself open to such a charge. For *fact* is really a rather more complicated concept than the question "What are facts?" can begin to suggest.

We will stand fast against the *equation* of *facts* with *objects*, *events*, *states of affairs*, or *true statements*, therefore, by refusing the bait offered by the question "What are facts?"

Now what?

Now I think we must have recourse to a move that is often very helpful in science and philosophy. Since it is difficult and perhaps even impossible to supply a general and airtight answer to the question "What are facts?" we might try coming to grips with the concept by inquiring how that particular word is used in the ordinary business of saying things about the world. This is the same recommendation I made in Chapter 2 regarding *definition*. Many are inclined to plow into certain philosophical questions about the nature of science brandishing definitions as if they were scimitars. And I dare say that one can make short work of the questions in consequence, for there is no quicker way to bring a game to an end than to refuse to play it. Doubtless Alexander the Great would have been a master in the use of definition; he was the chap, remember, who when set the puzzle of untying the Gordian knot sliced it through the middle with his sword. Similar moves are encountered regularly in scientific courses when poor Professor Snooks is suddenly charged to define "mass" or to define "momentum," "force," "space," or "time"; indeed undergraduates can quickly get a reputation for acuity by embarrassing their science teachers with essentially unanswerable questions like, "Come, sir, define what you mean by 'particle,' " or "Come, sir, let's have a clear and crisp definition of 'life' before we take another step." For the meanings of such conceptions are scattered through the whole of a scientific language, and to answer such questions at all adequately requires *exploration*, not *legislation*. It requires a patient familiarity with the language and the concepts of a discipline, not a sergeant-major's statement about how such terms will be understood for the purposes of a given discussion. In a similar way, to answer the "What are facts?" question with any crisp one-word answer is to have failed to see the subtle ambiguities and philosophical troublemakers in the concept; and worse, it is to have refused to take seriously the large problem (equally important to science and philosophy) about the nature of data, of evidence, in short, *of the facts*.

I could take you uphill and downdale over the logical countryside called "facts." We could explore in detail the sorts of logical job that reference to the facts performs. We could revel in the differences between the types of things that will rate as *facts* within quantum theory and relativistic mechanics and the types of things that will be called *facts* within electromicroscopy or bird ecology. And we might also notice some striking similarities in the uses of this concept in such different fields. We could examine the stylistic force of phrases like "As a matter of fact . . . ," "In point of fact . . . ," "The fact is that . . . ," etc., seeing what sorts of work such expressions are made to do in the writing of research reports for *Science* or *Nature* or *The Journal of Microbiology*. We could even consider some of the several exciting situations in science where two workers begin (apparently) from exactly the *same* facts, only to draw conclusions diametrically opposed to each other; we will brush against this matter later.

All this would be profitable, and (I think) interesting. But it would also be time-consuming, and I must forgo this particular excursion, for our discussion has many topics still to cover.

I will, however, come down hard on what I think would be the overall impact of such a logical sight-seeing tour. I will come straight out with the contention that what we call *facts* are linked logically with what we have come to know as that-clauses, and they are thus linked in the most intimate way possible. As has been said by the Oxford philosopher P. F. Strawson, facts are wedded to that-clauses.[3]

This is not to say that facts *are* that-clauses. If I said *that*, philosophical critics would be all over me like a swarm of gadflies. *That-clauses* are bits of language. Facts, we feel, are something more. But saying that facts are wedded to that-clauses no more entails that facts *are* that-clauses than saying that I am a married man entails that I *am* my wife. Nonetheless, if in my nature I were to seem rather elusive to you, you might find it illuminating to converse with my wife about me. In some such way the elusive nature of facts might seem less perplexing if we hold converse with *that-clauses*, the linguistic medium in which every so-called fact can be expressed, and indeed *must* be expressible.

Suppose I just say straight out that there is no more to any fact than that it can be expressed in a *that-clause*. For example, to the question "What are the facts about the chordates?" we might very well answer

   1. that they possess a notochord,
   2. that they have (at some stage) pharyngeal gill-slits,
   3. that they have a dorsal, tubular nerve cord,
   4. that they have a mesoderm, are bilaterally symmetrical, possess a one-way digestive tract and a blood vascular system,
   5. that they are segmented,
   6. that they probably evolved from some ancestral non-chordate stock which also gave rise to the echinoderms . . . and so forth.

Are not these the facts about the chordates? I think they are: And I did not state the facts, I gave them to you direct. These were not statements of the facts, for I made no statement. A that-clause is not a statement; a that-clause is neither true nor false. (This is an at least initial improvement over the equation of *facts* with *true statements*. For it is always a little queer to speak of *true facts*. So to speak might incline us

---

[3] P. F. Strawson, *Individuals* (London: Methuen and Co., 1959), p. 232.

to speak of *false facts*, a weird form of expression which has actually been perpetuated in German law.)

When asked what the facts are about the chordates, I just supplied them. I only presented the facts. I made no statements as I would have done had my answer been expressed as follows:

1. The chordates possess a notochord.
2. The chordates have (at some time) pharyngeal gill-slits.
3. The chordates have a dorsal, tubular nerve cord.

Now these *are* statements of fact. They are statements which state what the facts are. It would make sense to consider what the world would be like if these statements were false. But notice how, when we are asked *what* exactly it is that these statements state, we answer

1. *that* the chordates possess a notochord,
2. *that* the chordates have pharyngeal gill-slits, and so forth.

Now discovering whether or not such statements *are* statements of fact obliges us to undertake some scientific inquiry. We must first establish that chordates possess a notochord before we can state truly, "The chordates possess a notochord." This means *establishing it as a fact* that chordates possess a notochord.

This involves no shadowy undertaking, however. To find out whether chordates have a notochord we just zip a few of them down the back with a scalpel and take a look. What do we see? Notochords! All shapes, sizes, varieties, and in all shapes, sizes, and varieties of chordate. We do not see any facts, do we?

When asked by a lab instructor what it was that we discovered about the chordates we do not simply throw a sundry selection of notochords at our questioner. That would be neither polite nor strictly what you discovered. What you discovered was *that chordates possess a notochord;* or perhaps you would say you discovered *the fact that chordates possess a notochord*. There is nothing under the skin of the dogfish, the frog, the pigeon, or the rat that goes by the name "the fact that chordates possess a notochord." So there is nothing under their skins to be discovered corresponding to such a name. And yet, undoubtedly, you do no more than open up such sorts of animals one after the other to establish it as a fact that the chordates have a notochord, or to discover the fact that chordates have a notochord. We are now in a position to state facts, speak truly, about the chordates, even though we say no object, no event, no situation, and no true statement worthy of the name "the fact that the chordates possess a notochord."

"But wait a moment," comes the retort, "The *fact* that chordates have a notochord is what makes the statement 'Chordates have a notochord' true. It is the situation of the chordates' *having* notochords that allows us to speak the truth in such an assertion, and that situation is indisputably something in the world."

Initially, nothing makes a statement true. It *is* true or it *isn't* true. It states what is the case, or it doesn't do this; and that is the whole story. I do not make mirror reflections of myself adequate, they just are adequate. Getting puzzled about what makes statements true is like getting puzzled about how baby makes three—as if the little chap had a secret method of making three. But baby just does make three, in the same way that statements just are true or false. So, in a way, there is nothing *between* the furniture of the world and our statements about the world. Nothing tangible anyhow. The world is a congeries of objects, events, relations, situations, and states of affairs. These are the sorts of things *statements are about*. But statements *are not about* facts. *They state facts;* at least, some statements state facts. Some statements state facts, but in doing so they do not characterize the facts they state. Identifying the fact a statement states with the object, event, or situation the statement is about is a pure howler.

So, if you will allow a slightly artificial way of putting it, there is nothing tangible between the furniture of the world and our statements about the world, *only that-clauses and facts.* Statements (some of them) state facts about the world. Whether they do so or not is a matter of experimental inquiry. But when we ask what *are* the facts they state, we only get a string of *that-clauses*. It is these that-clauses which, so to speak, throw logical shadows on the world. We do not merely confront the world *en bloc,* as a kaleidoscope of shapes, colors, and objects, but we face it as *related* in this way or in that way; from our earlier discussions we have seen that the world is usually experienced as connected. And it is these connections *simpliciter* that it is the office of that-clauses to express linguistically. To consider such that-clauses is not necessarily to make statements about these connections, but only to appreciate these connections in their linguistic aspect. And it is these kinds of considerations that incline me at least to think of *facts* as constituted of no more than those aspects of the world that are expressible in that-clauses. There is no more to the fact that chordates have notochords than that the world is such that an aspect of it is expressible in the phrase "that chordates possess notochords."

My reasons for saying this are both negative and positive. Putting the matter in this way does not require me to answer the "What *are* facts?" question with anything like the too-simple answers we encountered in the terminology of "objects," "events," "situations," "states of affairs," or "true statements." None of these conceptions occupies the same logical space as *fact,* however much they may overlap. And (more positively) the consideration of facts as those features of the world we can express in *that-clauses*—or as the shadows of that-clauses thrown on the world—allows me to take the suggestion of the last chapter much more seriously. That suggestion consisted, you will remember, in remarking how our conceptions of the facts vary as does the language in which or with which we have those conceptions. The possibilities of conception, I argued, were only linguistic and logical possibilities. As Kant remarked, people who think in Latin cannot have certain conceptions the Germans have; as Wittgenstein remarked, there could be people who thought much more, or much less, definitely than we; and as I remarked (if you will pardon the sequence), what more is there to the world as known by science than the conceptions scientists have of it? *That-clauses* project, as it were, the possibilities of linguistic expression into the world. They sort the raw, chaotic data into what might be appreciated as conceivable or as intelligible. They provide the logical matrices through which we *see* the world. (*"Seeing that,"* you will remember, was an idiom wholly dependent on that-clause construction.) I think this must be pretty close to what Aristotle meant by his word *logos:* the world as it is made intelligible in speech and in perception.

Of course not just any old that-clause will express a fact: that pigs fly is not a fact; that triangles are quadrilateral is not—indeed, could not be—a fact.

Another way of putting this is to say that such clauses could not be incorporated in *true statements* of the form "It is a fact that pigs fly" or "It is true that triangles are quadrilateral." Hence, any *that-clause* which does express a *fact* could legitimately be incorporated into true sentences of the form "It is a fact that . . . ," "It is true that . . . ," "The fact is that . . . ," and so forth.

The word "fact," therefore, is a kind of linguistic wrapping paper. It is like the word "thing" in this respect. Whereas it is quite all right to say something like "There are three things on his bench," as an alternative to saying "There are a galvanometer, an electroscope, and an amber rod on his bench," there is something queer in saying, "The galvanometer is a thing." This expresses either an empty tautology or a redundancy. In somewhat the same way it would be queer to say, "It is true that the chordates possess a notochord, that they have pharyngeal gill-slits, and that they have a dorsal, tubular nerve cord—and these are all facts." Again, this is redundant. That these are all to be understood as facts is built into the very form of expression "It is true that . . . ," just as that a galvanometer *is* a thing is built into the form of expression "A galvanometer, an electroscope, and an amber rod were on his bench."

This is why the question "What is a fact?" is ridiculous. It is in principle just like the question "What sort of thing is a thing?" Words like "fact," "thing," and "object" are category-words embracing large classes of entities. "Thing" and "object" are just general and indefinite words or concepts used in reference to entities like galvanometers, electroscopes, and amber rods. "Event" is just a general word or concept embracing explosions, falls, growths, births, and deaths. And there is the whole story. So, too, "fact" is just a general and remarkably indefinite word or concept appropriate to the characterization of such a thing as that the sun is up, that oxygen is lighter than americium, that chordates possess a notochord. And that is the whole story. The hunt in the physical world for that which corresponds to expressions like

"that the sun is up,"
"that oxygen O is lighter than Am,"
"that chordates possess a notochord"

is a pseudo-hunt for pseudo-objects. There is no more to the characterization of these as *facts* than that they could be used in sentences of the form

"It is true that . . . ,"
"It is a fact that . . . ,"
"It is established that . . . ,"
"We have proved that . . . ,"

and so forth. Whether it is legitimate to use those phrases about the sun, oxygen, and chordates in these sentences depends on the usual well-known techniques of experimenting and observing to find out how the world is.

Fact, therefore, is a kind of umbrella-concept, a concept like *mind, intelligence, seeing, event, thing,* or *object.* Though facts do not take up space in the physical world, reference to them very often has quite a point, much the same point as has reference to the data, to the evidence, to the truth, or to what is the case. And that the facts are almost wholly articulated in terms of *that*-constructions is really very important. For it is through the office of *that*-constructions that most of the concepts important to scientific observation and scientific knowledge crisscross. All of which is to say that, like most of the really perplexing concepts in science and in philosophy, "fact" is a *hollow* word. It has no specific content, it is almost purely formal. It provides no easy and obvious answer to the question "What is a fact?" Neither are we provided with easy and obvious answers to the questions "What is matter?", "What is force?", "What is mass?", "What is life?", "What is organization?" "Matter," "force," "mass," "life," "organization," and, particularly, "fact" are chameleon words: They draw much of their color and their force from the particular scientific contexts in which they figure and do their work. Rather than to ask in general, "What are facts?" we should do much better to note to what use the word "fact" is put in this context or in that context.

So there is nothing, I feel, in a reference to the hard, objective *facts* of science that makes me want to qualify anything I said before. "Fact" is a category-word, a wrapping-paper word, a hollow word, a chameleon word. Given the thesis that changes in our language (and, *ipso facto,* changes in our conceptual framework) can change our appreciation of the nature of the world, the notion of *fact* is no antithesis, antidote, or corrective to this thesis. For *fact* is just one of the concepts that *transmit* such changes. What pass as the *facts* of a science reflect all the modifications in our pictures of the world, our modes of expression, and the character of our perceptions.

Part III | PERPLEXITY: THE PROCESS OF EXPERIMENTAL RESEARCH

12. Waves, Particles, and Facts
13. *Hypotheses Facta Fingunt*
14. Scientific Simplicity and Crucial Experiments
15. The Systematic Side of Science
16. Discovering Causes and Becauses
17. What Happens as a Rule
18. Theory-Laden Language
19. The Scientists' Toolbox
20. Laws, Truths, and Hypotheses
21. Principles as Platitudes

BIBLIOGRAPHY
PART III

Campbell, Norman. *What Is Science?* London: Methuen & Co., 1920.
Caws, Peter. *The Philosophy of Science.* Princeton: Van Nostrand, 1965.
Duhem, Pierre. *The Aim and Structure of Physical Theory.* Translated by P. P. Wiener. Princeton: Princeton University Press, 1954. (Originally published in French in 1904.)
Hall, A. R. *The Scientific Revolution.* Revised ed. Boston: Beacon Press, 1966.
Hanson, Norwood Russell. *Concept of the Positron: A Philosophical Analysis.* Cambridge: Cambridge University Press, 1963.
Hempel, Carl G. *Aspects of Scientific Explanation.* New York: Free Press, 1965.
_____. *Philosophy of Natural Science.* New York: Prentice-Hall, 1966.
Hume, David. *An Enquiry Concerning Human Understanding.* La Salle, Ill.: Open Court, 1958. (Reprint.)
Jammer, Max. *The Conceptual Development of Quantum Mechanics.* New York: McGraw-Hill, 1966.
Körner, Stephan (ed.). *Observation and Interpretation.* New York: Academic Press, 1957.
Kuhn, Thomas. *Structure of Scientific Revolutions.* Chicago: University of Chicago Press, 1962.
Mill, John Stuart. *A System of Logic Ratiocinative and Inductive.* London: Longmans, Green and Co., 1965. (Reprint.)
Nagel, Ernest. *The Structure of Science.* New York: Harcourt, Brace & World, 1961.
Ryle, Gilbert. *The Concept of Mind.* London: Hutchinson's, 1949.
Scheffler, Israel. *The Anatomy of Inquiry.* New York: Alfred Knopf, 1963.
Toulmin, Stephen. *The Philosophy of Science.* London: Hutchinson, 1953.

12 | Waves, Particles, and Facts

THE LOGIC of notions like *see, observe, witness, notice, data, evidence,* and *facts* has, for my purposes at least, been thoroughly explored. For the next few chapters I should like to see how the morals drawn from our work so far can be applied in cases of actual scientific perplexity. The present chapter, therefore, as well as the two to follow, will be addressed, not to logical *exploration* (as the previous chapters were), but to the *application* of the fruits of our study to typical cases of experimental research.

We will discuss *the facts* of the controversy over the nature of light. We will consider the apparent logical clash between the undulatory and corpuscular theories. Next, the general question of *hypotheses* will be raised. The work that hypotheses perform relative to any experiment or observation will be remarked and appraised. Finally, that queer phenomenon the *crucial experiment* will be cross-questioned.

In all this I shall not hesitate to refer back to, and even repeat, points made in previous chapters. Sometimes I will call your attention to features of an argument upon which I have already dwelt in detail. Important points can always bear repetition, particularly when they are made in a new context.

It will be recalled that the phenomena of interference are among the most distinguishing characteristics of wave motion. Hence the appearance of an interference phenomenon is reliable evidence for the existence of waves. Why this should be so is in itself a difficult philosophical question. An interference phenomenon on a fairly large scale may be produced as follows: run together two identical wave trains traveling in nearly the same direction; viewed from 45° above

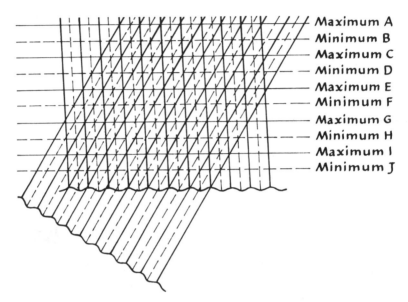

Two wave trains interfering with one another. Troughs of the waves are indicated by dotted lines, crests by solid lines. The solid and dotted horizontal lines show where the two wave trains reinforce and counteract one another. Thus, the horizontal solid lines can be seen to trace out the path of overlap of the crests of the two wave trains. The horizontal dotted lines trace out the path of interference, viz. the points at which the troughs of one wave train cross the crests of the other. Maxima occur at A, C, E, G, and I. Minima occur at B, D, F, H, and J.

Along the lines marked "maximum" the crests of the two waves coincide. On the lines marked "O" a crest of one wave coincides with a trough of the other, the effect being that each is neutralized. If these were water waves they would lap up and down to a maximum extent along stakes placed at A, C, and E, but they would remain at rest alongside stakes placed at B and D. Were they simple sound waves, a loud sound would be heard at A, C, and E, but silence would prevail along B and D.

All this was fairly well understood at the beginning of the 19th Century. It was then that Young and Fresnel addressed themselves to the problem of determining what was the physical difference between the different parts of the continuous spectrum of white light. This is as opposed to mere difference in physiological effect, i.e., the difference between what we call "colors." That is, after the Newtonian discovery that white light was composite, it was still an open question as to what were the differences (besides our physiological reactions to them) among the several components of the spectrum.

The theory was that if light were to be shown to have either a wave nature or a corpuscular nature (the great assumption being that it must have one or the other), an experiment had to be devised which would place two narrow beams of light into the situation sketched above. If interference phenomena are observed, their light must be propagated in waves. If no such phenomenon is forthcoming, then light is at least not wavelike, and probably corpuscular. An interesting logical aspect of this situation is the centrality of the assumption that light must be *either* undulatory or corpuscular *but at least one of these and not both*. Without this large assumption having been fed into the deductive machinery one could never conclude that light was wavelike merely from the fact that interference phenomena were exhibited. That the Fresnel and Young experiments could be thought to be, in a way, crucial to the decision between the corpuscular and undulatory theories of light only makes sense when considered against this presupposition.

Fresnel and Young, of course, both devised effective experiments for the exhibition of interference. But because Young's experiment involves *diffraction* we will consider only Fresnel's.[1]

Now, the best microscopes enable us to see details of somewhere between 10^{-4} and 10^{-5} cm. in size. Hence it is obvious that light, if it is wavelike at all, cannot have waves very much larger than this; otherwise the microscopic specimen would not reflect the waves and would therefore remain invisible. To demonstrate the interference of light waves, then, the angle between the interfering beams must be exceedingly small. Furthermore both beams must derive from the same source. This is the only way of insuring that they are identical in wave lengths and in phase. Hence Fresnel's experiment.

---

[1] For a fuller discussion of Young and Fresnel's work see Ernst Mach, *The Principles of Physical Optics*, tr. J. S. Anderson and A. F. A. Young (New York: Dover Publications, Inc., 1926), Chapters VIII–IX.

204 | Chapter 12

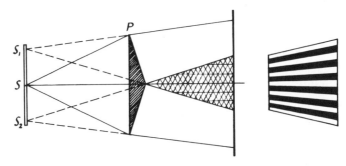

The Fresnel double prism experiment. Homogeneous light enters through the very narrow slit at S and is refracted through the double prism P. The effect is that of two virtual sources radiating from the original baffle (indicated by the dotted lines). The shaded area to the right of the prism shows the area where the two virtual beams interfere with one another. An interference pattern, indicated schematically at the far right, is thrown onto the vertical screen.

Two halves of a beam of light, coming from the very narrow slit S, are made to overlap by means of a double prism (this is equivalent to two prisms placed base to base). The prism angles are very small, less than one degree. Hence the angle between the overlapping beams is very small. Visible fringes appear on the screen at this place of mutual overlap. The bright fringes correspond to the A, C, and E of the earlier diagrams, the dark fringes to B and D. The explanation is exactly the same.

→ This experiment presents us with facts that provide irrefutable evidence for the existence of light waves. These facts are obvious. Such a pattern of fringes would not be possible were their source other than wavelike. If anyone were to ask you why you were so sure that light had a wave nature, you would probably take him into a laboratory, set up this simple experiment, and show him the facts: the light and dark fringes which can result only from the interference of something essentially wavelike. All that could keep him from appreciating these facts would be imperfect vision. He would then be unable to see the facts.

But, as you already suspect, this language is to come under fire. Some questions will be raised about it in a moment. First, however, let us consider briefly these other experiments which bear directly upon our appreciation of Fresnel's discoveries.

In 1887 Heinrich Hertz observed that a spark between two metal electrodes occurs at lower than normal voltages if ultraviolet light is shining on the electrodes.[2] Eleven years later J. J. Thomson demonstrated that when a metal surface was bathed in ultraviolet light the metal emitted negative charges.[3] Thus a negatively charged electroscope connected with a piece of zinc slowly loses its charge as light falls on the zinc plate. When the experiment is repeated with a positively charged electroscope, the leaves do not fall. Conclusion? Under the action of the light, negative charges leave the zinc plate, and hence the electroscope. (If, of course, the zinc plate were positive, it would retain the negative charges seeking to leave.)

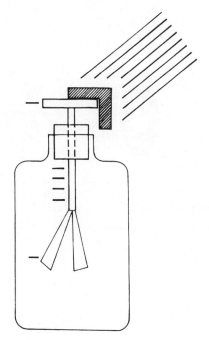

This is the photoelectric effect. The emitted charges are called photoelectrons.

A long and systematic survey of photoelectricity, the details of

---

[2] Heinrich Hertz, *Electric Waves*, tr. D. E. Jones (London: Macmillan, 1900), p. 63. Originally published in *Sitzungsberichte der Berliner Akademie der Wissenschaft*, June 9, 1887.

[3] J. J. Thomson in *Philosophical Magazine*, Vol. 48 (1899), p. 547.

which you will be spared, disclosed several facts of great importance. For one thing, *all* elements exhibit the photoelectric effect. But each element has its own characteristic photoelectric threshold. That is, for any given substance there is a minimum light frequency below which photoelectrons will not be emitted. Next, for a given surface, the photoelectric current—that is, the number of photoelectrons emitted per second—is directly proportional to the intensity of the light. Finally, and most spectacularly, for a given surface the kinetic energy of the emitted photoelectrons depends *only* on the light frequency; *it is independent of light intensity*. Consider this experiment:[4]

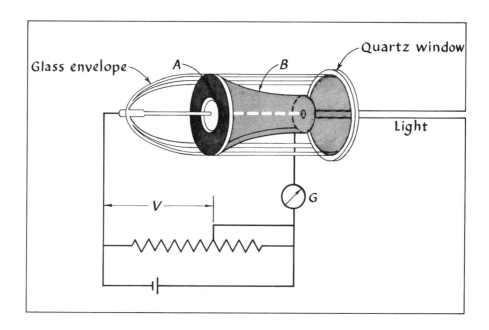

The incident monochromatic light from the right is chosen with a wave length considerably shorter than the photoelectric threshold value for the metal plate A. It is discovered that even if the potential

---

[4] See, for instance, David Halliday and Robert Resnick, *Physics for Students of Science and Engineering*, 2nd ed. (New York: John Wiley and Sons, 1962), Vol. II, p. 1088.

difference between the metal surface A and the collecting cup B is made zero there is still a current registered on the galvanometer G. Ergo, electrons are being emitted with some kinetic energy. But since the potential difference is zero the current cannot be the result of thermionic emission of electrons from the metal cathode. It must therefore be a photoelectric current. To learn how much kinetic energy the emitted electrons possess, a "retarding" potential is applied; the metal is made slightly *positive* relative to the collecting chamber. As this positive potential is increased a point is reached where the galvanometer indicates that no current is flowing. If five volts of potential difference are required to hold the electrons back, then they must have had five electronvolts of energy. By this method the amount of energy given to the electron by the incident light can be measured as both the intensity and the frequency of the light are varied.

The electron energy is found to be *independent* of light intensity. It may be noted now that the wave theory of light is completely unable to account for this. If light is depicted as a wave motion, increasing the intensity of the light must increase the amplitude of the wave and hence its energy. The sound waves from a sharp clap of my hands sets your eardrums in motion. The energy received by the eardrum will depend upon the amplitude of the sound wave. The louder or more intense the handclap the greater the amplitude of vibration of your eardrums. But increasing light intensity *fails* to give the photoelectrons more kinetic energy. It only causes the ejection of more electrons. Higher photoelectron energies can be produced only by increasing the light frequency. Consider now how this links up with A. H. Compton's experiment of 1923.

Compton was studying the scattering properties of x-rays. His idea was to incorporate x-ray scattering into the scheme of electromagnetic theory. Many physicists were similarly engaged at the time, but Compton alone employed a Bragg x-ray spectrometer to measure the wave lengths in the scattered beam. He discovered that x-ray frequency could be changed in the scattering process. Here schematically is the set-up:[5]

---

[5] A. H. Compton, *X-rays and Electrons* (New York: D. Van Nostrand Co., 1926), p. 266.

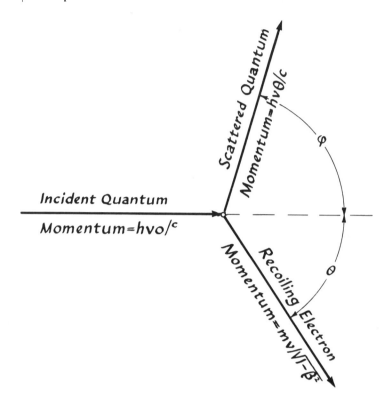

Compton found that part of the scattered beam had a longer but quite definite wave length. Now there is nothing in all of wave theory which allows for the possibility of a wave changing its frequency in this manner. Compton proposed that this long wave-length line was due to a "photon" collision with electrons in the scattering material. The electrons were regarded as being at rest before the collisions. And they were supposed to be free, that is, unbound to the carbon or other atoms in the scatterer. Collisions were conceived to be completely elastic. Compton interpreted the photon and the electron as comprising a two-particle system in the strict classical sense obeying the conservation of energy and momentum.

The mathematics involved in calculating the alteration of the wave length λ as a result of such a collision is somewhat complicated. It will not be presented here. (Cf. appendix to this chapter.)

At ø = 180° we encounter back-scattering. Here the value of

the shift in λ reaches a maximum of 2 times h/mc (sometimes called the "Compton wave length").

→ Clearly, then, experiment and theory confirm that light and x-ray photons exist as particles. They are localized in space and have momentum. In collisions they behave purely as particles; their wave properties play no apparent part. And yet they are peculiar particles, for obviously they possess no rest mass. Despite our difficulty of picturing a particle without rest mass undergoing an elastic collision, those are the facts. Dr. Compton could have displayed these facts to any doubting Thomas. He actually designed a cloud-chamber experiment in which the collision between electron and photon was traced and observed. The facts are there to be seen by anyone with eyes! When the scattering angle is 90° the scattered beam registers intensities of both the .710 Å of the incident beam *and* a longer .734 Å component—and that's a fact! At $\emptyset = 180°$ the back-scattering value is $2h/mc$, and that's a fact! And these facts are evidence that light is a particulate-discontinuous phenomenon, *just as the facts of Fresnel's experiment were, and are, evidence that light is a wavelike continuous phenomenon.*

Those are the facts. Open your eyes and look.

But we must now bring forward the morals of previous chapters and reflect again for a moment on the queerness of speaking of seeing facts. Have you ever *seen* a fact? I never have, or at least I have never *spoken* of having done so. Perhaps someone tries to tell us what the last fact he saw looked like. We shall immediately ask: Was it tall or short? Stout or thin? Black or white? Would heat affect it, or damp; might it be photographed, or reflected in a mirror, or set into a beam balance? These questions are enough to show that there is a logical fish here somewhere.

After my account of the Fresnel experiment I referred you to the facts thereof. I referred you to the light and dark fringes on the target screen. To persuade a dissenter that light really is wavelike, I suggested that you might show him the facts, in this case the fringe-pattern.

Clearly, however, a fringe-pattern is merely a tiny patch of light and dark. It could be photographed, or reflected; it might be narrow or broad, tall or short—all things that a fact cannot be. But a fringe-pattern cannot confirm or deny anything about the nature of light. What persuades us that light is wavelike is *the fact* that when two light beams of identical wave lengths overlap (in phase) they interfere to produce a fringe-pattern. It is this fact that confirms the hypothesis that light is wavelike, and not the fringe-pattern itself. The fringe-pattern is not of the right logical type to refute or deny any hypothesis or theory. Fringe-patterns just *are*, like rocks and blades of grass. They are not assertive, hence they cannot confirm or deny. If a mouse were to creep into a physics laboratory at night and flick a switch turning on the light sources in this Fresnel apparatus, would we say next morning that the diffraction pattern had been busily refuting the corpuscular theory of light since 2:00 a.m.? We might refer to *the fact* that a light beam incident upon such a biprism will produce such a pattern, and we might appeal to this fact in advancing our reasons for saying that light has a wave nature. But this fact did not come into existence at 3:00 a.m. It did not even come into existence at the beginning of the 19th Century when Fresnel and Young were hard at it. Facts do not *come into* existence at all, as you can easily persuade yourself by asking of facts a few birth-day-questions like, "How old is it now?", "What is its life expectancy?", etc. Facts may be discovered but they are never born. Or are they? And yet it is facts and only facts that support or defeat our hypotheses and theories. Facts and only facts are what make us say of propositions that they are true. What proposition would a fringe-pattern verify? What propositions *could* it verify?

Perhaps this purely logical point will become even clearer if we reconsider Compton's conclusions in the same terms. Two vapor trails in a Wilson cloud chamber could never, of themselves, confirm that light is granular and discontinuous, not even if the trails traced the most representative of all two-particle collisions. It is *the facts that* zinc bathed in ultraviolet light will lose electrons, *that* the energy of these electrons is independent of the intensity of the light, and *that* an x-ray photoelectron will bounce off a carbon electron in billiard-ball manner that force such an important theoretical conclusion on us.

Notice again how these conclusions are expressed. They are always put forward in sentence form, or in a "that"-clause construction. Fresnel might have said, "Light is wavelike in nature," and to support this he would have directed you to the fact that when, with a biprism, a light beam is split and made to overlap, fringe-patterns appear on the target. But there is nothing to be recorded on the retina that corresponds to this statement of fact. There is nothing three-dimensional that goes round labeled "the fact that when, with a biprism, a light beam is split and made to overlap, fringe-patterns appear on the target." But it is just this fact that makes us say that Fresnel's conclusions were true and those of his opponents false. A fringe-pattern *per se* could not make us say of anyone's conclusions that they were true or false. All it can do is make us blink with eyestrain.

The general logical point is this: If we have a theory or an hypothesis that we wish to test by experiment we must usually make certain preliminary deductions before we can even get matters down to an experimental stage.

Fresnel wished to learn whether light was wavelike or corpuscular. By way of the quite general principle or presupposition *that anything wavelike will exhibit interference properties* (and nothing particulate will exhibit these properties) he designed a way of finding an answer to the question, "Were two light beams, identical in wave length and in phase, to cross at a very small angle, would interference phenomena be manifested?" An answer to such a question, or any question for that matter, will always be expressed as a statement of some sort. Had Fresnel set out this question in his report to the French Academy he would have answered it thus: "After the experiment was conducted as outlined a fringe-pattern was observed." And this would be a true statement because of the *fact that* after conducting the experiment as outlined Fresnel did observe a fringe-pattern. Fresnel would not merely have posed the question and then gestured (with a grunt) towards a fringe-pattern. A fringe-pattern is not an answer to a question; a statement is. And if a statement is a true answer to a question, it is because it states what are the facts. And this particular statement of fact was taken to be the main part of an answer to the controversy over the nature of light because of the brace of assumptions (1) that interference phenomena in X was conclusive proof of the wave nature of X, and (2) that X could either be wavelike or particulate, but at least one or the other, and in no circumstances could it be *both* wavelike and particulate.

Compton *could* have given names to all the discernible elements in his ingenious experiment. He might have called the anode "Moe" and the scattering block "Garb," he might have called the reflector "Chris" and the scattered photoelectron "Ray." But then nothing that he names, not even the cloud-chamber collision, will be *what confirms the hypothesis* that light is granular and particulate. It is only the conclusion of a long and complicated deduction which can do this. But the conclusions are not namable because they are always expressed as statements, and statements are not namable. Scientific papers are not made bulky with bits of glass and metal glued to the pages and surrounded by inverted commas. They *are* made bulky of course, but for other reasons.

So again the big point to get straight when regarding the facts is that they are not to be found within and alongside other bits of apparatus on a laboratory bench. They are invisible, untouchable, odorless, tasteless, and very silent, but not because they are especially marvelous sorts of material objects. They are all these things rather just because *it makes no sense* to speak of seeing, touching, sniffing, tasting, or hearing a fact. (It is this that makes Professor Braithwaite's much-used phrase "observable fact" so puzzling.[6] For how else except by seeing, touching, sniffing, etc. might anything be observed?)

The second thing to see is that, just as people and objects are the counterparts of names, and properties and processes are the counterparts of adjectives and verbs, so facts are the counterparts of statements. The guarantee that a name is used correctly is the existence of some object (appropriately situated) which bears that name. The guarantee that a verb is used correctly is the occurrence of some process which is specified by that verb. But the guarantee that a statement is correctly and truly used is the existence of the fact it states. Yet there is no *place* to look for such a fact: The matter is logically more complex (as we have been trying to show). Hence discovering whether a statement is true or false is not just like seeing whether that chap in the distance is Bertie or Willie. For on closer inspection one can see *whether* it is Bertie or Willie. But with facts there is nothing to inspect more closely. Yet nonetheless it is facts that are the backbone of science and not objects, events, or situations. If one *had* to say what science consisted of it would be better to say that it consisted of facts, or of statements of fact organized in certain characteristic ways, than to say that it was an assemblage of objects, events, or situations. The Cavendish, Brookhaven, and Oak Ridge laboratories, with all their equipment and activities, do not constitute science, although the advance of science is aided immeasurably by the existence of such institutions. So if one *had* to make a mistake about the nature of science it would be better to say that it was an assemblage of facts rather than an assemblage of objects, or even an assemblage of observations. More particularly it is an assemblage

---

[6] R. B. Braithwaite, *Scientific Explanation, op. cit.*, p. ix *et passim.*

of specific facts like those that were discovered by Fresnel and Compton. Indeed, it is such facts as these and the interrelations between them that lead us to our rather uncomfortable set of conclusions about the nature of light. For the facts are that light is wavelike *and* that it is corpuscular. And yet the Fresnel-Young assumption that it must be one or the other but not both is lurking in the woods somewhere.

How can we get out of this logical thicket? And amidst all this chopping what is there to be learned about facts, about light theory, and about the nature of science from our efforts? Let us see.

I have been striving to present a case for there being a very powerful linguistic bias in our perceptual knowledge, in our thinking, and in our science. Our knowledge consists largely of statements which we believe to be true. *Knowledge* of any subject matter is not just a staccato review of the principal names and verbs which might be found in that discipline. We show that we know the "ins" and "outs" of a subject when we can state some relevant facts about it. We show that we are perceiving something when we can state some facts about the nature of our perceptions. But remarks like "red now" or "oscillating" or "buzz buzz" do not state facts about our perceptions, though they may perhaps have important roles to play in perceptual situations. Such remarks do not state facts about anything. And it is only *statements* about the character of our perceptions that can be incorporated into a wider system of our knowledge. It is only statements about what we perceive during experiments that can be incorporated into the body of our science.

Before our perceptions can be given any sense relative to what we already know they must be molded into propositions. Because only so can we mark an observation as relevant or irrelevant, significant or insignificant. Because only so can we understand two perceptions as being complementary or contradictory. Only so can we detect that the conclusions of the Fresnel-Young experiments support each other. A fringe-pattern and a diffraction pattern *per se* cannot support or contradict each other. What would it be like for two such patterns, plane or solid, to contradict each other or to support each other? No, it is only relative to the propositions that characterize them that such things as interference patterns can lend support to the contention that light has a wave nature. And it is only relative to the rather more complicated experiments of Compton that the sight of two ionization trails tracing an elastic impact lends support to the contention that light has a particle nature. If there is a contradiction here it is as always between propositions and not between events or perceptions. Diffraction patterns and ionization tracks cannot contradict each other, they cannot be maneuvered closely enough together for that. The contradiction, or apparent contradiction, derives solely from our habit of thinking so well of the proposition: "Propagation of motion in a medium can be *either* by particle action *or* by wave action *but not both*." This proposition certainly expresses what we encounter in our everyday molar-magnitude experience. Hence with this as a background the Fresnel-Young conclusion seems logically to exclude the Compton conclusion. Had the Compton conclusion been established first, however, and against the background of this presupposition, it would have seemed to exclude the Fresnel-Young conclusion. But the FACTS support both contentions. The propositions which express the "Fresnel-Young" perceptions and those which support the "Compton" perceptions are true. So, clearly, our classical and everyday proposition about waves-or-particles-but-not-both is a little wobbly. But it is *logically* wobbly. It is propositions that are crowding one another and not events or perceptions. We have stretched a notion beyond its limits without explicit warrant. The amazing thing, of course, is that we get away with this so often in science, and not that we are caught out occasionally.

Before pursuing this matter to the death, however, I should like to consider another aspect of facts. This is an epistemological consideration and not a merely logical one of the sort to which we have been attending. I wish now to inquire about the factors involved in our coming to know something as a fact, whereas until now I have been concerned with the logical status of facts relative to perceptions on the one hand and language or propositions on the other.

Charles Darwin once remarked on how odd it was that people should not see that every fact is a fact for or against some theory. I shall try to squeeze just a little of the juice out of this ripe remark.

Even after arguing that facts are not observable we were still content to say that on the target screen of the Fresnel experiment interference fringes might be observed. With the Young experiment we speak of a diffraction pattern and we say that we *see* the diffraction pattern. And indeed we *do* see this. But, as we have already noted so often and in such detail, there is a genuine difference between what we *see* and what strikes our retinas. Granted that we do not see any *facts* about interference fringes or diffraction patterns, can it even be said without qualification that we see interference fringes and diffraction patterns?

What strikes the retina are alternate bands of light and dark. To see these and to speak of these as interference fringes or diffraction patterns is not a low grade visual undertaking. Not everyone *could* see the bands in such a way, or speak intelligently of them in such a way. The significance of observational phenomena like interference and diffraction can be appreciated only against a background of at least some elementary wave theory, certain general principles like that of the rectilinear propagation of light, and probably a good deal of experience with the characteristics of water waves and sound waves. A very young child, whose vision is every bit as good as ours, will not see interference fringes or diffraction patterns. He will see alternate bands of light and dark— and that is all. And that is the substance of our own visual impression too, though for us it is a sophisticated visual experience. We *see* interference and diffraction. The raw visual datum of Compton's cloud-chamber experiment consisted only of two divergent fuzzy lines. Only a very great deal of training in the theory and practice of physics equipped him to see these as the ionization tracks of an elastic collision between a photon and an electron.

The significance we will attach to an observation is pretty largely a reflection of what we have been trained to regard as significant, which is just a way of saying that we see every new experience only through the lens of the knowledge we already possess. You and I can see diffraction patterns and hence can appreciate the facts in favor of the wave theory of light. But a totally uneducated person, a child, a savage, or a philosopher, would not see a diffraction pattern, but only strips of light and dark. And an understanding of the facts about the wave nature of light might be beyond such a person. He would not even have the capacity to appreciate what *are* the facts, much less make inferences from them. And he could certainly not be expected to see that the Fresnel facts and the Compton facts clash violently.

Moreover, brilliant men that they were, Fresnel and Young would have been in exactly this same position had they been at Compton's elbow in 1923. If a hundred years could somehow have telescoped together, bringing Fresnel and Young with their early 19th Century knowledge of optics face to face with Compton's x-ray spectrometer and ionization equipment, they simply would not have been prepared, in virtue of their stage of scientific advancement, to appreciate that the fuzzy divergent lines in the Wilson chamber could be understood as constituents of facts in favor of the contention that light is granular. Doubtless, a man as patient as Compton could have explained the significance of these tracks in such a context, and the two men would soon be in a position to see the facts as we see them. But getting them to that stage, notice, would be tantamount to briefing them on what has gone on in physical science during the last century and a quarter. It would be to supplement their knowledge by that much, for nothing less will allow an appreciation of the facts of such an experiment as the one performed by Compton.

As Darwin said, every fact is a fact for or against some theory. I will put it even more strongly: Nothing can constitute a fact *unless understood in terms of some theory*. And if you think that garden-variety facts like the fact that you are now reading, or the fact that you are now sitting down are exceptions and wholly innocent of theoretical substance, I invite you to inspect even them more closely at your leisure.

Seen in this way the *facts* of the Fresnel-Young experiments *do not crowd the facts of the Thomson-Compton experiments at all*. And, I suppose, to get us to see that the facts of experiments taken from different periods in the history of science do not clash (as we might hastily suppose they would) is just the task to which historians of science have often addressed themselves. The experimental contexts and theoretical backgrounds in terms of which these facts have significance are so vastly different as to put our understanding of them on virtually two distinct intellectual planes. If we say that light is wavelike and make no further comment on the differences between light waves, water waves, and sound waves, then *of course* the facts of Compton's experiment will jar us. But if we have learned all we should have learned by the time the granular nature of light is made apparent to us, we will have come to expect such spectacular discrepancies in regions of experience so unlike that of our everyday, middle-sized existence. In terms of the happy distinctions that were made in classical physics and that we make every day, the idea of a wedding between waves and particles is unthinkable. Still, this perplexity is possible only for someone who has not come to reckon with light *on its own terms*—for one who is prepared to see sound wave and water wave analogies *upset* when it comes to light radiation. For the *facts* are that light is wavelike *and* particulate, by which is meant that a complete theoretical and practical understanding of the nature of light will embrace and harmonize these two notions. As Dr. Ashmead of the Cavendish might say, "That's the way light *is*." What Compton's experiment showed was not that nature was self-contradictory, but that we are sometimes too eager to transplant the theoretical roots of old, familiar experience beneath new experiences which, though superficially similar, are really quite different. Light waves, water waves, and sound waves are superficially similar. That is all we were *ever* entitled to say. Compton shocked only those who did not look before they leaped, namely most of us. And if this lesson has led us to a discovery of new facts it has also led us to a new appreciation of the old ones.

The facts, then, are nothing we can kick. They are nothing we can see. And yet they are indispensable to the advance of science. This is because science is at least a body of organized knowledge, an enormously elaborate system of true propositions. The only things that can bear upon our knowledge must therefore be propositional in their logical form: enter *facts*. But besides deriving their *form* from the language in which we express the knowledge we already possess, "the facts" derive a good deal of their *significance* from that knowledge as well. The advance of science is a step-by-step affair. Problems are dealt with as they come up. There are no seven-league boots with which we can jump to the solution of the next century's problems. The step to Compton's conclusions, Compton's observations, Compton's facts would have been too wide for Fresnel and Young, not because they were deficient in intellect or visual perception, but because they did not have Compton's problems. The corpus of their knowledge was characteristically different from Compton's, or, for that matter, from ours. And if that be a truism, well and good. I hope that the nature of facts will henceforth be just as clear to you as that truism, and that you will only speak of seeing them, touching them, or kicking them under duress from your less cautious peers and teachers.

In the next chapter we will examine hypotheses, and I leave this thought for your tender mercies: *The facts are what our hypotheses call to our attention*. Our questions determine, to a large extent, what will count as answers.

218 | Chapter 12

APPENDIX *

THE COMPTON LONG WAVE-LENGTH SHIFT CALCULATED FOR A SCATTERING ANGLE (ϕ) of 90°:

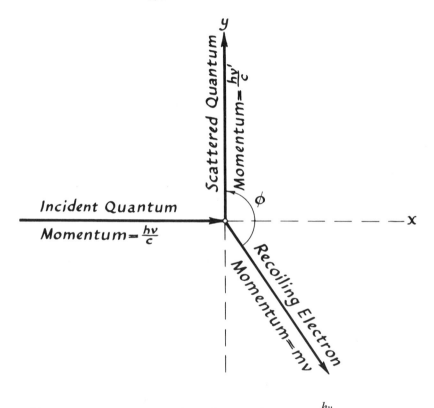

X-component momentum obeys the equation $mv_x = \dfrac{h\nu}{c}$

and for the Y-component momentum $mv_y = -\dfrac{h\nu'}{c}$

where ν' is the frequency of the scattered quantum.

* For calculations of the general case see, for example, Louis de Broglie, *An Introduction to the Study of Wave Mechanics,* tr. H. T. Flint (New York: Dutton & Co., no date), pp. 146–8 (non-relativistic) and Compton, *op. cit.,* pp. 265–8 and Appendix 6 (relativistic).

Energy conservation is given by the formula (neglecting relativistic effects)

$$h\nu = h\nu' + \tfrac{1}{2}m(v_x^2 + v_y^2).$$

Now we have from the momentum formulae that

$$v_x^2 + v_y^2 = \frac{h^2}{m^2c^2}(\nu^2 + \nu'^2).$$

Substituting this into the energy formula we obtain

$$\frac{\nu - \nu'}{\nu^2 + \nu'^2} = \frac{1}{2}\frac{h}{mc^2}$$

and putting the result in terms of the x-ray wave lengths $\lambda = \dfrac{c}{\nu}$ and $\lambda' = \dfrac{c}{\nu'}$, we get

$$\frac{\lambda\lambda'(\lambda' - \lambda)}{\lambda^2 + \lambda'^2} = \frac{1}{2}\frac{h}{mc}.$$

Since the shift in wave length $(\lambda' - \lambda)$ is quite small in comparison with the wave length itself, $\lambda^2 + \lambda'^2 = \lambda\lambda'$ approximately and

$$\lambda' - \lambda = \frac{h}{mc}.$$

This is exactly the wave-length shift as found by experiment for 90° scattering. At smaller scattering angles the shift is less, becoming zero at $\phi = \phi°$; and at $\phi = 180°$ it attains the value of $2\,h/mc$.

13 | Hypotheses Facta Fingunt

"Let the facts speak for themselves"; so runs an old but not very wise saying. For whoever *just* lets the facts speak for themselves will either be enveloped in silence or be deafened with the noise. That, at least, will be the thesis of this chapter.

I concluded the last chapter with the oracular pronouncement, "The facts are what our hypotheses call to our attention; our questions determine, to a large extent, what will count as answers." Let us explore this further. Let us see how hypotheses were used by two great scientists in spotlighting for the first time facts that are now quite familiar to us all.

Consider William Harvey's magnificent *An Anatomical Disquisition on the Motion of the Heart and Blood in Animals,* first published in 1628. Here follows a long quotation, the elements of which we shall subsequently examine at close range:

Harvey wrote:

> . . . when I surveyed my mass of evidence, whether derived from vivisections, and my various reflections on them, or from the ventricles of the heart and the vessels that enter into and issue from them, the symmetry and size of these conduits—for nature doing nothing in vain, would never have given them so large a relative size without a purpose— or from the arrangement and intimate structure of the valves in particular, and of the other parts of the heart in general, with many things besides, I frequently and seriously bethought me, and long revolved in my mind, what might be the quantity of the blood which was transmitted,

in how short a time its passage might be effected, and the like; and not finding it possible that this could be supplied by the pieces of ingested aliment without the veins on the one hand becoming drained, and the arteries on the other getting ruptured through the excessive charge of the blood, unless the blood should somehow find its way from the arteries into the veins, and so return to the right side of the heart: I began to think whether there might not be a *motion, as it were, in a circle.* . . .

The blood is incessantly transmitted by the action of the heart from the vena cava to the arteries in such quantity, that it cannot be supplied from the ingesta, and in such wise that the whole mass must very quickly pass through the organ. . . .

Let us assume either arbitrarily or from experiment, the quantity of blood which the left ventricle of the heart will contain when distended to be, say two ounces, three ounces, one ounce and a half—in the dead body I have found it to hold upwards of two ounces. Let us assume further, how much less the heart will hold in the contracted than in the dilated state; and how much blood it will project into the aorta upon each contraction—and all the world allows that with the systole something is always projected . . . and let us suppose as approaching the truth that the fourth, or fifth, or sixth, or even but the eighth part of its charge is thrown into the artery at each contraction; this would give either half an ounce, or three drachms, or one drachm of blood as propelled by the heart at each pulse into the aorta; which quantity, by reason of the valves at the root of the vessel, can by no means return into the ventricle. Now in the course of half an hour, the heart will have made more than one thousand beats, in some as many as two, three, or even four thousand. Multiplying the number of drachms by the number of pulses, we shall have either one thousand half ounces, or one thousand times three drachms, or a like proportional quantity of blood, according to the amount which we assume as propelled with each stroke of the heart, sent from this organ into the artery; a larger quantity in every case than is contained in the whole body. . . .

Upon this supposition, therefore, assumed merely as a ground for reasoning, we see the whole mass of blood passing through the heart, from the veins to the arteries, and in like manner through the lungs.[1]

---

[1] William Harvey, *An Anatomical Disquisition on the Motion of the Heart and Blood in Animals,* tr. Robert Willis in *Great Books of the Western World,* ed. Robert Maynard Hutchins (Chicago: Encyclopaedia Britannica, Inc., 1952), Vol. 28, pp. 285–7.

This use of hypotheses by Harvey must rank among the most brilliant in the history of science. Now what precisely was he up to?

Several things relevant to the circulation of the blood were well known in Harvey's time and before. It was known, for example, that there were two distinct sets of tubes in connection with the heart, namely the arteries and the veins. It was known also that in one of these sets, the veins, there were valves which would allow the passage of blood in one direction only.

These were just brute facts for many. Not only were physicians and anatomists aware of them, farmers, butchers, and cooks knew these things as well. The connection of these facts with other equally familiar facts, however, was for most people totally unsuspected.

This suggests in still another way how superficial is the view that science might advance if scientists merely studied the facts, merely let the facts speak for themselves, merely opened their eyes and looked. It is superficial because no inquiry can even get under way until, and unless, *some difficulty* is felt in a practical or theoretical situation. Only such a difficulty or problem can guide our search for order among the facts. And, reciprocally, it is only the discovery of such an order among the facts that will resolve the difficulty or problem. Most 17th Century cooks, butchers, farmers, anatomists, and physicians were completely undisturbed by the fact that there were two sets of circulatory conduits (one of which was equipped with one-way valves). Harvey could not possibly have discovered the *reasons* for there being such different sorts of tubing in the circulatory system had he not first recognized in this difference a problem demanding solution, a problem about the nature of the circulation of the blood.

What specifically were Harvey's problems?

Previous to the time of Harvey, the very vaguest notions prevailed regarding the use and movements of the blood. Some people thought that the arteries contained air. Others thought that they contained a subtle essence called "animal spirits." The animal spirits were supposed to have been generated in the ventricles of the brain; they were controlled by the soul, which was situated in the pineal gland. How the animal spirits got from the brain into the arteries was an anatomical detail which was bridged across by the imagination. (Credit for that jokelet goes to the physiologist, Professor W. D. Halliburton.)[2]

[2] W. D. Halliburton and R. J. S. McDowall, *Handbook of Physiology and Biochemistry*, 36th ed. (London: J. Murray, 1939), p. 219.

Another opinion held by some of Harvey's contemporaries was the one he criticizes in the passage quoted. This view was that the blood was produced in the stomach and upper intestine directly from the food we ingest at mealtime. The liquid thus fabricated was then pumped into the heart, and thence to all parts of the body, never to return but to be assimilated immediately at the extremities.

Harvey's *problem* consisted in seeing that neither of these accounts squared with the anatomical facts: the sizes of the cardiac chambers, the symmetry and relative sizes of the conduits, the structure of the valves, etc. It was his basic dissatisfaction with the traditional and contemporary accounts of the circulation of the blood, a dissatisfaction arising from a discrepancy between those accounts of the facts and what Harvey took to *be* the facts, that beset this great man with this great problem.

Of course the ability to perceive in some brute experience the *occasion* for a problem, and especially a problem the solution of which has a bearing on the solution of other related problems, is not a common talent among men. No rules can be given by means of which men will learn to ask significant questions. It is the mark of a scientific genius like Harvey to be sensitive to difficulties where others pass by untroubled. It takes a Harvey or a Galileo to feel the necessity of finding an order amongst phenomena in which most other investigators fail to notice that order is lacking.

How does a search for order among phenomena proceed? Naturally, there are thousands of ways, each peculiar to specific problems within particular disciplines. Sometimes the order attained is of a most refined mathematical variety. At other times it is, as sometimes in physics and chemistry, "merely empirical." (It is in the latter way that Boyle's discovery of the relation between the pressure and volume of a gas must have been understood in 1662; this is also the way in which Ampere's discovery of the relation between an electric current and its magnetic field was understood in 1820. Only very much later did a comprehensive theoretical understanding of these phenomena become possible.) Again, the ordering may be classificatory, as when we group whales, porpoises, and bats with the mammals on the basis of structural similarities, or when we consider collectively the wings of bats, birds, and insects on the basis of certain functional similarities despite the enormous anatomical differences among them.

Just how Harvey proceeded we will discuss in a moment. But it must be clear by now that before he could even state his problem he had to have a real familiarity with the subject matter that presented the problem. Scientific education today accepts the fact that familiarity with the subject matter presenting the problem is essential before any problem can be stated. It takes a lot of training in electrical circuit theory before one can raise genuine questions about hot-cathode electron tubes and half-cycle rectifiers. It took a great deal of training and experience in anatomical matters for Harvey to have been able to pose the wonderfully systematic questions that he did. And yet how curiously inconsistent all this is with the view that science proceeds by letting the facts speak for themselves; just keep your eyes open and wait.

Harvey certainly did not just wait. He might have stared at the arteries, auricles, veins, and valves on his bench until he turned blue in the face, and yet nothing remotely like what he *did* discover would have disclosed itself. For, while his previous knowledge of the circulatory system was great—he had studied the anatomy of the heart and tributary vessels with uncommon care and devotion to detail—Harvey *interfered* with the ordinary course of natural events by the most ingenious experiments.[3]

> a) He noted that when an artery of a laboratory animal was opened, the blood spurted with great force and in a jerky manner corresponding to the beats of the heart.
>
> b) When he tied the large veins near the heart, the latter organ became pale, flaccid, and bloodless. When he removed this ligature, the blood again flowed into the heart.
>
> c) When the aorta was tied, the heart became distended with blood. It could not empty itself until the ligature was removed.
>
> d) These experiments were performed on animals; Harvey also drew a ligature tightly around a human limb. Blood being unable to enter, the limb grew pale and cold. When the ligature was somewhat relaxed, allowing blood to enter but not to leave, the limb swelled. The ligature was removed and the limb soon gained its normal appearance.
>
> e) Harvey also noted a general constitutional disturbance resulting from the introduction of poison at a single point.

This latter observation is what gave Harvey his most significant clue, for it led him to conjecture that there might "be a motion, as it were, in a circle . . ." Indeed, only on this hypothesis could such a

---

[3] He did not merely *inspect*, he *interacted:* he was not content to *appreciate* the heart passively, he wished to tinker with it in every conceivable way. (Hanson.)

general disturbance be explained as a result of a local injection of poison.

Here, then, was the hypothetical suggestion that organized Harvey's research. All of his experiments show the influence of this leading query. Only such an hypothesis as that there is a fluid circulating all over the human body could have given Harvey's inquiry its direction, its economy, and its conclusiveness. In a sense this controlling hypothesis brought to Harvey's attention things that he might never have noticed otherwise.

As it happens his hypothesis was correct. But it need not have been correct in order to have fulfilled properly some of the essential methodological functions of hypotheses. Of course, whether or not an hypothesis *is* correct is for experiment to decide. Certain preliminary services, however, may be expected of hypotheses even before experiment. It is here that a well-constructed hypothesis can give our research maximum clarity and direction, and eliminate muddling ambiguities in the form of our question.

In any case, without an hypothesis Harvey's inquiry, or any other inquiry, could never have begun. Let the facts speak for themselves, indeed! For what *are* the facts, and *which* facts should we study? How long and how hard would Harvey have had to observe the differences between arteries and veins, and the internal structure of the heart, before a solution to his problem dawned upon him? Without an *hypothesis* that sought to relate all these phenomena within some sort of connected system of explanation Harvey might still be looking for the answer to his problem. For his task was to relate the facts about the cardiovascular anatomy to certain other facts that would fill out the details of his explanatory account. But *what* other facts? Not just *any* other facts, surely. The number of facts about the mammalian anatomy is endless. A merely undirected observation of the elements of the cardiovascular system would never reveal just those other facts that Harvey required to construct an interconnected account of the movement of the blood. Facts must be selected for study on the basis of an hypothesis. Or, more strongly, the facts draw what we regard as their most distinctive characteristics only by way of the hypotheses through which we apprehend them.

In directing an inquiry, an hypothesis must by its very nature incline us to regard certain facts as relevant and others as not. The fact that he was educated in Padua and Cambridge was irrelevant to Harvey's circulation hypothesis. The fact that Cambridge is 78 miles from Oxford and 93,000,000 miles from the sun was also irrelevant. The optical and hydraulic properties of the blood were likewise irrelevant, as were the chemical and thermal properties of the cardiac muscle. It would have been humanly impossible for Harvey to examine the movement of the blood relative to *every other* class of events. And even were he so much a God as to be able to consider the behavior of the blood relative to *every other* class of events, had he ever wanted to *explain* that phenomenon to us poor mortals he would have had to present it to us exactly as he did: in terms of a circumscribed experiment governed throughout by a logically well-defined hypothesis. Only so could we have understood the point of his remarks.

Now just what is meant by saying of some hypotheses that they express relevant connections between phenomena, while other hypotheses do not? Some of Harvey's predecessors considered the physical necessity of breathing in air and the state of the empty arteries (as revealed in *post-mortem* examinations) to be most relevant to an understanding of blood motion. Hence their hypothesis that the arteries were conduits for the passage of air from one part of the body to another. And others of Harvey's contemporaries took the physical necessity of ingesting food, and the further requirement that all members of the body should receive the nourishment of that food, as highly relevant to an understanding of blood motion. Hence their hypothesis that the blood vessels were conduits for liquid food, i.e., blood, manufactured in the stomach and upper bowel and pumped by the heart to the extremities once and for all. But Harvey, seeing that these accounts did not square with the things he had discovered before about the relative sizes of the blood vessels and cardiac chambers, dismissed them and formulated an hypothesis in terms of which these relative sizes and capacities were construed to be facts of paramount importance.

It was on the basis of his *previous knowledge* (which was, of course, considerable) that Harvey conjectured that nature, doing nothing in vain, would never have given the cardiac ventricles and vessels so large a relative size without a purpose. His experience had been that animal conduits always reflected in their sizes and shapes the character of the substances they transported.

Thus, relative to Harvey's hypothesis, the place of his education and the distance of the sun were irrelevant. No relation was known to exist between such phenomena and the motion of the blood. And there is no reason to expect a change in our current theories with respect to this. An hypothesis is understood to be relevant to a problem if it expresses determinate modes of connections between a whole range of events including the event in question; it is irrelevant otherwise. *In other words, hypotheses are specific reflections of theories we may have about the world. Hypotheses are theory-loaded conjectures.*

No recipes are possible for hitting upon relevant hypotheses. Indeed, a great deal of caution must be shown in the dismissal of an hypothesis as irrelevant. For, while we would all agree that our distance from the sun is irrelevant to an inquiry into the nature of the cardiovascular system, we must remember that there was a time when everyone agreed that our distance from the sun was irrelevant also to an inquiry into the nature of the cycle of the tides and the phases of the moon. And, while in 1823 an hypothesis that postulated a corpuscular character for light would have been put down as futile, in 1923 this very postulation assumed the utmost importance, as we saw in the last chapter. In the absence of detailed knowledge of a subject matter we can make no well-founded judgments concerning the relevance of hypotheses. The hypotheses which do occur to an investigator are, therefore, a function, in part at least, of his previous knowledge. Hypotheses mediate, so to speak, between what we already know and what we are about to learn. They mold, color, and relate the facts that in a later age might be conceived only in terms of some highly organized system of knowledge. One might even go so far as to say that what we take to *be* the facts *here and now* are only the things we have discovered through asking, or being forced to ask, questions of a strictly hypothetical nature. Random observation uncontrolled by hypotheses never resulted in a science. Letting the facts speak for themselves can only result in a deafening confusion or a primordial silence.

But equally as wonderful as the way in which Harvey tortured the few facts at his disposal with his probing hypothesis was the way in which he developed that hypothesis, as well as those of his opponents. Very often two hypotheses which seem fairly compatible can be shown to issue (deductively) in propositions that are diametrically opposed to each other. Seeing, then, which of these two contradictory propositions is supported by experiment provides a way of selecting between hypotheses and the theories behind them.

Harvey's performance in this connection is a masterpiece of scientific inference. Attacking the received views of the circulation of the blood, Harvey compares the relative infrequency of our ingestion of food with the regular and rapid pumping action of the heart. From this he concludes that on such a view the veins would become and remain drained for much of the time, and that the smaller arteries would follow the relentless cardiac action. But neither of these things is ever observed to happen in the normal man. His veins never go dry, even during long periods of fasting, and his arteries certainly do not burst after every meal. So there are two marks against his rival theory before Harvey is even warmed up to his task.

In a magnificent *reductio ad absurdum* argument Harvey begins by granting his opponent's hypothesis, and on the most favorable terms, too. He concludes from that hypothesis to a physically impossible state of affairs. He lets the opposition choose which liquid capacity they will assign to the left ventricle. What will it be? Three ounces? One and a half? Two? In *fact* the capacity is three ounces, but Harvey can prove his point while allowing only half that. Now how much blood is ejected during each contraction? For, as Harvey says, "All the world allows that with the systole something is always projected." What will it be? One-fourth the capacity of the left ventricle? One-fifth? One-sixth? Or even one-eighth? Have it as you will! If only one-eighth of the ventricle's capacity is pumped into the aorta at each contraction of the heart, the situation would *still* be that in one half-hour more blood should have passed into the aorta than is found in the whole body at any one time. And, of course, on the account offered by some of Harvey's opponents this would leave the heart without blood. That is, nothing would be ejected at the systole for the better part of one day. But this is absurd, for "all the world allows that with the systole something is always projected." Hence the opposition is reduced to the position that at the systole something is *always* projected and yet (by their hypothesis) that there are times when at the systole nothing is projected —a logically awkward position, you will agree.

This was the logical death blow to those who disbelieved in the complete circulation of the blood. It also provided a strong impetus to Harvey to continue with his own positive experiments, some of the details of which I set out earlier. *For what could be more encouraging to the establishment of an hypothesis than to discover that alternative hypotheses result in contradiction the moment they are developed de-*

ductively? What could be stronger confirmation of the thesis that the blood circulates than the proof that if the blood is assumed *not* to circulate a complete absurdity is the result?

Moreover, in a similar way, Harvey was able to show how the assumption that the blood *does* circulate not only does not result in this absurdity, but succeeds in relating in a most comprehensive manner all those facts that had proved so devastating to the opposition. He was able to argue so as to show why the cardiac and vascular valves are constructed as they are, why the aorta and vena cava are of such great dimensions, and why the pulmonary artery makes the unusual route it does. He was even able to suggest, if incorrectly, how the blood gets from the arteries back to the veins. For if the blood does indeed circulate, this exchange must take place. So, in a way, Harvey's hypothesis *drew him on* to the consideration of this further important problem.

Of course, Harvey was unable to follow this part of the blood cycle. He had no lenses powerful enough for that. *He* had considered that the tissues at the extremities were sponge-like. The arteries (after considerable branching) were thought by him merely to empty into these tissues. Later this blood was collected by the veins in much the way that drainage pipes collect water from a swamp. (And notice how, when experimental complexities or unrefined techniques leave us with a vague, unclear feeling about the nature of the phenomena before us, we seize on models suggesting in themselves just such a vagueness. Thus the tissues for Harvey were like swamps. Electrons are like clouds.) The discovery that the ends of the arteries are connected to the commencements of the veins by a definite system of small tubes, the capillaries, was made by Malpighi some 30 years later. Malpighi observed them in the tail of a tadpole. Seven years later still, Leeuwenhoek observed the circulation in the lung of a frog.[4] But that Harvey was in error here in no way detracts from the importance of his conjecture, as we will see in a moment. Indeed, without the circulation hypothesis to guide them, Malpighi and Leeuwenhoek would not even have understood the nature of what was before their eyes. For it was only as a consequence of the circulation hypothesis that a proposition about the existence of capillaries could have any sense to it at all.

---

[4] For further discussion of these and related developments see Charles Singer, *A History of Biology*, rev. ed. (New York: Henry Schuman, 1950), pp. 151–5, 164–8.

Thus by tracing the consequences of his opponent's hypothesis Harvey showed it to be absurd. And by tracing the consequences of his own hypothesis he found himself able to explain a whole range of hitherto inexplicable phenomena. And, moreover, this deductive elaboration led directly to the realization that there was a further problem demanding solution. It led ultimately to the discovery of the capillaries.

Such a deductive elaboration *must* follow the formulation of every hypothesis. For we can only discover the full meaning of an hypothesis —whether and to what it is relevant, and whether it offers a satisfactory solution of the problem at hand—by discovering *what it implies*.

The technique of developing an hypothesis deductively is indispensable to scientific procedure. Only by seeing what follows from a conjecture can one determine what are its theoretical foundations and what is its experimental significance. We have had a classical example of this from the early chapters of modern biological science. Let us consider another equally classical experiment from the first pages of modern physical science.[5]

Galileo had shown that, if we neglect the resistance of the air, the velocity with which bodies fall to earth does not depend upon their weight. And it was known that bodies accelerated during a free fall. It was not known, however, what the relation was between the velocity, the space traversed, and the time required for a free fall. What general account might serve to explain this situation?

Galileo entertained two hypotheses. According to the first, the increase in the velocity of a freely falling body was proportional to the *space* traversed. But ultimately Galileo rejected this by the following argument: He considered that one of the deductive consequences of such an hypothesis was that a body should travel *instantaneously* through a *portion* of its path, something Galileo held to be impossible.[6] This particular argument is now known to be erroneous. Nonetheless, as Harvey was to do later, Galileo tried to trap his potential opposition in an absurdity. That we now know he did not succeed in this as well as he had hoped does not obscure the fact that it was his intention, one of which he thought highly as a technique of scientific argument.

---

[5] See also A. R. Hall, *The Scientific Revolution*, rev. ed. (Boston: Beacon Press, 1966), Chapter III.

[6] Galileo Galilei, *Dialogues Concerning Two New Sciences*, tr. Henry Crew and Alfonso de Salvio (Evanston, Ill.: Northwestern University Press, 1950), pp. 160–2.

He considered next the hypothesis that the change in velocity of a freely falling body during an interval of time is proportional to that time interval; i.e., $v = at$ (where $v =$ velocity, $a =$ velocity increase per second, and $t =$ the number of seconds the body has fallen). That is, the acceleration of a falling body is constant.

But if ever there was an assumption that could not be put to the test directly, it is this one that the acceleration of a falling body is constant. Galileo had to deduce other consequences from the constant-acceleration hypothesis and then test them directly, just as Harvey had to deduce from the circulation hypothesis the consequence that a local injection of poison would lead to a general constitutional disturbance.

From his constant-acceleration hypothesis, then, Galileo deduced the proposition: "The distances freely falling bodies traverse are proportional to the squares of the times of their fall." [7] Of course, even this sub-hypothesis presented Galileo with some formidable experimental problems. With his own pulse beat and a primitive water-clock as his only timing devices, certain modifications were essential if anything like accuracy was to be forthcoming. So, he made still other inferences from this sub-hypothesis to even lower-level hypotheses having to do, not with freely falling bodies, but with freely rolling bodies; not with vertical descents, but with descents down inclined planes. And from all this he was able to establish that a body descending for two seconds will travel four times as far as a body descending for only one second, and a body descending for three seconds will travel nine times as far as it would have in but one second.[8] The hypothesis that bodies fall so that their acceleration is constant was thus strengthened greatly, albeit by these indirect but deductively related means.

Further implications of this hypothesis were deduced and verified with precision. Hence the constant-acceleration hypothesis gained universal acceptance largely because the wide circle of its *implications* was verified experimentally.

Considered in general terms, Galileo's procedure might be characterized somewhat as follows:

First, he selected but a certain portion of his experiences for study. Neither did he just let the facts speak for themselves nor did he stroll through the world of facts, wide-eyed and with notebook in hand, ob-

---

[7] Paraphrase of Galileo's Theorem II, Proposition II, *ibid.*, p. 167.
[8] *Ibid.*, pp. 171–2.

serving at random. By simply dropping different weights from a standard height—from the Tower of Pisa, runs an old but improbable story—Galileo assured himself that the velocity of falling bodies was independent of their weight. This resolved some of his problems, but it only raised others in their stead. If the behavior of freely falling bodies was not dependent on weight, then upon what did it depend? The ancients had already determined that certain properties of bodies were *irrelevant* to their dynamical behavior. The temperature, the odor, the colors, and the shapes of bodies were disregarded; they were held to be insignificant to any inquiry into gravitational acceleration. But some of the ancients had also regarded the distance and the duration of the fall as irrelevant. This was an assumption Galileo refused to make. He proceeded, therefore, to formulate hypotheses in which these properties of bodies were the determining factors in their behavior when falling.

So to select the relevant factors was in large part a reflection of Galileo's previous knowledge. Like the ancients, he neglected the colors and odors of bodies because general experience seemed to indicate that their colors and smells could vary greatly without affecting their gravitational behavior in the slightest. But the selection was also based partly upon Galileo's tentative guess that other properties sometimes regarded as unimportant, viz., the distance and duration of the fall, were in fact very relevant indeed. Galileo had already made in physics successful researches in which the quantitative relations explored by the mathematics of his day played a fundamental role. He was well read in ancient philosophy and had an unbounded confidence in the thesis that the story of nature was written in mathematical language; he read Euclid as a prolegomena to the Book of Genesis. It was not, therefore, with an unbiased mind, a mind empty of strong convictions and interesting suggestions, that Galileo set out to solve the problems of motion. He did not keep an open mind on the subject of acceleration in the sense that *any* physical fact whatever might claim his attention. His mind was open more in the sense of being a trap with which to ensnare all and any information relevant to falling-body motion, the better to extort every last clue leading to the answer to his question. This relevant information, Galileo was persuaded, was restricted only to the velocity, time, and distance of fall of freely falling bodies, plus certain constant proportions.

There were thus two sets of ideas controlling Galileo's study of the motions of bodies. The more comprehensive set consisted in the mathematical, physical, and philosophical convictions which determined his choice of data, and what were, for him, their relevant properties. The other set consisted in the special hypotheses he invented in order to discover the relations between the relevant factors of the data. The first set was a relatively stable collection of beliefs and convictions. Galileo would probably have held on to these even if neither of his two hypotheses on the behavior of falling bodies had been confirmed in experiment. The other set, especially in Galileo's day, was a much more unsettled collection of suggestions, hunches, and unestablished theories. He might easily have sacrificed his very simple equations between velocity, time, distance, and acceleration for more complex formulations had his experiments demanded he do so.

In a similar way, Harvey might have sacrificed his circulation hypothesis for something much more intricate and elaborate. But this would not for a moment have shaken the corpus of his stable knowledge about the human anatomy, and about the effects of certain sorts of systematic disturbances, like the inhaling of smoke or of water, or the ingestion of liquids to the exclusion of solids.

And in a similar way again, large areas of relatively modern physical theory have already become part of our stable scientific beliefs. While certain highly specific hypotheses about the nature of cosmic radiation or electron spinning may be sacrificed tomorrow if need be, the corpus of relativity and quantum theory is in nothing like this wobbly position. It is only against the background of our acceptance of relativity and quantum theory that these specific hypotheses make sense. Just as it is only against the acceptance of Snell's law and the principle of the rectilinear propagation of light that specific hypotheses about the refractive index of a given substance make sense. Many philosophers, and some scientists, fail to appreciate this, imagining as they do that all scientific knowledge has exactly the same logical claim on us. They suppose it all to be hypothetical, merely probable, and in danger of collapse at the first sign of an unfavorable experiment. Nothing could be more unrealistic.

Thus Harvey's circulation hypothesis and Galileo's constant-acceleration hypothesis make sense only against the considerable background of their respective collections of stable knowledge. To one who lacked the background knowledge of either of these great investigators, letting the facts about blood motion or falling-body motion merely speak for themselves would have issued in an unearthly silence. And for one who *possessed* the background knowledge of a Harvey or a Galileo, but who refused to pinpoint his inquiry with sharp hypotheses constructed out of specific questions, letting the facts speak for themselves would have resulted in a deafening confusion. For, clearly, the facts about the blood or about falling bodies are unlimited.

So it is our *special* assumptions about a class of phenomena that we formulate as hypotheses. When these are well formulated they give us an insight into the nature of our investigation and the nature of the world which no amount of passive observation could begin to match.

But when is an hypothesis well formulated? Are there any formal conditions which an hypothesis ought to meet before it can be said to be "well formulated"? In my opinion there are. Of course, to some extent these will vary in content and in detail from discipline to discipline and from problem to problem. A really complete account of such formal requirements would have to deal with a great range of particular hypotheses in a most elaborate way. For, clearly, some of the formal requirements of an hypothesis about the nature of x-radiation will differ widely from those of an hypothesis about the effect of x-radiation on human tissues. Nonetheless, some general statements can be made.

Initially, an hypothesis ought to be formulated in such a manner that deductions can be made from it. Consequently, a decision ought to be reached as to whether the hypothesis does or does not explain the facts considered. Let us look at this condition first from one, then from another, point of view.

It is often the case, indeed most usually the case, that an hypothesis cannot be verified directly. We cannot establish directly that light is wavelike, or that it is granular. It was not open to Harvey's direct inspection to determine whether or not the blood circulates. Galileo had to do more than just *look* to recognize in temporal duration the operative concept of gravitational acceleration. There was no simple observation by which Newton could settle whether or not bodies attracted each other inversely as the square of their distances. In each of these cases the hypothesis was stated so that by means of the well-established techniques of logic and mathematics its implications could be traced clearly and then subjected to experimental testing.

Thus the hypothesis that the sun and Mars attract each other proportionally to their masses, but inversely as the square of their distances, could not possibly be amenable to observation, not even the observation of the great Newton. But one set of consequences from this hypothesis, namely that the orbit of Mars is an ellipse with the sun at one of the foci, and that, therefore, given certain initial conditions, Mars should be observable at different points of the ellipse on stated occasions, is capable of being verified, as Kepler had shown earlier.

Consider this from another point of view. Unless each of the constituent terms of an hypothesis denotes, or draws its significance from, some determinate experimental procedure, it is impossible to put the hypothesis to an experimental test. The hypothesis that the universe is shrinking, but in such a fashion that all lengths, including our standard measures of length, are shrinking in the same ratio, is empirically meaningless. It can have no verifiable consequences. The animal-spirits hypothesis that preceded Harvey's account of blood motion was of this type; no deductions were forthcoming. There was no experimental process for detecting the presence, or the absence, of this very subtle fluid.

Another obvious condition for an hypothesis is this: it should provide the answer, or at least *an* answer, to the problem generating the inquiry.

Thus the theory that the blood circulates accounts for the relatively large sizes of the cardiac chambers and conduits. It accounts for the peculiar valves in the veins, and it leads to an explanation of how the blood is aerated in the lungs and how it nourishes the tissues of the body. And Galileo's constant-acceleration hypothesis accounts for the known behavior of bodies near the surface of the earth.

Nonetheless, we must not suppose that hypotheses whose consequences are not all in agreement with observation are always useless. False hypotheses (e.g. Harvey's, that the arteries just spilled the blood into spongy tissues, the fluid being drained back into the veins) may direct attention to the existence of wholly unsuspected phenomena (the capillaries in this case) and hence increase the evidence in support of other theories. The history of science is replete with hypotheses which, though false, have proven to be most useful. The phlogiston hypothesis, the theory of caloric, or specific, heat substance, the one-fluid theory in electricity, Lamarck's theory that acquired characters are inherited, and the hypothesis of spontaneous generation—all of these, though false, have left an indelible mark on the advance of modern science.

An even more striking illustration is the following: In ancient Babylon the number seven was held to possess magical properties. Because Babylonian astronomers believed that the planets had to be seven in number, they were led to look for, and to discover, the rarely seen planet Mercury.

The English logician De Morgan once remarked, "Wrong hypotheses rightly worked from, have produced more useful results than unguided observation."[9] We ought not to attend to the truth or the falsity of propositions to the exclusion of the methods by which they were suggested in the first place and by which they were established.

A further condition of hypotheses is most important. Galileo's hypothesis of constant acceleration enabled him, not only to account for what he already knew when he formulated it, but also to *predict* the occurrence of phenomena undiscovered at the time of his prediction. He was able to show, for example, that if the acceleration of a freely falling body was constant, then the path of projectiles fired from a gun inclined to the horizon would have to be parabolic. Harvey's circulation hypothesis enabled him to predict a general bodily disturbance resulting from the introduction of a poison at a single point. He *could* have predicted, but did not, that an artery hemorrhage may be stopped by a ligature placed between the heart and the wound, and that a vein hemorrhage requires ligature on the opposite side of the wound from the heart. An hypothesis becomes verified, though not of course proven beyond every doubt, through the successful predictions it makes.

And if an hypothesis expresses a universal connection among phenomena, it must maintain itself through its predictions and not be maintained in the face of *every possible* attempt at falsification. When one considers how the phlogiston, caloric, and ether hypotheses were saved and rarefied whenever an unfavorable observation presented itself, the force of this condition is clear enough. It is to the everlasting credit of Albert Einstein to have proved that classical mechanics is limited in just this way. For in such classical equations as that $S = V_0 t + \frac{at^2}{2}$ the "t" has no experimental significance whatever when we are dealing with bodies moving with speeds comparable to c, the velocity of light. What timepiece could possibly supply the required values?

---

[9] Augustus De Morgan, *A Budget of Paradoxes*, ed. D. E. Smith (Chicago: Open Court, 1915), Vol. I, p. 87.

Another way of putting this is that an hypothesis must be capable of being *refuted* if it specifies one order of connection rather than another. An hypothesis cannot pretend to explain *no matter what* may happen. The hypothesis that everything that happens happens through the will of God is deficient in just this respect. For nothing that *does* happen can possibly score against it. Again, the echoes of the phlogiston, caloric, and ether theories, in their later much modified forms, may be heard here. The enormous strength of the hypotheses of men like Kepler, Galileo, Harvey, Newton, and Darwin is that their conjectures *could* have been proven false by observations unfavorable to them, but that this situation very rarely came about. In modern physics, chemistry, and biology, one of the first things looked for in an hypothesis is what could serve to eliminate it. Clearly, if nothing can eliminate it, it will explain nothing at all.

A last general condition for hypotheses is that they should be simple; of two hypotheses that are about equal in explanatory power, the one with the fewer intricacies and least amount of logical plumbing will be preferable. But as this will require considerable development I will treat of it when we discuss so-called "crucial experiments."

In any case I hope that you will agree with me that *the facts* are not just lying around like pebbles on the beach waiting to impress the first retina that comes along. Metaphorically the situation is more like this: Our hypotheses behave like fine filter lenses; by their formulation they will focus upon and make prominent only certain facts, leaving others blurred or completely out of focus. They will leave still others wholly invisible. The more acute our *hypothesis* is as an experimental instrument, being ground and polished with ever more precise logical and mathematical manipulation, the more select and specialized will the relevant facts appear to be.

14 | Scientific Simplicity and Crucial Experiments

BEFORE trying to get us tied up in new knots, I should like to tidy up some loose ends left by the knots of the last two chapters.

What is it to say that of two hypotheses about equal in explanatory capacity the simpler will probably win our acceptance? Is it not clear that science consists largely in discoveries that the world is more complicated than we had imagined?

How, then, can we say that the tendency is to accept the simpler hypothesis whenever possible? Let us explore this a little.

Consider first two unlikely hypotheses from lunar astronomy. One scientist says that the orbit of the moon relative to earth is a closed curve such that the distances from points on that curve to the geometrical center of the earth are all equal. Another investigator describes the curve as being such that the spatial area enclosed by it is the largest one that can encompass a curve of that length.

In point of fact both hypotheses are incorrect. But what is important here is that all the logical consequences of the first hypothesis are the same as those of the second. Indeed, from a logical point of view the two hypotheses are not different in any essential respect. The evidence that falsifies one of them also falsifies the other without any amplification or modification whatever. Were the two scientists to quarrel about their theories of the lunar orbit they would be quarreling only about the use of words, or about aesthetic preferences for their different formulations of what is fundamentally the same hypothesis. These two hypotheses are logically equivalent.

This, however, is not the case to which we must attend. I cite it only to remove it from further consideration. The situation to which we will now address ourselves is one in which two hypotheses are *not* logically equivalent although their consequences are incapable of being distinguished experimentally. This may be the case whenever our techniques of observation are not sufficiently sensitive to cleave apart the logically distinct consequences. For example, Coulomb discovered that the repulsion force between two positively charged bodies, A and B, depended on the charge on A, the charge on B, and the square of their distance of separation. Thus, $F = C\, qq'/S^2$.

But an alternative theory may assert that the repulsive force between bodies of like charge is inversely proportional, not to the second power of their distance, as with Coulomb, but to the 2.00008th power of their distance. Clearly, no experiment at our disposal will be capable of distinguishing the two hypotheses. These two hypotheses are indistinguishable experimentally although they are not logically equivalent.

What further condition may therefore be imposed so that we might become able to decide in such cases as these between rival hypotheses?

The answer which I put forward tentatively is this: We choose the *simpler* of two hypotheses in all such situations. This is advanced tentatively because there are several *prima facie* objections to it which will occur immediately. But without considering these now I will contrive to elucidate and strengthen this contention.

Consider the famous heliocentric theory of solar astronomy. This theory was first advanced by Aristarchus, the ancient Greek. Copernicus revived and greatly strengthened this old idea in his tract *De Revolutionibus Orbium Coelestium,* published in 1543. It was designed to give a comprehensive account of solar, lunar, and planetary apparent motion. But, of course, the theory of Hipparchus and Ptolemy had been formulated for exactly the same purpose. Both theories enable us to account for these motions. And, even as late as the 16th Century, neither theory permitted a prediction which could not be made by the other, the only exceptions to this being the question about the phases of Venus and the apparent diameter of the moon. Indeed, for many applications the two theories are mathematically equivalent and therefore *logically equivalent,* just as were our two theories about the moon's orbit. In other applications the theories, though not logically equivalent, are *indistinguishable on experimental grounds*. This was also the nature of the hypothetical rivalry between Coulomb's law and our alternative account of electrostatic force.

Furthermore, the Ptolemaic theory had a distinct advantage over the Copernican account in that it did not conflict with the testimony of the senses. Men could see the sun rise in the east and sink in the west. From the point of view of what is there to impress the retina, the heliocentric view is very sophisticated and tenuous indeed.

Nonetheless, Copernicus and the Copernicans found the heliocentric theory *simpler* than the ancient theory of Ptolemy. But what sort of simplicity is this? In what sense of "simple" is heliocentrism simpler than geocentrism?

"Simplicity" is sometimes to be understood as "familiarity." To one unfamiliar with the landmarks of modern science and mathematics, a geocentric theory of the heavens will doubtless seem simpler than a heliocentric theory. In the latter case we must actively revise our "natural" interpretation of what is supposed to meet our eyes.

In the same way, the theory that the earth is flat is more simple in the sense being considered than the theory that the earth is round. Thus, Lactantius wrote in the 4th Century A.D.:

> Is it possible that men can be so absurd as to believe that there are crops and trees on the other side of the earth that hang downward and that men have their feet higher than their heads? If you ask them how they defend these monstrosities: how things do not fall away from the earth on that side, they reply that the nature of things is such that heavy bodies tend toward the centre like spokes of a wheel, while light bodies as clouds, smoke, fire tend from the centre to the heavens on all sides. Now I am really at a loss what to say of those who, when they have once gone wrong, steadily persevere in their folly, and defend one absurd opinion by another.[1]

Lactantius surely appealed to the *simpler* of two hypotheses. But "simplicity" in this sense could never serve as a reliable guide in choosing between rival hypotheses in modern science. Because a new,

---

[1] Lactantius, *On the Heretical Doctrine of the Globular Form of the Earth* (Hanson's translation). See Lactantius, *The Divine Institutes,* tr. Sister Mary F. McDonald, O.P. (Washington: Catholic University Press, 1964), Book 3, Chapter 24, pp. 228–30.

Scientific Simplicity and Crucial Experiments | 241

and therefore unfamiliar, hypothesis could not be simple in this way and yet it might be chosen for its simplicity. What is simple in this sense to Lactantius, say, may not be so simple to another; what is familiar and easy to one may be strange and difficult for another. It would be absurd to say that it is in this sense of "simple" that relativity and quantum physics are simpler than classical physics, or that heliocentricity is simpler than geocentricity, or that the notion of a globular earth is simpler than a flat-as-a-pancake earth. So this first sense of simplicity is ruled out.

Let us say provisionally as a second try at the meaning of "simplicity" that one hypothesis is simpler than a rival hypothesis if the number of *independent elements* in the first is smaller than in the second; if, that is, the number of indispensable and irreducible assumptions involved in accepting the first is smaller than the number involved in the acceptance of the second.

Thus, plane geometry is simpler than solid geometry, not merely because the first is usually easier to master than the second, but also because and primarily because it is a study of configurations in two dimensions while solid geometry is addressed to the study of entities in three dimensions. Likewise, plane projective geometry is simpler (in this sense) than plane metric geometry, because the first attends only to transformations which leave invariant the colinearity of points and the concurrence of lines, while in plane metric geometry there is added the study of transformations which leave invariant the congruence of segments, angles, and areas all of which are highly variable when projected. And for the same reason topology, though very much more difficult to master than plane projective geometry, is simpler in that it is concerned only with those properties of figures that are invariant through *any* transformation, distortion, or projection.

For exactly the same reason a logic which requires only the notions of *negation* and *disjunction* may be said to be simpler than arithmetic, which requires also the notions of "number," "zero," and "successor" (though there is a powerful argument to the effect that negation and disjunction are themselves sufficient to generate even these purely arithmetic conceptions). So, too, mathematics as a whole is simpler than physics, which requires the additional notions of "mass" and "force" and "momentum," or at the very least "energy"; and physics is simpler than chemistry with its valence theory—even though it may be argued that the latter can be collapsed into one of the departments of quantum theory. Chemistry is simpler than biology, the latter being concerned as it is with what might be called "life" and its component processes, organic growth, ingestion, circulation, assimilation, respiration, organic oxidation, excretion, reproduction, etc. And this account of the matter might still be tenable even after every such organic process were shown to be at bottom physicochemical in nature.

However, there are theories which, though they *apparently* rest on fewer independent elements than rival theories, do not in fact do so, and for a special and interesting reason. For example, a theory of animal behavior which postulates but a single unlearned impulse, self-preservation perhaps, or maybe the urge to reproduce, is sometimes understood to be simpler in the sense being considered than theories which rest upon the assumption of several independent unlearned impulses. But *this* simplicity may be only superficial (indeed it is seldom otherwise). With theories of the first type it is often necessary to introduce *special* assumptions or qualifications of the single impulse in order to account for the observed variety of types of animal behavior. Stalking prey, hiding and storing food might easily be understood as cases of self-preservation. But what about basking in the sun, playing and frolicking and over-sleeping, things which certain animals are known to do and which seem quite indifferent to the concept of self-preservation? The theorist's task then becomes one of arguing that these really are cases of self-preservation despite appearances. Likewise, mating and home-building are clearly cases concerned with reproduction. But the theorist who would build all of animal behavior upon just this urge will have to jockey into place activities like hibernating, scavenging, and fighting with additional presuppositions about the *real* function of such behavior.

Unless, therefore, *all* the assumptions and presuppositions that a theorist requires to make his hypothesis work are stated clearly and in such a way that the logical relations between them are evident, it is impossible to say whether a theory or an hypothesis is in fact simpler than another alternative theory or hypothesis.

Hence a third sense of simplicity: Consider two hypotheses, both the conclusions of theories which possess (roughly) the same number of independent elements and which are equally capable of introducing order into a certain domain of inquiry. But one theory may be able to reveal various facts *as related* on the basis of the systematic implications of its assumptions. Diverse phenomena will show themselves to be related in terms of these assumptions much as diverse theorems in logic or mathematics will show themselves to be related in terms of certain appropriate primitive terms, axioms, and principles of inference. Whereas the second theory does not do this. The second theory can order the phenomena in question only by the introduction of special assumptions formulated *ad hoc* and unconnected in any systematic fashion.

The first theory is, then, simpler than the second. Simplicity in this sense is a *simplicity of system*. An hypothesis that is simple in this sense is characterized by its generality. One hypothesis, therefore, will be said to be more simple or more general than another if it can, while the second cannot, exhibit the connections it is investigating as being special instances of the relations it takes as fundamental.

Now the heliocentric theory of Copernicus, especially as it was developed by Newton, is *systematically* simpler than the theory of Ptolemy. In terms of the fundamental ideas of the heliocentric theory we can account for the succession of day and night, the seasons, the solar and lunar eclipses, the phases of the moon and the interior planets, the flattening of the earth at the poles and the oblate spheroidal shape of our globe, the precession of the equinoxes, and many other events within our solar system. A Ptolemaic astronomy can also account for all these things. But it requires *special* assumptions in order to explain some of them, assumptions that are not related systematically to the type of relation which Ptolemaic astronomy takes as fundamental, namely geocentricity.

The simplicity sought in the advanced stages of a science are of these two varieties, then. The attempt to reduce the number of independent elements required for hypotheses has infused many old sciences with new life. The new physics, modern chemistry, and several of the biological disciplines are in part the result of scientists wishing to reduce the number of isolated assumptions required to give sense to an hypothesis. But these disciplines display a comprehensive systematic simplicity besides. They are characterized by a remarkable increase in the number of logical interconnections between their several parts and aspects. *Simplicity in modern science, then, consists in a) reducing the number of independent assumptions and b) increasing the number of logical interconnections based on these assumptions.*

Unless we remember this, some of the changes taking place within modern science today will indeed seem arbitrary. For quite fundamental changes in a scientific theory are frequently invoked for the sole purpose of finding a more general theory which will explain what previously required two different and independent theories to explain. Maxwell, Einstein, Darwin, Bohm, Vigier, and Pavlov are just a few of the names appropriate in this connection.

When we are advised to adopt the simpler one of two theories, it is usually *systematic* simplicity that is meant. In this sense, the general theory of relativity, though mathematically far more difficult than the Newtonian theory of gravitation, is really simpler than the latter. Unlike classical theory, relativity physics does not require the introduction of *ad hoc* assumptions like attractive *forces,* absolute *spaces,* and instantaneous *simultaneities*.

Needless to say, however, at an advanced stage of science distinguishing between the two theories (and their component hypotheses) on the basis of their relative *systematic simplicity* can be extremely difficult. And the whole nature of the choice may be greatly complicated by such factors as the direction and emphasis of past research, the already accumulated knowledge about the subject matter of theory, and, indeed, the aesthetic preferences of the scientist himself. Despite these qualifications, however, the history of science seems to record a series of dramatic theoretical decisions in all of which systematic simplicity excels.

So much for the nature of theoretical simplicity. More will be said about its consequences presently. For of course we do not choose the most simple hypothesis *no matter what!* It would be simpler in *every* sense of simplicity to have the earth as the fixed center of a spherical universe all the celestial bodies of which made a complete circuit of our world 364 times a year, whatever "a year" would mean in such a context. Our mathematics would be supremely simple: just a few independent notions like *point, body, line,* and *dimension,* all welded within a rigorous geometrical system of explanation, and all eminently intelligible.

But the world is *not* that way. Arguments about the simplicity of hypotheses or theories are always relative to our observations of *what there is*. Scientists do not just adopt the simplest hypothesis without further question, for if that were so we should never have got beyond Thales. No, we show our tendency to select the simpler of two hypotheses only in cases where both hypotheses are known to square with our observations. This may be the situation where the rivalry is such that our experiments cannot distinguish between hypotheses, as, for example, with the opposition between Coulomb's inverse second-power law and its hypothetical competitor which involved not the second power, but the 2.00008th power. Or it might arise where two different theories are identical in one, several, or all of their consequences. The hunt for the simplest of a group of hypotheses arises only when all of them square with our observations. Those that do not do so are not even eligible for the contest.

Observation, however, is not so simple a matter as it is sometimes believed to be.

Earlier chapters raised the question of the differences between what a man of the 20th Century sees in the east at dawn and what a man of the 13th Century saw in the east at dawn. Apparently the retinas of these two men are similarly affected. Had there been cameras in the early 13th Century, they would have recorded much the same spectacle at dawn as would a 20th Century camera. All the visual impressions are the same. Yet the 13th Century chap sees a brilliant orb rising majestically through the sky above a stationary horizon. One of us would see, I hope, the horizon dipping away, or, rather, turning away, from the line between the sun and our eyes. But were we suddenly transported back to the 13th Century (as a result, perhaps, of a typographical error in one of Professor Einstein's papers) how could we get the 13th Century man to see what we see? By asking him to look harder and longer? By asking him to recite the details of what is happening before his eyes? No. *We must teach him what we know before we can show him what we see.*

The most random observation is charged with theoretical considerations drawn from our own experience and training. This is patently obvious the moment you ask a man *what* it is that he sees. He says he sees the pole star, perhaps, or the earth eclipsing the moon, or a hawk hovering over a hot haystack, or a storm approaching. But would a 13th Century man have seen any of these things in just the same way? That he would not is apparent the moment we consider how comparatively recent in the history of science are explanations of what we see.

This point was advanced earlier as follows: There is a little bit of *seeing that* in every bit of seeing. The gentleman says he sees the pole star, but at least part of what he means is that he sees that Polaris, being so close to the axis of the earth's rotation, will remain relatively fixed, while the other constellations Ursa Major, Cassiopeia, Orion, etc., will appear to turn around Polaris once every twenty-four hours. He sees that Polaris can serve to indicate geographic north to a traveler in the northern hemisphere. He sees that at the north pole Polaris will shine almost at zenith, while near the equator it will appear on the horizon. And even if it never occurred to him to ask why a certain star was called "Polaris" instead of any of a million alternative names, even if he just traced the pattern of the heavens on his memory and assigned the proper names to each star thus recorded (just as a child might deliberately commit to memory the street plan of his town or village), it would still be the case that he *saw that* Polaris is situated in such and such a way relative to Ursa Major and to Orion and to Cassiopeia.

Unless *observation* is merely equated with immediate, ineffable experiences like the buzzing in one's ears or the color patches in one's visual field—and you would be surprised to discover how many philosophers and scientists have carried out just such an equation—it must be recognized how very theory-loaded our observations of the world are. In short, we can observe intelligently or unintelligently. But there is no more in the east at dawn for the scientist to see than there is for the lunatic to see. And yet the scientist sees incomparably more. The objects of our seeing, hearing, touching, tasting, and smelling acquire meaning for us only when we can link up what is directly given in experience with what is not. A brilliant white spot of light against a deep-blue background has an ineffable, incommunicable, and very personal quality. After all, the impression itself is not different in type from what I see after colliding with a football or with someone's fist. But in normal circumstances such a spot of light can be seen as a star, can *mean* a star situated within a certain region of the heavens many light-years away.

When we see, we classify the objects of our visual perception as *stars, eclipses, hawks,* and *thunderstorms* partly on the basis of noted similarities between things, partly on the basis of our theories about the physical nature of the things in question, and partly on the basis of our predictions and expectations of what the things in question will do in certain specifiable circumstances. And all this is *there* in the seeing. It is not something tacked on to this visual impression. It is not an afterthought. Thus, despite superficial resemblances we see bats not as birds but as flying mammals, and we see porpoises not as fish but as aquatic mammals. We see the sun not as a shining balloon floating daily from horizon to horizon but as a relatively fixed local star around which we make our spinning revolutions.

Indeed, as I have argued, the consideration of raw sense experience (like visual patches and bursts of sound), in isolation from the theoretically fuller experiences we encounter in our everyday observation, is a highly sophisticated process. There is something basically wrong with the idea that our ordinary observation of the material objects of the material world is a psychological compound out of this color patch, that sound, those tactile sensations, etc. For on such a view we *would* be imprisoned within our immediate sensations and never able to break out into an understanding of a world external to our sensations, a conclusion which is in fact reluctantly accepted by many philosophers and scientists. Those who have held to such a view were probably just misinterpreting the remarks of others like Russell and Price who (far from giving an historical account of baby's psychological growth from the apperception of colors and sounds to the identification of stars and storms) were actually asking after the *logical guarantee* of what we say we see in our ordinary observations. Russell and Price were concerned with the cash value in actual experience of any observation statement like "That is the sun." They found such a cash value only in logically more primitive statements like "There is a brilliant yellow-white disc in my visual field," etc. They put it that statements about the sun were logical constructions out of statements about shining discs in my, or your, visual fields. Or, more extremely, they sometimes went so far as to say that *the sun* was a logical construction out of one's perceptions of it. All material objects were but logical constructions out of sense-data.

But if my argument has been right, if it is only a fully interpreted experience that has significance for us and the already accumulated knowledge we have of the world, and if sense-data like color patches and sounds are just partial aspects of our epistemologically prior experiences of tables, chairs, and stars, then it must be clear that the situation is the reverse of what Russell and Price claim. It is our everyday experience of material objects that puts a guarantee on our sense-datum experience. We say that this privately experienced red patch is red just as are the roses, the stop lights, the fire engines, and the robins' breasts we encounter every day, that privately experienced sound is shrill just like the whistling of the kettle, the sound of a jet airplane, or the yapping of a puppy. It is our ordinary perceptual life which provides the matrices required for identifying and recognizing even our private sense-datum experiences.

In short, material objects are not logical constructions from sense-data; sense-data are logical destructions of material objects. When our theory-laden observations of things and events in the material world are drawn and quartered by the logician, it is possible to distinguish several sorts of elements, some of which are present for everyone with normal sensory powers and others of which are present for some and absent for others depending on their theoretical outlook. At dawn any normal eye can discern a spatial gap opening between the horizon and the sun, whether it be in the 13th Century or in the 20th Century. But a great deal more than normal vision is required to see the sun as the earth's satellite or, more recently, to see the sun as the center of our planetary system. And yet, due to man's insatiable search for intelligibility, it is always as the earth's satellite *or as* the center of our planetary system or as something-or-other that the sun is seen. Who could satisfy a child's question about the nature of daylight merely by referring him to the brilliant disc and the widening gap between it and the horizon that greet every normal eye at dawn? Mythology and magic, astrology and alchemy were just the attempt to put the *seeing as* and *seeing that* into our everyday seeing. Without these elements our daily experience would be a chaotic nightmare of flashes, bumps, and rumbles. In the letter if not exactly the spirit of Kantian philosophy, "experience would be a meaningless rhapsody." To extract from our ordinary observations the essence of sensation is a highly sophisticated, strictly academic undertaking, not without some value perhaps. But whether undertaken for psychological or purely logical reasons such a process of intellectual abstraction only makes sense when seen against the normal case, i.e., our observations of stars and street-sweepings *as* stars and street-sweepings.

Another complication about the nature of our observations is this: The hypothesis which *directs* observation also determines in large measure what factors in the subject matter are noted. The human heart, e.g., must have an infinite number of properties. Its color will vary indefinitely at various stages in its life and death, its shape will vary indefinitely as the angle from which it is seen is changed, and so forth with unnumbered other properties of the heart. But William Harvey's hypothesis, remember, let all these pass unnoticed. Or if they were noticed they were set aside as irrelevant to the inquiries at hand. Nonetheless, other hypotheses about the human heart, hypotheses developed by men other than Harvey and about aspects of the heart other than those that held Harvey's attention, may treat these properties as very relevant indeed.

So unless the conditions under which an observation is made are known, the observation will be very unreliable, if not altogether worthless. Of what value is an observation that a certain liquid boils at 80°C. if we do not also note its density and its atmospheric pressure? But again, only a theory will lead us to observe all the relevant factors, only a theory can give any sense to the notion of *relevance;* only a theory will indicate whether atmospheric pressure is a single factor or whether it may be resolved into other factors, as force is resolved into magnitude and direction. And what would have been the worth of an observation that a chamber of the heart held 2.5 ounces of liquid if we did not know whether the organ was in systole or diastole, or for that matter whether the blood moved through the heart at all? An alternative theory might have taken the organ to be a kind of churn, its purpose being just to swill the blood through the tissues. But in any case what will be relevant under what conditions is for some comprehensive theory to suggest. Mere looking, unguided by theoretical consideration of any sort, is from most scientific points of view wholly insignificant and without consequence. And this is another reason why the attitude of some philosophers who think the business of scientists is primarily just to observe at large, collecting a great body of facts before sitting down to think about their relations and interconnections, is so myth-eaten. The man who really suceeds in going into his experimental work with a completely open and unbiased mind is likely to present to the scientific world only hollow and indirect bits of odd information. This is not to say that a man should enter the laboratory with preconceptions about what *must* be the outcome of experiment. It is to say that experiment will have a direction and force only when it is designed to bear on some aspect of some theory about the world, and only when the experimenter has some pretty definite theoretical expectations about the nature of the phenomenon he is investigating.

Yet another snag in the too-simple "open your eyes and look" view of observation is this: All but the most primitive observations are conducted with the aid of specially devised instruments. The nature and limitations of the instruments must be completely understood before any reliable observations are possible. The readings of such instruments must be corrected continually, and they must be interpreted and reinterpreted in the light of some comprehensive theoretical system. Without a thorough understanding of the principles of geometrical and physical optics, of the laws of refraction and diffraction, how could a man use astronomical equipment to any good effect? He would not even be in a position to observe celestial phenomena. And so it is with 99% of all contemporary science. Opening one's eyes and looking is not nearly enough. As I have been urging, one must know what he is looking for, which is another way of saying that he must have some theories about the phenomena confronting him. But he must also know *how* to look for what he is looking for. This requires considerable training. Thus he must not only train his eyes, he must train himself to use the instruments that increase the power of his eyes and other senses. All this is required before one intelligent observation is forthcoming.

From all this, from the considerations I raised in previous chapters about *facts* and *hypotheses*, and from the complications apparent in our notions of observation, is it not clear that the sharp distinctions customarily made between things and theories, facts and hypotheses, observations and interpretations ought to be re-examined and critically reappraised? When it is seen that our scientific knowledge is not merely a description of what strikes the retinas of people who have B.Sc.'s, but is, rather, a vast system of beliefs gathered and molded relative to the specific problems of scientists and the specific theories and hypotheses that are designed as provisional answers to those problems (and relative, too, to all the formal and aesthetic considerations that we gesture at with words like "simplicity," "intelligibility," "predictability," "explanatory capacity," and so forth), only then can one come to appreciate in natural science the most profound intellectual achievement of the western world.

The growth of scientific knowledge is like the growth of a great city. As new roads and highways are laid connecting the docks to the railway station, or the market to the nearest farm center, suburbs spring up quite haphazardly. A new interest or a new discovery in the outlying districts may cause a great development there while other once-busy districts languish. New York did not grow in concentric circles under the rationale of the city planning committee. The best laid plans of our contemporary architects and engineers would deliquesce if gold were discovered in downtown Manhattan or if a morbid disease attacked the flora in Central Park. And to a less dramatic degree it is phenomena rather like this that have led to the often lumpy and distorted growth that we call New York.

The development of science is likewise a pursuit of problems often local, often more widespread. The suburbs of biophysics and biochemistry sprang up when problems about the theoretical boundaries of physics, chemistry, and biology arose. And it is well known how some of the established families of the older disciplines view the rapid growth of these new areas of research.

But one thing is clear. Just as suburbs do not spring up only from the city planner's recommendations about where there is a lovely site most pleasing to the eye, so scientific knowledge does not grow from a random curiosity about what there is. Suburbs are answers to problems of a commercial, domestic, and social nature. Inter-scientific problems function in much the same way. And it is in an important analogous way that every significant observation in science carries with it a lot more than merely meets the eye. More than normal vision is required when the development of a science is at stake. Potentially, the development of every science is at stake in every and any observation we make.

Which brings me at long last to the point of this chapter. It is a common belief, one held no less by scientists than by laymen, that a *single crucial experiment* may often decide between rival theories. If one theory entails a proposition which contradicts a proposition entailed by a second theory, then by conducting an experiment which tests the truth or falsity of one of these propositions we can eliminate one of the two theories.

Examples of this situation are numerous in science. You all remember Sir Humphry Davy's destruction of the theory of caloric, or specific, heat substances.[2] Before the 18th Century heat was regarded as an im-

---

[2] See, for instance, Stephen F. Mason, *A History of the Sciences*, rev. ed. (New York: Abelard-Schuman, 1962), pp. 486–7, and for a more detailed account of the overthrow of the caloric theory, Duane Roller's "The Early Development of the Concepts of Temperature and Heat," No. 3 in *Harvard Case Histories in Experimental Science*, ed. J. B. Conant (Cambridge, Mass.: Harvard University Press, 1966).

ponderable substance. It lodged in the pores of substances. Accordingly, when an object cooled, the caloric fluid flowed out of it. It flowed into the substance when the object got warmer. This theory accounted for all the (then-) known facts about heat. An alternative theory was soon propounded, however, one which regarded heat as a form of motion. This theory also explained the (then-) known facts. Davy performed the experiment which was allegedly crucial between the two theories. He rubbed together two pieces of ice which were isolated from all sources of heat. They melted. According to the caloric theory the ice must have combined with the caloric fluid to produce water. But the source of this caloric could not be explained by the caloric theory in terms of Davy's experiment. On the other hand, the melting of the ice was easily explained on the kinetic theory of heat. Hence the experiment was (and usually still is) regarded as crucial.

Consider two other hypotheses, two that we considered earlier in a somewhat different form. Hypothesis I asserts that light consists of minute particles traveling at tremendous speed. Hypothesis II holds that light is a form of wave motion. Both hypotheses explain a certain fundamental class of optical phenomena: the rectilinear propagation of light, the reflection of light, the refraction of light, etc. Hypothesis I, however, entails that the velocity of light is greater in water than in air, while Hypothesis II entails that the velocity of light in water is *less* than in air. Both of these entailments cannot be true, obviously. So here, as with the caloric versus kinetic quarrel, we have an ideal case for a crucial experiment. In 1850 Foucault was able to show that light travels faster in air than in water.[3] According to all doctrines of crucial experiments the corpuscular hypothesis should have been banished forever.

But, as we saw in detail, this has not happened. Certain optical effects, e.g., the photoelectric effect and the Compton effect, can only be explained on a corpuscular theory of the nature of light. What is wrong with the impeccable logic of crucial experiments? The answer is the same as before, and it again calls attention to how intimately related are facts and theories, knowledge and hypotheses.

In order to deduce from Hypothesis I that light will travel with a greater velocity in water than in air, and in order that Foucault's experiment should be performed at all, *many other highly theoretical assumptions must be made about the nature of light* and about the instruments

---

[3] Mason, *op. cit.*, p. 473.

we employ in measuring its velocity. Consequently, it is not Hypothesis I *above* that is put to the test by Foucault's experiment, it is Hypothesis I and all the assumptions upon which it and the experiment rest. It is not the simple optical registration of what Foucault did that matters here. It is his *observations* that are important, and important only against the assumptions, theories, and hypotheses that all hung in the balance before the Foucault experiment.

The logic of the crucial experiment, then, is as follows: If we adopt Hypothesis I, and all the assumptions it requires, then the velocity of light in water should be greater than its velocity in air. But this is not the case. Therefore, *either* Hypothesis I is false, i.e., light does *not* consist of high-speed particles, or the assumptions required to give Hypothesis I teeth are (in part or completely) false. One of these assumptions would be, of course, that light must be *either* wavelike or corpuscular *but not both*, the same assumption made by Fresnel and Young. Now if we have very good grounds for believing our assumptions to be true, we speak of Hypothesis I as having been refuted by experiment. Nonetheless, Foucault's experiment is really a test of Hypothesis I *and* all of its assumptions. Just as an ordinary scientific observation is a registration of a pure sensation *plus* all the assumptions necessary to give those sensations meaning. If in the interests of making our scientific knowledge formally more systematic we are forced to revise our assumptions, then the crucial experiment must be reinterpreted. It need not then decide against Hypothesis I. And of course some of the most profound revolutions in modern science have consisted, not so much in the criticism of old hypotheses, but in the criticism of the assumptions underlying these hypotheses. Specifically, many of the hypotheses advanced by Newton yield predictions of phenomena that Einstein also predicts, and with no greater accuracy. The real difference lies in the system of assumptions that supports these superficially similar predictions, and, of course, in the experiments and observations that would be taken as appropriate to each case.

I leave it to you to scrutinize the Davy experiment. What more than just the caloric theory of heat, or the kinetic theory of heat, was at stake in Sir Humphry's experiment? In other words, what assumptions had Davy to make before he could even conceive of his experiment as being relevant to arguments about the nature of heat?

Every experiment tests, not just an isolated hypothesis, but the whole body of relevant knowledge that is involved by the logic of the problem, the experiment, and the hypotheses (or tentative answers to the problem). If it is claimed that an experiment does refute an isolated hypothesis, this does not mean that the hypothesis rests on no assumptions whatever, but only that it rests upon assumptions that we are quite unwilling to sacrifice. If an unsupported body fails to fall we may consider every possible suggestion other than that there is something wrong with our laws of terrestrial gravitation, and it was in just this way that Compton's experiment was scrutinized in 1923. A change in a wave length of light was of the same importance to optical theory as a levitating body would be to classical mechanics.

This point is sufficiently important to deserve another, final illustration. Suppose we were concerned to discover whether our "space" was really Euclidean, a problem that once beset the great mathematician Gauss. We need a physical triangle for this. If its interior angles are equal to two right angles, our space is Euclidean. If this is not so, then our space is not Euclidean. The vertices of such a triangle are taken to be three fixed stars; the legs of the figure are the paths of rays traveling from vertex to vertex. Suppose, however, that after measurement and calculation we discover that the sum of the interior angles of this celestial triangle is less than two right angles! Must we conclude that Euclidean geometry is not true? Not at all! Two alternatives (at least) are open to us.

1. The discrepancy between the Euclidean and the observed values of the angle sum may be said to be due to errors in measurement. Or

2. We may conclude that the "lines" joining the vertices of the triangle with our instruments are not *straight lines* in the Euclidean sense. Euclidean geometry may then still be said to be physically true but with the proviso that light does not travel along Euclidean straight lines in interstellar space.

To think that such an experiment would disprove Euclidean geometry is just to confess one's inflexible adherence to the principle of the rectilinear propagation of light, an assumption which, though highly supported by local evidence, is neither indubitable nor unrestricted in its application. We might accept the last alternative either because we have independent evidence for denying the rectilinear propagation of light or because greater systematic coherence is introduced into the body of our physical knowledge as a consequence of this denial.

Crucial experiments, therefore, are crucial against some hypothesis *only* in terms of a relatively stable set of assumptions which we do not wish to abandon. But, of course, no set of assumptions is permanently impregnable.

Crucial experiments are out of the same logical bag as pure observation and hard, uninterpreted facts. They are philosophers' myths that many scientists spend a good deal of time telling and retelling, not because they properly characterize the real situation in experimental science, but because they trip lightly off the tongue and make laboratory work seem a frightfully objective business. But talking in this manner is rather like an Englishman eating spaghetti. Slicing and cutting things up in this way may make for easy oral manipulation, but it is only a travesty of the genuine scientific situation in which theories, hypotheses, experiments, and facts are interlocked and intertwined in a fearfully complicated way. It is the task of philosophy of science, not to scissor all these elements into tidy and discrete packets, but to try to give an account of how, why, and where these aspects of science conflate and mingle.

15 | The Systematic Side of Science

IN Chapter Fourteen I spoke at length about crucial experiments. I argued that crucial experiments are crucial only against the background of a relatively stable set of assumptions, assumptions we are not prepared to abandon lightly. Crucial experiments, therefore, are tests not only of isolated hypotheses but of an hypothesis plus all those assumptions that underlie the enunciation of that hypothesis. The Fresnel, Young, and Foucault experiments were crucial in the controversy over the nature of light only when viewed against the assumption that *light must be wavelike or particulate, but at least one or the other and in no case both*. So these famous optical experiments were crucial to the question of the nature of light *and crucial also* to underlying hypotheses, including this one. What are called "crucial experiments" are only the surface ripples manifested by deep and complex crosscurrents in scientific thinking. They certainly constitute much more than the superficially simple decisions they are sometimes represented as exhibiting.

At the close of my argument in favor of this way of regarding crucial experiments, however, I made a reference to Euclidean geometry and its interpretation as a theory about physical space. Some readers will no doubt feel that I overstated, understated, or misstated the case. In order to meet such objections it will be necessary to digress briefly in this chapter to a discussion of the nature of formal systems like logic, algebra, and geometry, a discussion that might have been overlooked otherwise.

The contentious claim from the last chapter was that if, after measurement and calculation, we discover that the sum of the interior angles of a celestial triangle is less than two right angles, we need not conclude that Euclidean geometry is false. Two alternatives (at least) were said to be open to us.

1. The discrepancy between the Euclidean and the observed values of the angle sum may be said to be due to errors in measurement. Or
2. We may conclude that the "lines" joining the vertices of the triangle with our instruments are not *straight lines* in the Euclidean sense. Euclidean geometry then may still be said to be physically true, but with the proviso that light does not travel along Euclidean straight lines in interstellar space.

Those were my contentions and they may lead some to say that I only succeeded in revealing my antiquated Euclidean conceptions in putting the matter as I did. For surely, it might be urged, the paths traced by interstellar light rays *are straight lines* and that is all there is to it. The only complication is that Euclidean geometry does not work in describing spaces of such magnitude. For me to contrast a non-Euclidean straight line with a Euclidean straight line is only to reveal my "classical" adherence to traditional conceptions of the nature of space. It is for me to put the light ray into a Euclidean framework, discovering it there to curve. But so what? Why should *this* comparison be in order at all? If we're going to be non-Euclidean, then let's be non-Euclidean and stop referring back continually to the properties of terrestrial space. The only meaning "straight line" has in stellar astronomy is "light ray." So light rays describe straight lines in interstellar space, Q.E.D. Nothing else could be a straight line, so with what are we contrasting the light ray? In such a space particularly there is *no* Euclidean straight line with which to contrast the paths of the light rays, so why make these fanciful and speculative comparisons? Paths of light are always straight lines. What could be straighter? And just because one sort of geometry is appropriate to one context while another sort is appropriate to other contexts, this does not change the issue in the slightest. The only thing that *could* count as a physical straight line is a light ray, so let's not be changing names every time we remark that not all spaces are Euclidean. We knew that already.

This is a formidable criticism, and I think that it is, in some ways, correct. But part of its apparent force derives, I think, from a misconception about the nature of a formal system. And to draw this out I must make an excursion now into the theory of formal systems.

To begin with, let us note that no set of isolated propositions (however informative they may be) can constitute a science. Telephone directories, dictionaries, cookbooks, and inventories may be supremely accurate, and well organized too, but they are not works of science. In a proper work of science propositions form, more or less, a logical *system* of propositions. That is, all the propositions stand to one another in a logical relation of some sort, be it equivalence, opposition, entailment, inclusion, etc. The character of this systematic side of a science requires some attention, rather more than we have accorded it so far.

No proposition, of course, can be *demonstrated* (in the logical sense) by any experimental method. Most of us have at one time or another *proved* or demonstrated the Pythagorean theorem that in a right triangle the square on the hypotenuse is equal to the sum of the squares on the legs. And yet it is safe to say that there will still be someone who, when asked how the theorem may be proved, will suggest protractors, carefully drawn triangles, and finely graduated rulers. But such a person would only indicate by that approach how little he had advanced upon the methods of ancient Egyptian surveyors.

Suppose that person were to attempt to prove the Pythagorean theorem by actually drawing on a piece of sheet metal the squares on the three sides of a right triangle. Suppose then that he cut these three squares out of the sheet of metal and then set out to weigh one against the other two, thinking that if they balance old Pythagoras spoke the truth, while if they do not he lied.

Would this be a proof? Of course not! We can never be certain that the sheet metal is absolutely uniform in its density, nor can we be perfectly sure that our pieces are really square. Hence, even if a series of experiments were to fail to disclose a perfect balance in the weights, we should not take *that* as evidence against the view that there *would* be perfect equilibrium *if* our lines were perfectly straight, the angles of the square perfect right angles, and the mass of the metal uniformly distributed through the sheet. (There, by the way, is a perfect simulacrum of the spirit of mathematical physics.) A logical proof or demonstration consists only in exhibiting a proposition as the necessary consequence of other propositions (whether materially true or false does not matter).

The demonstration asserts nothing about the factual truth or falsity of either the premises or their logical consequences.

"But hold on," you may protest, "Do we not prove that the theorems of geometry are *really* true? Is not mathematics the most certain of all sciences in which some property is shown to hold for all objects of a certain type, once and for all? Is not the Pythagorean theorem true of *all* right triangles? How can you admit that something is in fact proved true of all triangles while refusing to grant that we are establishing the material or factual truth of such a theorem? Come, sir, either you mean *all* right triangles or you do not mean *all* right triangles. Which is it to be?"

This protest, however, only ignores what has already been noted, namely that a logical proof is a pointing out or a showing of the implications between a set of propositions called "axioms" and another set called "theorems." That is the whole story about formal proof, that from one set of propositions (true or false) another set of propositions (true or false) follows. The axioms themselves are never demonstrated, nor are they necessarily established experimentally—they are only accepted. And they are accepted only for the purpose of tracing their implications.

"Hold on again," comes the reply to this. "The axioms are not proved because they need no proof. Their truth is self-evident. We do not just accept as axioms *any* old set of propositions, but only those which are obviously true, like 'The whole is greater than its parts,' 'Equals plus equals produce equals,' and 'Through two points only one straight line may be drawn.' Such propositions as these are indeed the only satisfactory basis for geometry, because only by their means can we establish the truth of propositions not so obvious or self-evident."

This really represents the traditional, classical view of the nature of formal systems. Almost until the end of the 19th Century it was generally believed that geometrical and mathematical axioms are materially true of the physical world, and that the very cogency of formal demonstrations depends on these axioms being materially or factually true.

But the answer to the question of whether we just accept as axioms *any* old set of propositions can be caricatured in this way, "Yes, we do just accept as axioms *any* old set of propositions." This is not quite right, of course. But it is far less wrong than the traditional view, a view which confuses these three different issues:

1. How is the material or factual truth of axioms established?
2. Are the axioms in fact materially true?
3. Are the theorems the logical consequences of the explicitly stated axioms?

Let us consider these issues one at a time.

1. The answer often given to the first question, "How is the material or factual truth of axioms established?", is the traditional one, that axioms are self-evident truths.

But this is no more than complacency on the part of those who answer in this way, for the answer cloaks some genuine difficulties. Just exactly what is *self-evidence*? Is it obviousness in the psychological sense, or a kind of irresistible impulse to assert some proposition, or the inconceivability of any contrary proposition? If it is any of these then it is clearly unreliable as a criterion of truth. Consider these once self-evident propositions: that nature abhors a vacuum, that bodies fall at a speed proportional to their weights, that light is either wavelike or corpuscular but not both, that every surface has two sides, and so forth without end. Indeed, contradictory propositions about every variety of subject matter have each been declared (at different times) to be fundamental intuitions and therefore self-evidently true. But whether or not a proposition is psychologically obvious to a man depends to some extent on his individual training and the cultural context in which it was undertaken. A proposition for one person self-evidently true (in this sense) may not be so appreciated by another person.

(There is the story told of Professor Hardy, the eminent mathematician, who, in the process of setting out a proof for his students, commented on one step by saying that it was obvious. One of his students did not think it was obvious and asked the professor to elucidate. Hardy was taken aback slightly. He pondered the step in question, then left the lecture-room for ten minutes, returning finally with several sheets of scrawled-on note paper. Saying, "Yes, it's quite obvious," he continued with the proof.)

This self-evident view assumes us all to have the capacity to establish the truth of universal-general propositions dealing with matters of fact simply by examining the meanings of the propositions. But again, mark the important and fundamental difference between understanding the meaning of a proposition and knowing its truth. I can understand the meaning of propositions about unicorns and centaurs without commenting on their truth. Nor is a case of establishing the truth of a proposition necessarily a semantic inquiry into the meaning of that proposition, though of course one must be relatively clear about the meaning of a proposition before one can estimate its truth. In pure, uninterpreted mathematics these two aspects move rather more closely together. The question of the assessment of the truth of a proposition is intimately associated with the inspection of its meaning, even though the two are still separable. In applied mathematics the two are widely separated.

A fundamental and enormously important reason, therefore, for denying that the axioms of geometry, or any other branch of mathematics, are *self-evidently true* is that each of the axioms has at least one significant contrary. It is always possible to consider a mathematical axiom outside of its "home system" and to meaningfully entertain its denial. This need not result in absurdity. What does result in absurdity is to accept certain premises and certain rules of transformation only then to deny the theorems following thereon. This is nonsense. But it is not nonsense to deny an axiom (even though in so doing you are no longer playing the game in which the axiom figures).

"But," comes the objection again, "Does not the mathematician discover his axioms by observing the behavior of matter in space and time? Could it not be argued that mathematics is really the formal science of physical space and time, or at the very least the science of our human conceptions of space and time? Is not our mathematics just a kind of higher-order physics, a physics dealing with the more general characteristics of spatiality and temporality? And for that reason are not mathematical axioms more certain than theorems, and *ipso facto* more certain than any physical interpretation of those theorems?"

Our backbencher begins now to sound as if he had read Kant or Eddington, or both.

To reply to this I need only to appeal to an old distinction made by Aristotle, a distinction between the *temporal order* in which the logical dependence and interdependence of propositions are discovered and the *logical order* of implications among propositions.

Doubtless many of the axioms of mathematics express what we believe to be the truth concerning selected parts of nature, e.g., geometrical optics, terrestrial counting, etc. Doubtless, too, many advances in pure mathematics were inspired by suggestions arising in natural science. Nor is there any doubt that mathematics as an inquiry—that is, as a kind of fact-collecting enterprise—did not begin historically with a number of axioms from which theorems like that of Pythagoras were subsequently derived. Euclid's axioms and other propositions of his geometry were well known hundreds of years before his birth. Very likely they were held to be factually or materially true as well. It has been said of Greek geometers that they regarded themselves as describing physical space in its most abstract character. Euclid's chief contribution, of course, consisted, not in discovering traditional theorems, but in exhibiting them all as integral parts of a system of connected truths. Euclid's questions must have run rather like this:

> Given the theorems about the angle sum of a triangle, about similar triangles, the Pythagorean theorem, etc., what are the minimum number of assumptions or axioms from which these theorems might be inferred?

Thanks to Euclid, instead of containing what were believed to be but a cluster of independent propositions, geometry became the first known example of a deductive system. The axioms were *discovered later* than the theorems, although the former are *logically prior* to the latter.

It is common belief, I almost say *prejudice*, that the logically prior propositions are somehow "better known" or "more certain" than the theorems. Moreover, many people have the vague notion that the logical priority of some propositions to others is connected somehow with their being true.

Mathematical axioms, however, are simply logical assumptions, formal hypotheses, used for the purpose of systematizing and sometimes even discovering what those axioms imply. They are propositions merely entertained to see what they logically contain, much in the way that we can consider the disposition of pieces in someone else's chess problems just to see what the possibilities are. From this it follows that axioms *need not* be known to be true before theorems are discovered. In general the axioms of a science are psychologically much less obvious than the theorems. The material or factual truth of theorems is rarely established in natural science by first showing the material or factual truth of the axioms. On the contrary, it is usual that the material truth of axioms is made increasingly probable by establishing empirically the truth, or the probability, of the theorems.

This, I think, is all that is required to deal with the first issue confused by the traditional doctrine of axioms as self-evident. The first issue was, remember, "How is the material truth of the axioms established?" The answer is that *qua* axioms their material truth is not a vital consideration. They could, after all, be *false* axioms and nonetheless still be the axioms of a consistent formal system. This is not to say that the material truth of an axiom is established in the way that the material truth of any proposition whatever is established. It is only to say that questions about the material truth of an axiom are quite independent of questions about either the meaning or the formal implications of that axiom.

2. The second issue confused by the traditional pronouncement that mathematical axioms are self-evident was raised in the question, "Are the axioms materially true?" But clearly, from the arguments advanced up until this moment, an answer to the question of whether a given axiom is or is not materially true cannot be given on purely logical grounds alone. A significant answer to such a question must be in large part determined by the special natural science whose business it is to investigate experimentally the subject matter of such an axiom, i.e., what piece of the world the axiom is *about*. For, as we saw, the logical importance of an axiom can be assessed in complete independence of considerations regarding what that axiom is *about*.

To put it baldly, the *material* or *factual* truth or falsity of an axiom is of no professional concern to logicians or mathematicians. These latter gentlemen are interested only in the fact that certain theorems are or are not implied by certain axioms.

And it is on these grounds that we may distinguish between *pure* mathematics, which is interested only in matters of formal implication, and *applied* mathematics (in certain contexts indistinguishable from *natural science*), which is interested both in formal implication and in questions of material and factual truth. To answer such questions, one requires not just logic and reasoning but experiment and observation.

3. The third issue raised earlier was whether or not the theorems are the strictly logical consequences of the explicitly stated axioms. From all that has been said, it must be clear that whether or not the theorems and axioms are so related must be determined by logical methods alone. The question lies beyond the power of experiment and observation to

answer. Of course, the determination of whether theorems and axioms are related logically is not always an easy matter. For centuries Euclid's proofs were accepted as valid, although they made free use of assumptions other than those he stated explicitly. A very steady growth in our conceptions of what constitutes logical rigor in mathematical demonstrations has brought us to a point where considerable logical maturity as well as a special technical competence is a prerequisite for deciding questions of formal validity. And of course in certain branches of mathematics the cogency of certain demonstrations has not yet been established.

We are now in a position to summarize our first conclusions about the nature of a formal system. Propositions can be demonstrated, and can *only* be demonstrated, by exhibiting the relations between them and other propositions. But, of course, not *all* the propositions in a system can be demonstrated. This would lead to circularity. We may note, however, that propositions which are axiomatic in one system may be demonstrable in another system.

What we call pure mathematics, therefore, is a *hypothetico-deductive* system, to use a frightful bit of philosopher's jargon. The axioms of pure mathematics serve as hypotheses or assumptions, which are entertained, considered, explored, just for the purpose of discovering what other propositions they imply. In general, the logical relation of axioms and theorems is just the relation of premise and conclusion on a grand scale. If the whole of Euclidean geometry, or any other mathematical system, were compressed into one single complex proposition, the proposition itself would be hypothetical. It would be of the form "If x then y." The axioms would be the antecedent of this proposition while the theorems would be the consequent. Or, again, the axioms would characterize the whole formal structure of the proposition while the theorems would just be the elements unpacked, so to speak, from that structure. Of course the formal validity of a hypothetical proposition of the form "If x then y" can be assessed without any reference whatever to the truth of either the x or the y component. (E.g., "If x were a body moving without the influence of external forces then y would be observed to happen." It makes sense so to speak even though the material truth of this proposition cannot be determined.) And in the same way the formal validity of any mathematical system expressible in the form of a complex hypothetical proposition can be assessed without any reference to the truth or falsity of either axioms or theorems.

Perhaps we can now appreciate this in rather less general terms.

Consider the simple arithmetic equation "$2 + 2 = 4$." What does it mean? How do we assess its truth or its validity? Let me appeal again to the analogy of the game of chess. Our chess tokens have names, of course. We speak of *kings, queens, bishops*, etc. But such names, and the standard meanings which attach to them, are obviously quite unnecessary to the whole business of playing chess. All that is required is a statement of the *rules* which govern the uses of the various pieces. Indeed, when I think of a *knight* or a *rook* in chess, only the typical moves of such tokens come to my mind and not any of the medieval associations that may have attended my learning of the game. Likewise, all we require to develop the game of mathematics is an array of symbols and a compendium of the rules governing the manipulation of those symbols. It is utterly unnecessary to assign any meaning to the symbols themselves other than the meaning that accrues to them by virtue of their being manipulable as laid down in the rules. How does this attitude deal with "$2 + 2 = 4$"?

The *meaning* of "2," all the meaning it can have, is given in the equation "$2 = 1 + 1$." It supplies no more information and is required to supply no more information about the *meaning* of "2" than does the dictionary statement "*Frère* means *brother*" to a Russian-speaking person, or "*Rouge* means in French what *rot* means in German" to one who knows neither French nor German, or "*Right line* means *straight angle*" to a non-geometer. There is no more than this to "$2 = 1 + 1$." If a *verbal* definition is to explain the meaning of a term, the terms used to define must themselves have a meaning for the interpreter. The symbol "1" would have to be given a material reference like an apple or a stick before "$2 = 1 + 1$" could convey any meaning in this sense. But this is totally unnecessary. The meaning of "$2 = 1 + 1$" is not something assessable by reference to apples or sticks. In order to play the game of mathematics, such as proving that $2 + 2 = 4$ from the definitions "$2 = 1 + 1$" and "$4 = 1 + 1 + 1 + 1$," one need not assume that the symbols designate anything at all.

To put it another way, definitions in mathematics are *syntactic* rules, not *semantic* rules. They instruct you about how certain symbols are related to certain other symbols, not how symbols are related to objects. They are concerned only with the business of *formal substitutions* and not at all with *material application*.

Knowing that "2" is substitutable for "1 + 1" does not provide the slightest clue about what "2" may be properly applied to. In this way a mathematical definition resembles a dictionary translation as understood by a person ignorant of either of the inter-translated languages. The question of the application of mathematical symbols in factual contexts is not really a mathematical question at all. It is a question for the natural scientist, who, as we saw when we discussed types of measurement, seeks in the world for those series of events which can be described by the use of mathematical symbolization.

A mathematical definition, then, like "$x^2 = x \cdot x$" is not an *application rule*, it is a *transformation rule*. It shows how one expression may be transformed into another expression in exactly the way that the rules of chess tell you how one configuration of tokens may be transformed into another configuration.

This is the essence, I feel, of the language of formal science. But now I must add a proviso.

Not all definitions of mathematical symbols are *explicit* in the sense of making it possible to eliminate one symbol or symbol cluster by a simple replacement. Some definitions, of course, amount to *postulates* or *axioms* which define implicitly. Thus, if "2 = 1 + 1" is to function as a transformation rule, we must know how to operate with symbols like "=" and "+." But this information is not set out in explicit *one-for-one* definitions. It only comes across to us through a set of axioms which are pretty largely devoted to the exhibition of all possible operations with such symbols. Thus the five axioms of Euclidean geometry are really implicit definitions of notions like "point," "line," "between," "plane," etc. The axioms of arithmetic implicitly define "=," "+," "·," etc. (It is in a sense roughly analogous to this that the three laws of motion are said to define implicitly "mass," "force," and "momentum.") For example, the arithmetic symbol for addition is defined in the so-called commutative, associative, and distributive laws:

$$x + y = y + x$$
$$(x + y) + z = x + (y + z)$$
$$(x + y) z = (xz) + (zy)$$

These axioms implicitly define the symbol "+."
"Equality" ("=") is defined by:

if x = y then y = x
if x = y and y = z, then x = z
x = x, for all x

The symbols which are, in a given language system, explicitly undefined, and recur only in postulates that exhibit how they may be used without fixing a definite interpretation for them, are called the *primitive* terms of the language. They are capable of various interpretations. Hence they really function as variables; they have no definite meaning. (In a most illuminating manner Professor Braithwaite in *Scientific Explanation* shows how words like "electron" and "gene" function in a very similar way within *their* languages; they lack definite meaning in so far as they are employed as theoretical concepts or primitive terms.)

Such symbols have a quite definite *formal* meaning, however. Relations expressed by the axioms strictly delimit the possibilities of interpreting those terms. Thus "=" may be said to designate, by virtue of the axioms set out above, any two-term relation having the formal properties of the axioms (e.g., numerical equality, logical equivalence of propositions, identity of individuals); and "+" would designate any operation having the formal properties of commutativity, associativity, distributivity, of which there are many examples. It is a good exercise to scan a chemistry text noting the great variety of uses accorded to "=" and "+," a variety made possible strictly by the latitude of the axioms.

So these axiom sets really give implicit definitions for a whole class of relations and operations, something that tends to become concealed by the use of specifically algebraic symbolization.

The process whereby we narrow the class of entities to which formal axioms are applicable is called *interpreting* an axiomatic system. We *could* have a very strict and rigid arithmetic which allowed only numerical equality to count as a proper interpretation of "=." But this would rob the system of much of its flexibility, for our ordinary arithmetic embraces many interpretations of the symbol for equality.

Or consider a case from Cartesian geometry. One of our familiar axioms of Euclidean plane geometry reads as follows: For any two points A and B there is just one straight line connecting both A and B. But if in this axiom we replace the word "point" with the words "ordered couple of real numbers," if we replace "straight line" by "linear equation in two variables," we obtain the quite familiar proposition from ordinary algebra: "For any ordered couple of real numbers there is just one linear equation in two variables." Adding to these transformations the further one that "x contains y" will be read as "x is a solution for y" we find the Euclidean proposition "*Two straight lines have either no point in common or else exactly one point in common*" translating into the algebraic form "*Two linear equations in two variables have either no common solution or else exactly one common solution.*"

There the system of algebraic propositions and the system of plane geometric propositions can be regarded as alternative interpretations of a common axiom system whose members contain only logical constants like "+," "=," and variables (which are little more than blank spaces between the constants). Such an axiom system, capable of a variety of interpretations, is a system of pure mathematics.

From all this the nature of a formal proof should stand out clearly. A proof is just a transformation of a sentence or class of sentences into another sentence in accordance with transformation rules. A purely formal calculus is a language (if indeed "language" is the right word) whose symbols are altogether meaningless, so that it is exhaustively described by a statement of rules for the formation of propositions and of rules for the derivation of propositions.

Out of such a purely formal calculus a normally interpreted calculus may be generated by the addition, in another language, sometimes called a *meta*-language, of *semantic rules* to accompany the *syntactic rules* of the calculus itself. The meta-language instructs us how to apply the symbols in question to objects, while the formal language itself is concerned only with the range of permissible operations on its own symbols.

In general, the question of how the non-descriptive symbols of a mathematical system acquire descriptive or factual meaning for a language user is a question for the psychologist and anthropologist, not for the logician or mathematician. The natural scientist must, of course, be all these people rolled into one, and that is why his ability to make these distinctions is of enormous importance to an appreciation of his own work.

All this is meant to throw suspicion on the view that mathematics has a special subject matter, call it *quantity* or *number*, or *space* or *time*, or what you will. So too we must be suspicious in another way of the Kantian-cum-Eddingtonian view that geometry is a system of necessary propositions about physical space.

In so far as a geometrical axiom belongs to pure mathematics it is not a proposition at all. It is not true and it is not false. It is a *postulate*, an axiom in which the terms "point," "straight line," and the like are undefined; they are really variable terms. Only when they are assigned specifically physical meanings, i.e., when the symbols of the formal system are *applied* to the entities in our physical space, do we obtain propositions.

Such a physically interpreted proposition, however, is an empirical hypothesis and not a timeless mathematical truth. In fact, the Euclidean theorem that the internal angles of a triangle equal two right angles does not hold for triangles of astronomical dimensions. Hence the Euclidean postulates (which include the famous parallel postulate, and which indeed implicitly define *straight line*) cannot be held to express the properties of interstellar space. The angle sum of celestial triangles formed by light rays has the value which one would expect it to have if space had the properties expressed by the axioms of Riemann's geometry, a formal system which includes a denial of Euclid's parallel postulate. And since Riemann's axioms can be interpreted as expressing the formal properties of a curved surface, most physicists follow Einstein in saying of interstellar space that it is curved. Euclidean geometry is said to be applicable only to small surfaces that may be idealized as "flat," but Euclidean straight lines are only segments of arc in infinite celestial circles, and Euclidean triangles are only small patches of a surface that is ideally spherical.

The proposition that any Euclidean triangle has an angle sum of two right angles is irrefutable because it is completely formal, decontentized, analytic, tautologous: A Euclidean triangle is a triangle that has the properties implied by Euclid's axioms. But such a proposition belongs to logic and pure mathematics and is no more about physical space than the proposition that all fathers are parents is about persons.

So the argument that in interstellar space light rays *are* straight lines either is a proposition of physics, in which case it is false, or signals the double determination (a) to use the expression "straight line" in a non-Euclidean sense and (b) to assert that physical light rays traverse paths describable by such a non-Euclidean geometry. But the second part of this determination is just an hypothesis about physical space which looks as though it is becoming well established. It is not a proposition of pure mathematics, it is an empirical application of pure mathematics. One could still point out that what is interpreted as a straight line in astronomy has no formal similarity to the straight lines of pure Euclidean geometry. This would be a legitimate comment, a remark of comparative logic, and not, I feel, an indication that the speaker is entrapped by old-fashioned Euclidean ways of thinking. It is only to say that, given the formal geometries of Euclid and Riemann, one of these will have an application on the astronomical scale when *straight lines* are interpreted as *light rays* while the other system will not. It is not senility which leads me to remark that the light from Arcturus does not travel a path describable as a Euclidean *straight line*, it is only an observation made within comparative astronomy.

As Einstein put it so well, "As far as the laws of mathematics refer to reality, they are not certain; and as far as they are certain they do not refer to reality." [1]

---

[1] Albert Einstein, "Geometry and Experience" in *Readings in the Philosophy of Science*, eds. Herbert Feigl and May Brodbeck (New York: Appleton-Century-Crofts, 1953), p. 189.

16 | Discovering Causes and Becauses

IN THE next three chapters *cause* and *purpose* will be examined. We are enormously versatile in our uses of both words. This occasionally makes us think that the *notions* behind the words are confused. The result? We become confused ourselves. First, some of the confusions.

We say that where there is boiling there is heat, where there is smoke there is fire, where there are swamps there are mosquitoes. But while we might say that the heat *causes* the water to boil, we need not say that the fire *causes* the smoke. (The brakes on my car usually smoke, but they have never yet broken into flames.) And would we say that the swamp *causes* the mosquitoes? Not always, surely. Still less would we say that the swamp *causes* our insect bites.

It will be our aim in these chapters to discover why it seems natural to say that the heat caused the water to boil, and why we hesitate to say that the swamp caused our insect bites.

It is often important to learn that an event A is regularly accompanied by another event B. To realize that the formation of swamps is accompanied by an increase in the number of mosquitoes was an essential step in our battle with malaria. It is important to learn that an outbreak of typhoid is usually accompanied by a sharp increase in the local death rate. But it may be a matter of some urgency to discover more than merely what *accompanies* the formation of swamps and an epidemic of typhoid; we often require to know what *causes* the swamp formation and what *causes* the epidemic.

What are we after when we ask for the cause of a toothache, an epidemic, a war? For what are we in search when we seek the break in the circuit that is causing the power failure, the leak in the gas insulator that is causing our meter readings to give unexpected results, the deficiency in the soil that is causing the poor growth of our crops?

In all these cases, notice, we are looking for something *by the control of which we could prevent the effect.* Remove the swamps and you remove the mosquitoes. Remove the decay in the tooth and you remove the toothache. Remove the contamination from the drinking water and you remove the typhoid. Find the circuit break, the insulation leak, and the soil deficiency and you have found where our apparatus requires mending.

Here, then, when we say that A causes B we mean that the *removal* of A would lead to the *removal* of B. B does not happen when A does not happen.

These are all cases where we wish to discover the cause in order to *prevent* the effect. Our aim is to control A in order to curtail B.

Very often, however, our intention to discover a cause is not directed to the curtailing of its effects. For we do not wish to prevent a good harvest, a decrease in the death rate, the success of a series of experiments. Yet we may wish to know the causes of these phenomena, if only to keep them with us, or to bring them back when they have left us temporarily.

In these cases, of course, we aim to discover a cause A, not to rectify it or stamp it out. Here we hope to *produce* A, in the hope of encouraging its effects—good harvest, increased longevity, successful experiments.

The pattern in *these* examples is this: We say that A causes B when the occurrence of A leads to the occurrence of B—when A happens, B also happens.

Now, these two senses of "cause" are profoundly different. The difference is sometimes signaled by saying that one is a "necessary" cause and the other is a "sufficient" cause. A "sufficient" cause of B is that event which produces B, or, as is more usual, those events all of which conspire to produce B.

The analysis of what we mean by *the cause* of B will be correspondingly different, depending on whether we wish to eliminate B or perpetuate it.

For the hero who is trying to save Little Nell (she is trapped in the dynamite warehouse), it is enough to discover the burning fuse that the villain set alight. Stamp out the fuse and Nell is saved—if the hero does not stamp too hard, that is. The burning of the fuse is a necessary cause of the intended explosion.

This simplicity will hardly do for the mining engineer. He wishes, not to prevent an explosion, but to bring one about, one that will blast just the right amount of rock away from the coal face. The cause of this explosion must be considered in more complex terms. For the desired effect is brought about only by a skillful combination of various factors; the charge must be of exactly the proper concentration; it must be set into the rock face at the right place, at the right angle, at the right distance from the tunnel supports; the charge must be dry and the fuses too; the fuse leads must be of the proper length; etc. Carelessness with respect to any one of these factors may result in a failure to bring about the desired explosion. We do not hunt here for a single event and call it the *cause* of B; rather, we contrive to cluster many special events together all to cause B.

Again, when we ask what caused the clock to *stop* running we are usually after something very unlike what we are after when we ask what caused the clock to *begin* running, what caused it to keep running. One or two events, e.g., a broken mainspring, a clogged escapement, a worn balancer, could have caused the clock to stop. But it is only the whole intricate mechanism of the clock—wheels, cogs, springs, and levers—which by an ingenious conspiracy causes the clock to begin running and keep running properly. One last example. Consider a newly born infant: The questions "What made it start breathing?", "What made it stop breathing?", and "What makes it continue to breathe?" all invite entirely different types of answers. "What made it start breathing?"—the doctor's slap! "What made it stop breathing?"—an obstruction in the bronchi! "What makes it continue to breathe?"—well, even supposing that I could answer that, if I began doing so there would be no space left for discussing causation.

Such a simple ambiguity, so obvious when pointed out, is often overlooked. We sometimes inquire about causality without having come to see that questions like "What caused the forest fire?" and "What caused the fire in the grate?" employ the word "cause" in entirely different ways. We are asking for the *necessary* cause of the forest fire, that which, had we eliminated it, would have eliminated the forest fire. But we are asking for the *sufficient* cause of the fire in the grate, that cluster of elusive circumstances which tends to encourage a warm, efficient, and cheery blaze in the grate.

Not all, not even most, of the important causes in science fall easily into either category. The causes of cancer or of sleep, of life and death, of the arrangement of the solar system and the galaxies—all of these are enormously complex, whether we would perpetuate or curtail them does not matter. For instance, the elimination of cancer—even lung cancer—probably requires far more than just the discovery of a necessary cause. Nonetheless, the distinction should be kept before one in all analyses of causality. It is very often ignored by philosophers, who are busy both affirming and denying such propositions as "Every event has a cause" or "Repeat the cause of X and X will occur repeatedly."

There are further elementary ambiguities in our notion of *cause* besides this one between *necessary* and *sufficient* causes.

An insurance company may report that the explosion in the warehouse was caused by Mr. Thing's cigarette butt. In this conclusion they pick out from a great range of cooperative factors, e.g., the inflammability of the wares, the wood shavings, the oil drums stored in the garage, the long drought, etc., just the one which human beings could have controlled. In the eyes of the law, too, Mr. Thing's negligence would have been taken as the cause of the fire.

But were Mr. Thing's house blown down in a hurricane, the insurance report would probably have said that the *hurricane* caused the demolition of the house. Hurricanes, of course, are things over which human beings have no control at all. They are often referred to as acts of God, something "caused by" God. God tends to be an important figure in discussions of causality. He is credited with being the cause of everything men cannot explain, or everything insurance companies can escape paying for, or everything theologians think will lend strength and force to their arguments.

Another ambiguity: You may say that the cause of what you are hearing is a vibrating of your tympanic membrane due to disturbances in the air which fills the room. Or you might say that the "true" cause is the action and reaction of the parts of someone's larynx where these disturbances originate. Then again, you may claim that the *real* cause was the failure of the speaker's parents to strangle him at birth. Which is correct?

All of these contentions are correct. "Cause" sometimes denotes a condition near to its effect in space and time, contiguous and propinquitous, as the philosophers might say, and sometimes it denotes a condition very remote in both these respects. The cause of the red ball hitting the black ball is that it was just nudged by the cue ball. But the cause of the earth's characteristic revolution and rotation may have been some original cosmic explosion, the literal *creation*.

We say that the cause of the wax melting was the increase in the temperature. We also say that the cause of the melting was the low melting point of the wax.

We say that the warehouse burned because Mr. Thing was careless with his cigarette. We also say that the cause of the existence of the Great Barrier Reef is the recurrent death of unnumbered polyzoa and the vast accumulation of their calcareous skeletons. We might even say that our delinquency problems are caused by the state of contemporary society. By that we clearly mean something, though it is far from easy to say exactly what.

Someone might say, I wouldn't, that the cause of the existence of Mr. Thing's cigarette is the tobacco from which it was made. Or perhaps it was the specific behavior of the machine which rolled out the cigarette in just the form it did. Maybe the cause of the cigarette was the idea which led to the design of the machine. The cause of the cigarette may be the demand of Mr. Thing and millions of other people for just such a lethal commodity, or it might have been Sir Walter Raleigh's early discovery and popularization of the uses of tobacco. Perhaps the cause is God, the first cause, or Mr. Thing's early death, the "last" cause.

The word "because" is even more worry-making. Consider: "I did it because I was told to do it." "This must be the answer because the premises allow no alternative conclusion." "It must be love because everything looks rosy." "He deserves an 'A' because he has studied 18 hours a day." "Churches and cinemas flourish on Sundays: Go to the pictures because it increases church attendance." And so on.

"Cause" is a logical labyrinth. It is not surprising that in analyzing this notion philosophers and scientists often go in circles, or get completely lost, or come out the way they went in. They get headlocks on themselves with puzzles like this: A cause must be different from its effect. Considered in terms of time, this means that there must be a time interval between the final moment during which the cause operates and the first moment during which the effect comes into being. Were this not so, cause and effect would be temporarily indistinguishable: It would be impossible to specify where a cause terminated and its effect began. Since it always *is* possible so to distinguish cause and effect, a time interval must be sandwiched in between them. However, on this account cause and effect could never be a possible relation if every cause ceased to operate before its effect came into being. If we allow small time intervals to split cause and effect apart, on what grounds can we refuse to allow larger time intervals the same office?

Hence it is impossible that there should be a time interval between cause and effect, and it is also impossible that there should not be such a time interval. A pretty state of affairs! "Pretty? Beautiful!" some philosophers would say.

There are many other puzzles concerning the logical status of this cause and effect relation: Do we learn of it purely by experience? Is the proposition "Every event has a cause" a contingent proposition? Or is it necessary; is it undeniable precisely because we are unable to conceive of a totally uncaused event? Perhaps that it is the effect of some prior cause is exactly what we mean by calling X "an event," or at least part of what we mean by "event."

Or is the proposition an article of faith, a desperately illogical prayer somehow built into the laboratory practice of every scientist?

Is the will free? Can there be a causally unconditioned choice? How do ideas in the mind cause motions in the body? Is God subject to the law of cause and effect? If not, how can He have effect upon the world in which we live? But if He *is* so subject, can He be said to be omnipotent, all powerful and subject to nothing?

Let us talk some common sense about these notions of "cause," "causality," and "because." I hope that I will be successful in the attempt to arouse in you a suspicion of 95% of all the highly advertised problems about causation. When you see how many of them stem from inattention to the features of our garden-variety uses of the words "cause" and "because" you may be able to resist joining certain members of the older academic generation in their tour of the logical labyrinth.

One thing is apparent straightaway. The notion of cause and effect as we understand it has become dominant, I would say overdominant, only in the last 300 years. It is not stressed much in Greek philosophy. Plato and Aristotle were not perplexed by the ambiguities of the concept.

The rise of our modern notion of cause and effect coincides completely, of course, with the rise of our modern notion of a scientific theory and the subsidiary notions of explanation and prediction. And, as I shall argue, the coincidence is far from being an accident. So it is that we hear that modern science is a search for causes. Non-causal talk is apparently "unscientific."

In all this we must distinguish what scientists *do* from what they *say* they do, and especially from what philosophers say they do. Scientists often say that they are searching for causes. Yet you will look a long time before you find the word "cause" in a scientific paper or even in a text. You will not hear the word very often in the workaday speech of scientists. It is usually only at lunch time, or at Mrs. Thing's cocktail party, that a scientist will proclaim himself to be a hunter of causes.

I would add that the very *idea* of cause and causation is used far less in actual laboratory science than most people, you and I included, would imagine.

Nevertheless, scientists get worried about "cause." This is particularly true in the biological sciences. For, if everything is the effect of prior causes, it seems to follow that there is no room in the world for purpose, purposeful action, deliberation, choice, strategy, not even for intelligent decision or goal-directed activity. Our most typically reflective behavior is what it is, apparently, only because we have been pushed and molded in a thousand different ways by nature, nurture, and Dr. Freud. We are all just bits and pieces of the great clockwork that is the universe.

Consider an avalanche. The course it takes down the side of a mountain may be explained in detail without any reference to a target. It veered to the left when it struck that outcropping of solid rock, it plunged straight down the middle of the shallow crevasse, it struck the shepherd's house and knocked it flat. It struck what it struck, but not *in order* to strike it. The avalanche did not *decide* to demolish the shepherd's house and then consult a road map for the shortest route to it.

Let us redesign the situation with respect to the question

"Why . . . ?" Often when we ask "Why?" of something, we expect an answer of the form "Because certain antecedents made it . . ." Thus: "Why did the avalanche occur?" "Because the rains washed away the earth and soft rock from beneath the massive boulders perched above. As these tumbled, other matter was shaken loose from the face of the mountain, all of it massing together in a great landslide."

Often, however, we ask "Why?" not to discover the antecedents of an event but to discover what it is in aid of. Question: "Why are you immersing that bit of muscle in wax?" Answer: "In order to facilitate slicing it into sections for examination under the microscope." Question: "Why are you swinging the container like that?" Answer: "To separate the light from the heavier particles of matter suspended in the fluid."

We note that the "why" in "Why did the avalanche occur?" is a "why" in the first sense only. It requires an answer in terms of antecedents, not in terms of the designs or purposes of the avalanche. And we note (with mixed attitudes) the "improper" questions of children: "Why did it want to rain?" "Are the clouds angry?" "Was the avalanche trying to hurt the shepherd?" (See Jean Piaget's *The Child's Concept of Causality*.)

Perhaps we even come to feel that "why" questions of the second sort (the "what for" questions) ought to be given up entirely in favor of "why" questions of the first sort (the "what from" questions). This is the Baconian-Cartesian dictum against "final causes" (so-called). Under pressure from all sides we might come to suspect every account of things that does not deal exclusively with descriptions of the inexorable pattern of cause and effect.

Hence gloom and despondency. Learned men inside and outside of science are led to ask medieval-sounding questions about the freedom of the will and the nature of the mind. For many of them biology becomes just a kind of promissory mechanics.

Here is a historical point I should like to make before settling down to engage some of these difficulties.

Philosophers like Francis Bacon, Descartes, and Spinoza ruled out final causes from science. Scientific questions are never "what for" questions, said they. Answers to scientific questions are never answers to "what for" questions, said they.

But the whole idea behind *final causes* has become exceedingly confused since the days of Aristotle. It was confused even by the time

Bacon and Descartes inherited it. Aristotle, speaking primarily of manmade things, argued that there are four different sorts of explanation that we may require.[1]

We may ask:

What is it made of? This Aristotle called the "material explanation." Thus *bronze* is the material explanation of the statue.

or *What is it made into?* This is Aristotle's "formal explanation." Thus *a discus thrower* is the formal explanation of the statue.

or *Who (or what) made it?* This is "efficient explanation" for Aristotle. "Myron, the great sculptor of Greece," provides an efficient explanation of the statue.

or *What is it made for?* Aristotle's "final explanation," or explanation in terms of purposes. That it was made to commemorate and glorify the athletes of Greece is the final explanation of this statue.

Thus the existence of Mr. Thing's cigarette is explained *materially* by saying that it is made of shredded tobacco leaf that has been dried, cured, and rolled in a piece of rice paper. The *formal* explanation is that the tobacco and the paper are made into this cylindrical entity that we call "cigarette," the tobacco being stuffed into the paper cylinder. An *efficient* explanation of the cigarette would be that the shredded tobacco, after having been placed in a feeder bin at the top of a rolling machine, is passed onto a conveyor belt in small heaps, each of which contains an amount of tobacco sufficient for one cigarette. The individual heaps are paired off with and rolled into small sheets of gummed paper. Then the loose edges are sliced to form clean edges, etc. The *final* explanation of Mr. Thing's cigarette is that it was made to be smoked by Mr. or Mrs. Thing or Mr. or Mrs. Anyone-at-all. Its purpose is to provide us with at least some of the things that cigarette manufacturers claim for it.

Now, Aristotle generalized all this. He went on to say that everything, whether man-made (like cigarettes) or natural (like tobacco), had four possible explanations. Thus there were material, formal, efficient, and final explanations of the sun and the moon, the stars, tides, birth, death—everything.

This suited Christian philosophy very well. God became the Su-

---

[1] Aristotle, *Physics*, II, 3, 194b17–195a3.

preme Artificer; the final explanation was in His purposes. We will not linger on this point, however.

The Greek word, αιτια, which I have translated (following Professor Ryle of Oxford) as "explanation" was latinized as *"causa."* Hence it was anglicized into "cause," a word that is very inappropriate, I think, for conveying Aristotle's meaning. "Explanation" is a much better word for this. Perhaps the word "reason" is better still. Causes of X are explanations or reasons why X happened.

Thus it is that translators and historians of philosophy saddle us with the altogether odd notion that the "what for" of a cigarette is its final *cause*. As though the virtues of a cigarette are what cause it to come into being; as though the maiden flight of the Comet aircraft was an operative cause of its construction.

This oddity is not Aristotle's creation, however. His fourfold distinction is proper and useful were it but understood for what it is and not for what its tortured translations suggest. Aristotle was talking about ways of accounting for natural events. He was not talking about the actual pushes and pulls which bring natural events into being. Small wonder that philosophers have had difficulty in imagining what kind of push or pull a final cause is. Small wonder that science is said not to be a hunt for final causes. The wonder *is* that enough sense could be attached to the expression "final causes" to permit one even to say that science is not concerned with them.

In fact, not one of the four members of Aristotle's distinction corresponds to the notion of *cause* that we use in scientific explanation (if, indeed, we use the notion at all). How wicked of philosophical scholars to perplex us with the language of "material, formal, efficient, and final *causes*" when "material, formal, efficient, and final *explanations* or *reasons*" would have been so much less troublesome and so much nearer to Aristotle's meaning!

Final *causes,* whatever they are, will not be dealt with in these chapters very much. The final explanations of which Aristotle spoke, however, may hold our attention later.

Let us return to the problem of reconciling cause and effect with purpose, or reconciling the questions "What from?" and "What for?"

Many people have this picture:

To be an effect is to be pushed from behind (in time).
To have a purpose is to be pulled from in front (in time).

It is from this picture that their despondency arises. For it seems too queer to say that what is still to be can make things happen here and now. So presumably—alas—everything is pushed from behind. Science can only be of causes and effects, not of purposes pulling in front. Scientists can only study billiard balls, not billiard players. Hence the uneasiness about whether or not psychology is a science, for, as G. K. Chesterton put it, "Geographers study maps, psychologists study chaps."

Later, I hope to persuade you not only that the pull-from-in-front idea is wrong but that the push-from-behind idea is wrong as well. Causal explanations have nothing to do with pushes.

For the moment, however, let us ask, "What are causes and effects?"

Clearly they are not material objects. Consider:

Were you and I looking out the window of a moving train, would either of us say, "You count up the causes (the cows) and I will count up the effects (the sheep)"? An inventory of the equipment at the Cavendish would not include a section on causes and effects. I can keep a great many things in my pockets, but not causes and effects. And this is not just because my pockets are too small.

It is because causes and effects are not objects at all. It would be better to say they are happenings, events, occurrences, even though that's not quite correct either. Billiard ball A collides with billiard ball B, and B moves off with a mathematically predictable velocity. The burner imparts heat to the vessel, the liquid in the vessel bubbles and froths. The circuit-breaking switch is closed, and the galvanometer needle begins to oscillate.

But it is not billiard ball A, the bunsen burner, and the switch that are causes, nor billiard ball B, bubbles, or the galvanometer needle that are effects. By themselves, these are just what they are, objects. They are just billiard balls, burners, switches, bubbles, and needles. They only take up space. It is what these things do, or what they are made to do, that are the specific causes in these cases.

It is the *colliding* and the impact of ball A with ball B that is the cause of B's velocity. It is the *combustion* of the gas emitting from the burner, and the subsequent *transmission* of heat to the vessel and the fluid therein, that is the cause of the boiling and frothing. It is the *closing* of the switch, and the subsequent *flow* of electric current, which actuates the galvanometer needle. Collisions, combustions, transmissions, closings, and flowings—these are the names of events, not of objects. Likewise with boilings and oscillations.

Causes and effects, then, are *doings* of certain specifiable sorts. They are not entities possessing some uniquely causal property. So one mistake *not* to make is to imagine that there is a class of objects, like cue balls, hammers, bunsen burners, and generators, that are specifically qualified for the title *causes*. These things, by themselves, are dead and inert. It is what they do and can be made to do, it is the events in which they figure as principals, that we call *causes*. And the same applies to effects, although in a way to be qualified later.

Roughly, we get to cause-and-effect talk when we get to verbs, e.g., "collide," "combust," "transmit," "close," "flow," etc. But not all verbs, perhaps; "resembles," "exists," "implies" do not sound like causal verbs, though cases can be thought of where we might qualify even this. (E.g., "an impurity exists in this ore, and it will cause us difficulty at the refining stage," "A entails B; this will cause us to reconsider our deductions.")

Still, it is safe to say that causes and effects are happenings, processes, events. So don't let the fact that the words "cause" and "effect" are nouns trap you into thinking that they are the names of two species of objects. Not all nouns are names of kickable things. They do not designate objects but transactions between objects. It will appear later that even this is not quite right, but we will limp along with it for the time being.

Not all processes, transactions, and events are as dramatic as those we have been considering, however. We usually have in mind the clicking of billiard balls, the driving of nails, the opening of valves, and the striking of detonators as our models of cause and effect. This is liable to give a Keith Robinson twist to our analyses of causality.

Nothing nearly so dramatic is involved in the question, "What caused the butter to melt?" Maybe the heat of the sun was the cause. But this is not like a push from behind, a shove from the side, or a blow with a hammer. More so with the question, "What caused the butter not to melt?" Perhaps it was because the butter dish was in the refrigerator. And putting the butter dish into the frig is not like hitting it with a hammer, not even with a *utility* frig.

It is interesting to speculate what our ideas of cause and effect might have been had melting butter been our model instead of billiard balls. As it is, the world may seem to us a succession of clicks, pushes, ticks, and tocks. Had the melting of butter or wax seized our imagination instead, the world would have appeared to us as a series of simmerings, drippings, meltings, and splashes, and our early studies in physics would have been mainly of things plastic and liquescent.

In any case, we tend to take causes and effects as *changes* or alterations to things or non-changes, non-alterations, to those things. Thus something may cause the tree to fall. And something causes it to remain standing. Something causes an increase in the number of mosquitoes, and something causes the number to remain constant. Clearly, what will count as a cause or an effect is largely a function of what we are looking for, what our problem is, what we consider noteworthy, where our interests lie, the way we express our questions.

The same point is put by saying that not all changes are sudden and drastic, many are very gradual. On the one hand we have wax melting, on the other we have a cartridge exploding. More extreme than either of these are such phenomena as the evolution of the ungulate's hoof (a process which is infinitely more gradual than the melting of wax) and the orbital shift of an electron (an event infinitely more rapid than the explosion of a cartridge). The first is a change which took, and is taking, so long and the latter a change which is so unbelievably quick that we hesitate to use the language of cause and effect in either case. A cause that has an effect through hundreds of millions of years or one that is spent in a hundred millionth of a second is so unlike our billiard-ball, hammer-and-nail notion of cause that we may well wonder whether it is even in the class of happenings.

Butter melting, natural uranium decaying, the ungulate hoof evolving—into how many causes and effects are these divisible? There is clearly no answer. In fact, the account of a remorseless tick-tock sequence of causes and effects is pure myth. Most of the processes for which we give, and expect, causal explanations are not in any way divisible into ticks and tocks. The origin of such a view probably derives from the earlier physical theories about the communication of motion by impact or collision.

Very often, however, the notion of something causing something else is conveyed not by the word "cause" but by expressions like "make," "produce," "lead to," "issue in," "bring about." And the conception of "effect" is often rendered by language like "due to," "because of," "produced by," "issue of," etc. Yet, few of these ways of paraphrasing the words "cause" and "effect" seem to encourage universal assimilation of all cases of causation to the model of motion by impact or movement by collision. And it is precisely this model that is responsible for a good many of the logical and philosophical difficulties which have made causality one of the most venerated puzzles in western thought. That much of the force of this puzzle drops away when the words "cause" and "effect" are rendered by less pompous and more useful expressions like "leads to" and "is due to" shows how few of our difficulties about causality arise from any actual complexity in nature and how many of them derive from our own failure to appreciate how we use our language in causal contexts.

I have tried to give a survey of the kinds of agony theorists have felt about causality. I have also made suggestions about how the difficulties got started and why they are perpetuated. My principal hope at the end of this chapter is that you will join me in refusing to be overimpressed by the venerable tradition which makes a kind of ceremonial rite of this cluster of problems.

Causation can be discussed clearly, simply, and above all intelligibly. And since scientists like most people *will* worry about this concept, we shall continue to sort out some more of its features in the next two chapters, when we consider the *logical* relations between what we call "causes" and "effects."

17 | What Happens as a Rule

How must a set of events be related before we are justified in saying of one of them that it is the cause of the others?

Here are some of the traditional requirements:

a) They must be different. Were two events no different, we could not distinguish which was cause and which effect. That seems obvious enough. Difficulties arise, however, when we try to say *how* different, and in *what respects* different. For, if we are even to speak of two events *as* two events, they must differ from each other in *some* respects. Each tick of the clock, or one pulse of a Geiger counter, is not the cause of any other ticks or any other pulses. A single chime of Big Ben is not itself the effect of any other chime. Let us agree, however, that a *cause* event and an *effect* event are in *some* respects different.

On the other hand, the difference cannot be too great, either. The difference cannot, for example, be one of logical type. That the number *two* is an aperiodic number makes it a very different sort of thing from the total eclipse of the sun over Southport on June 29, 1927. But the difference here is of *logical strata*. These two phenomena never *could* be the same. They could not be *un*different. (There is little point, therefore, in announcing that they are different.) And surely they would never be related to each other as cause and effect.

But let's leave it at that for the moment. Causes and effects must be different in some sense of "different," a sense that lies between the sense in which *any* two events at all are different if they are to be spoken of as two *events* and the sense in which logically dissimilar occurrences are different.

(The more one thinks about this traditional requirement, the more ludicrous it seems. Apparently the message is this: Indistinguishable events cannot be distinguished, and *a fortiori* cannot be distinguished as cause and effect; hence, any events that *do* stand to each other as cause and effect are distinguishable, i.e., different. This is a compelling argument, but not very startling and not at all informative. However, so be it: Causes and effects must be different from each other.)

b) Another requirement that the philosophical tradition has exacted of the cause-effect relation is as follows: Two events, if they are to be described as cause and effect, must stand to each other in terms of the relations "earlier" and "later." It won't do to speak of the explosion as causing the lighting of the fuse, or to speak of the stopping of the clock as the cause of the mainspring's breaking. In general, causes should precede their effects. And well-behaved effects should lag behind their causes.

But not by too much, of course. The golfer whose ball drops into the cup 20 minutes after he has made his putt is in for trouble when he tries to convince an opponent that the game is his by one stroke.

Still less do we allow million-year gaps between the time at which some "cause" terminated and the time at which its alleged effect began.

But though these questions about *how much* later and how much earlier are difficult, it is still true that we do apparently require causes to precede their effects. What kind of a cause-effect pattern would we have were this requirement controverted? I cannot imagine. Perhaps, then, it is really tautological to say of a cause that it must precede its effects. Perhaps that it is an event preceding another event called "effect" is precisely what we mean by referring to a happening as "a cause." That, however, is a sleeping dog past which we can afford to go on tiptoe. Let us just agree, for the moment, that causes precede their effects.

So we are now backed into the following position: Causes are different from their effects, and causes precede their effects.

Not without another reservation, however. For just as we saw that a mere *difference* between two events is insufficient to warrant marking them as causally related, so too the fact that one event precedes a second is not enough to guarantee that they are to each other as cause and effect. Monday invariably precedes Tuesday, Wednesday, and Thursday; it does not, however, *cause* Tuesday, Wednesday, or Thursday either altogether or one at a time.

c) The third traditional requirement was this: that cause and effect must be spatially continuous, by which is meant that there should be no spatial gaps between the area through which the cause operates and the area in which the effect manifests itself. This requirement was, apparently, contrived to make life difficult for dowsers, magicians, and telepathists. But the general condemnation of action-at-a-distance that is implied in this requirement led to the discomfiture of a few Newtonians as well; at least it made them more self-conscious about the terms "force" and "attraction."

The traditional expectations of the cause-effect relation, then, were these: i) the two events had to be different, in some as-yet-undecided sense of the word "different," ii) the two events had to be in a relationship of temporal succession, iii) the two events had to be spatially continuous; action-at-a-distance was not permissible with the event-pairs we call "cause and effect," and iv) . . .

What else? This was the big question Hume had inherited from the philosophical tradition of the 17th Century. For, clearly, there must be *something* else. Many pairs of events fulfill all three requirements and yet are not at all related as we feel cause and effect must be related.

Consider a story I have invented just for this discussion: It is a story of two little boys in Southport in the year 1927. They were both given chemistry sets as presents. Immediately Tommy took a strip of litmus paper and dipped it into the weak base solution that came with the set. At that moment Southport was plunged into the darkness of a total eclipse of the sun. Small wonder that he cried, "Johnny, don't put your blue paper into this bottle, it makes you blind!"

Now, we would say that the passage of the moon between Southport and the sun at the exact moment when Tommy dipped his litmus paper into the alkaline solution was a pure coincidence, a strictly fortuitous occurrence. (At least I hope we would say this. There is a periodic tendency on the part of philosophers, and especially students of philosophy, to deny the existence of fortuitous occurrences, coincidences. This strikes me as just silly. The conjunction of a complete blackout with Tommy's litmus-paper test is as fine a coincidence as could be hoped for. It is surely not a case of cause and effect.)

Nonetheless, all the requirements are fulfilled. Tommy's action of dipping the litmus paper into the base liquid was *different* from the complete darkness that followed, it *preceded* the darkness, and was *spatially continuous* with it. But it is still a coincidence.

Hence a favorite philosophical pastime, both before and after Hume and even at the present time to some extent, has consisted of suggesting the missing ingredient in this analysis of causality. For, after all, any account of the nature of the causal relation that lets a mere coincidence slip through as a case of cause and effect must be deficient. How, then, do truly causal relations differ from coincidences? What is the further element besides (1) difference, (2) temporal succession, and (3) contiguity that characterizes a sequence as "causal"?

This further element many philosophers found in what they called "necessary connection." They might have argued in this way:

It is the mark of a coincidence that either event *could* have happened without the other. Little Tommy could have performed his litmus-paper test without an eclipse having ensued. The eclipse of the sun could have occurred without Tommy's ever having been given a chemistry set, indeed, without Tommy's ever having been born.

But this is not the case with a genuine causal connection. You cannot remove the balancer from a clock without the clock stopping; the clock *has* to stop when the balancer is removed or damaged. Likewise, sucrose cannot be heated with mineral acids without giving up equal parts of glucose and fructose; it *must* do this, if it is to be called "sucrose."

There is a compulsion in these examples that is lacking in a coincidence of the sort expressed in the sentence, "Tommy dipped the litmus paper and all went black." The point is that Tommy could have dipped the paper without the blackout having occurred. But Tommy could not have bashed in the balancer of the clock with Daddy's hatchet without the clock failing. Sucrose cannot be treated as stated without behaving as stated. This compulsive and binding tie between two causally related events is the "necessary connection" between them. It is just what coincidences lack. Thus, "necessary connection" is the fourth major ingredient of truly causal relations. *So runs the traditional argument, anyhow*.

Now, what is the nature of this necessary connection, if indeed it *has* a nature?

Were I to have said that Tommy is *taller* than Johnny, *near* to Johnny or *to the left of* Johnny, the relations of which I spoke would be perceivable. Stand Tommy next to Johnny. You can *see* that he is taller than Johnny. Is he on Johnny's right or on his left? A glance will give the answer.

In the same way you can *see* that A is to the left of B in the complex symbol "ALB." However, you may stare until Harvard wins the boatrace but you will not *perceive* any necessary connections between objects or events.

And you may watch me remove the balancer of the clock, perceiving that the internal mechanism of the instrument ceases its movement straightaway. But then you will have seen everything that there is to be seen. There is nothing between the removal of the balancer and the deceleration of the timepiece's gear train that a skilled and careful observer will detect but that may be overlooked by the novice. The ability to perceive necessary connections is not something one acquires along with a degree in mechanical engineering or a certificate in horology.

Tommy was wrong in his account of the litmus-paper-eclipse incident. But not because he missed seeing or hearing or smelling something that we have been trained not to miss. There was nothing wrong with Tommy's perceptions, only with his explanations. I can say this even after recognizing that my earlier remarks would bring the notions of perception and explanation much closer together than my language here suggests.

So, whatever this "necessary connection" is, it is nothing perceivable, and *ipso facto* nothing we can perceive well or ill, carefully or carelessly (if, indeed, it is even permissible to speak of perceiving carefully or carelessly). There is no difference whatever, so far as perception goes, between the putting of litmus paper in an alkaline solution prior to a blackout and the removal of the balancer from a clock prior to its stopping.

How would you proceed to show Tommy that the blackout was *not* due to operations with his chemistry set? How would you persuade him that it was all just a great coincidence, just like the time the church-bell chimed when Johnny pressed the doorbell button?

First, you would probably try to prove to him that *everyone* in Southport had experienced the blackout, not just those who were playing at chemistry with litmus paper and base solutions.

Next, you would argue that children (and grownups, too) often play at chemistry, with litmus and alkaline, and blackouts are not always the result. You might even invite him to try his little test again, confident that Southport has had all the solar eclipses it is going to have for one day.

Finally, you would explain that the darkness in Southport was due to the moon passing directly between Southport and the sun. Besides some few remarks about solar astronomy you would be saying things like, "Whenever large objects get between us and the sun, it gets darker; this happens also when you walk behind a large building, or when you go into the basement or into the cinema, or when Southport is overcast with thick clouds. But in those cases a *little* light always leaks through. When the moon gets in the way, however, hardly any light gets through to us. That's why everything went so very dark."

That is, persuading Tommy of what was and what was not the cause of the sudden darkness in Southport involves telling him some very *general* sorts of things. Words like "probably," "everyone," "often," "whenever," and "everything" hold the elements of such an explanation together, the point being that we do not say, "This A caused that B," unless we are prepared to say, "*Whenever* an A, then a B."

The notion of a *causal law* thus arises very early in our attempts to distinguish coincidence from causation. And it is in this connection, I will argue, that our feeling arises that some necessary connection binds cause to effect.

Now, when Hume addressed himself to the question, "What besides *difference, temporal succession,* and *spatial continuity* objectively characterizes any given case of cause and effect?" he boldly answered, "Nothing!" And when asked how he then distinguished coincidence from causation, he replied that it was the continued and regular repetition of causal sequences of the form "A and then B" that eventually creates in our minds a lively expectation of a B whenever we are confronted with an A.

In its main outlines this answer is, I feel, correct. For it suggests that in single, discrete observations of event-pairs nothing is disclosed to us by perception that will enable us to mark a given sequence as causal rather than merely coincidental. Only the experience of a *range* of occurrences wherein B regularly follows A can justify such a distinction, the simple linguistic fact being that if B regularly follows A, then it is no coincidence that B follows A. When we refer to a particular pairing of events as a *coincidence* we mean exactly this, that the two events are not regularly paired. But whether or not two events *are* regularly paired requires a lot more than one observation of the events in tandem.

Hence part of Hume's argument is as follows: The difference between coincidence and causation is that a causal sequence is a regular succession of A by B. By definition, a coincidence is not a regular succession of A by B. But to decide whether a given B does regularly succeed an A requires an experience of a range of happenings, not just a single observation of this B following this A. Hence, so far as only single observations go, there is nothing to distinguish coincidence from causation.

Of course, two events may pair off regularly without being cause and effect. I regularly wind up the clock before going to sleep. It is thus no coincidence that I do so. Still, winding the clock does not *cause* my going to sleep, as the ringing of the alarm causes me to awake. But more of this later.

There are good reasons for trying to state Hume's argument without employing his archaic psychological language, however. The really sound point that he makes (viz., that "cause" and "effect" are titles that only the experienced are in a position to confer) can be made without the offside complications about what goes on in the mind—the repetitions of impressions, the existence of lively ideas and expectations. Once one gets into these latter brambles, getting clear again is task enough without the particular thorns causation contributes. Besides, Hume's contention is really logical, not psychological, in nature, and that is the way I intend to treat it henceforth.

Another way of putting this point is to remark that the "necessary" of the expression "necessary connection" has to do, not with nature, but with language. Necessity has to do, not with the nature of things, but with the nature of arguments, specifically, rule-governed arguments. At least this is what I am going to try to sell to you. If nothing else it may make you at least a little suspicious of the question, "What kind of push from behind is a cause?"

Let us look further at this expression "rule-governed." Our everyday actions are to a large extent rule-governed. And in so far as our scientific actions are rule-governed too, the word "law," though misleading in other respects, serves to bring out a certain similarity between them.

Many games—chess and bridge, for example—are also rule-governed. Suppose that we are teaching Tommy to play chess. We say approvingly, "He moved his bishop diagonally," as we would not say approvingly, "He moved his pawn backwards."

Or suppose that bridge is the game in question. To say "he revoked" is not to say just that Tommy *did* something, any old thing. It is to say that he did something of a certain sort, something that contravenes a rule covering this case and all similar cases. "Tommy revoked" means "Tommy did not follow suit though he could have, and it is a general rule of bridge that suit be followed."

There is, indeed, a compulsion, a necessity to move the bishop diagonally in chess and to follow suit in bridge. To play chess or bridge at all is to follow the rules of chess or bridge. You must follow those rules if you are going to refer to your activities as "chess playing" or "bridge playing." One is not free to improvise in games like this. So we can infer directly from 1, "He is moving his bishop according to the rules," to 2, "He is moving his bishop diagonally."

Now, I am not saying that inferences governed by what happens "as a rule" in nature are identical to inferences governed by what happens *as a rule* in chess. They are very unlike in many ways. But there are similarities, too. And as regards this eel-like notion "necessary connection," the similarities can be marked with profit, provided they are not overworked.

The major similarity is this: Words like "bishop" and "revoked" are not simple, "phenomenal" words like "red," or "smooth," or "tall." They are, as I will say, "theory-laden," and because of this "rule-governed." One must know the games of bridge and chess to use the words "revoke" and "bishop" with anything like understanding. Likewise, one must have experience of nature, have some theories about it (however primitive), know what happens "as a rule," before he can use causally-loaded words like "inhale," "dissolve," "charge," and "expand" with anything like understanding. That is, the compulsion we feel when we consider strictly causal connections between events is a logical compulsion built into the very language we use to describe the events. The language of causality is not *neutral,* it is impregnated with theories, *our* theories. It is from the logical nature of these theories that causality draws its distinctive characteristics. And it is from confusion about that logical nature that at least some of our venerable problems about causation derive. But as I will return to this thesis again and again, I need not club it to death now.

Let me approach this same point from another angle. Earlier we noticed how odd it would be to treat causal connections as if they were objects. Inventories are not drawn up as lists of causes and effects; research teams are not divided into those who count up the causes and those who count up the effects.

Nor are causal connections *properties* of objects. Consider: "The clock's balancer is made of mild steel; it is 14.6 grams in weight, 1.1 centimeters in diameter, 2.5 millimeters thick; it is inset with three adjustable weights and mounted on a shaft that rides in two jewelled bushings; and, oh yes, it is causal."

Causal connections, then, are not objects, and they are not properties of objects. As we said, they are relations between happenings.

But not so fast. "A preceded B" is a statement of a relation between A and B. Likewise with "A is the same color as B," "A is larger than B," and "A is next to B." Now, it is tempting to lump expressions like "made happen," "is due to," "is the cause of" and "issues from" with relational words like "precedes," "has the same color as," and "is larger than."

This temptation ought, however, to be resisted. For note one very big difference; to establish that A precedes B, or is the same color as B, or is larger than B *observation* of the relation in question is quite enough. To establish that B is due to A, or happens because A happens, observation (in this sense) is not enough. *A causal law must be established first:* Before we can say that this B is due to that A we must first be assured that *any* B is due to some A. That is Hume's contention and my thesis here.

There are several quite manageable categories of action about which we often speak in terms of observing and witnessing. "Kicking," "throwing," "slicing," "twisting," "covering," etc. are words that indicate activities for which witnessing the situation in question is quite sufficient for determining whether it is one in which kicking, throwing, slicing, twisting, or covering did actually take place.

But is *revoking* such an action? Is checkmating? Surely just witnessing is not enough. One must know the rules of bridge or of chess in order to say whether some witnessed situation is one that involves revoking or checkmating. The argument here is that causally-loaded words like "inhale," "dissolve," "charge," and "expand" resemble words like "revoke" and "checkmate" in precisely this respect; that not one of these is a mere verbal register of what is there to be seen by everyone with eyes, or what is there to be perceived by anyone with senses. They require, besides observation, knowledge of the rules, or of what happens "as a rule."

Hence one must be suspicious of pigeonhole words like "object," "property," "relation," etc. They are useful enough for making preliminary distinctions but they will seldom hold together under a heavy logical load.

Causal connections are obviously not objects and obviously not properties. So, according to a tradition, we feel they must be relations, as though it were perfectly clear what a relation is. And doubtless causal connections are more like relations than they are like objects and properties of objects. But causal connections are very unlike the standard sorts of relations signified by "is taller than," "is darker than," "is above," and "is below"; and these unlikenesses are just as important as the likenesses.

So, while it is all right to say that the sentence "The stopping of the clock followed the removal of the balancer" is a straightforward description of observations, it is not all right to say the same of the sentence "The stopping of the clock was due to (was caused by, was the effect of) the removal of the balancer." That is, the expressions "was due to," "was caused by," and "was the effect of" are not out of the same logical basket as the expressions "came after" and "preceded," a fact that is obscured by calling them all "relations."

A further point:

When I know B to have been the effect of A, then I could have inferred B on meeting an A or, as is sometimes said, I could have *predicted* B on meeting an A. Often, but not always, I can infer from B back to A as well.

Thus from the falling of rain I can infer to the wetness of the streets, and from the removal of the balancer I can infer to the stopping of the clock. And conversely, from the wetness of the streets I can (with some confidence) infer back to the prior falling of rain. But from the stopping of the clock I cannot, without further information, infer back to the removal of the balancer. The mainspring may have snapped, the escapement may have clogged, a bushing jewel may have become dislodged; these or any of a number of other mechanical failures may have resulted in the stopping of the clock. So the inference from effect to cause in this case is a little risky.

We express inferences like these in some such way as this: "The streets are wet, *so* it must have been raining." Or, as is more usual, we see that the streets are wet and say, "Oh, it's been raining."

We say, "The balancer has been removed; *therefore* the clock stops."

Or we say, "The sucrose is being heated in mineral acids, *so* we shall soon have equal parts of glucose and fructose."

Now, the point I am trying to secure is this one: The (so-called) necessary connection between causes and effects is, at least, an inference channel. The *necessity* consists in there being a legitimate inference from A to B, and in our feeling compelled to infer B whenever we are confronted with an A, and (with reservations) *vice versa,* given, of course, that we know the causal law "Whenever A, then B."

What precisely is an *inference,* then? To show precisely the nature of inference is one of the most complex undertakings of analytic philosophy. But it is enough to say here that one assertion linked with another assertion by a word like "so" or "therefore" constitutes an inference. Logically these are closely connected with the "because" and the "if-then" constructions.

Thus the following assertions share many logically important features in common:

"The balancer has been removed so the clock has stopped."
"The clock stopped because the balancer has been removed."
"If you remove the balancer then the clock will stop."
"The removal of the balancer is the cause of the clock's stopping."

These are really four different ways of saying the same thing.
But is this enough?
Is the sentence "B was due to A" equivalent to the sentence "The occurrence of B was inferrible from the observed occurrence of A in accordance with the known law that whenever an A, then a B"? Does the sentence "The sunshine melted the ice" simply mean that the observation of the sun shining on the ice provided us with a sufficient

premise for the conclusion, "therefore the ice will melt"? That is to say, does the word "cause" just mean "inferrible precursor"? Is the cause of the ice's melting just that from which the melting of the ice can be inferred? Is the cause of the clock's stopping just that from which the stopping of the clock can be inferred?

No. This is not nearly enough. For it leaves unmentioned two vital and closely connected points.

a) We can quite often infer correctly from B to A as well as from A to B. From the wetness of the streets I can infer to the prior rainfall. And from the current rainfall I can infer to the wetness of the streets. From the explosion of the cartridge I can infer to the pulling of the trigger, and from the pulling of the trigger I can infer to the explosion of the cartridge.

However, we certainly do not think that effects explain causes in the way in which causes explain effects. The melting of the ice does not explain the heat of the sun, but that the sun is hot does explain the melting of the ice. The stopping of the clock does not explain the removal of the balancer as the latter explains the former.

Thus, though cause and effect are mutually inferrible (or symmetrically inferrible), there is an asymmetry between them. The asymmetry is this: "Why" questions are answered only by reference to causes, not by reference to effects. The reason why the ice melts is because the sun is shining on it. The reason why the sun is shining on it is *not* because the ice will melt. Why did the clock stop? The balancer was removed. Why was the balance removed? Not because the clock stopped, surely.

b) "Red sky in the morning" does not explain "Rain in the afternoon," though this latter is inferrible from "Red sky in the morning."

A pain does not explain a scar. But we can infer from the scar that it must have hurt.

A red sky is a symptom of coming rain. But it does not "make it rain." What makes it rain is the same thing that had made the sky red earlier.

The *wound* explains both the pain and the scar. The wound and the pain are both inferrible precursors of the scar. But only one of them is the cause of the scar. From both the wound and the pain we can infer the scar. But the pain does not explain the scar as the wound does.

The removal of the escapement and balancer of a clock explains both the sudden whirring noise of the mainspring and the subsequent stopping of the clock. The removal of the escapement and the sudden noise of the mainspring are both inferrible precursors of the clock's stopping. From either of these the stopping of the clock can be inferred. But only the removal of the escapement is the cause of the stopping of the clock, for only it *explains* the stopping of the clock.

So "cause" means more than "inferrible precursor." There is more to the nature of the cause of a phenomenon than that the phenomenon is inferrible from that cause.

What, then, is this extra logical ingredient of "accounting for" that lurks in our explanations? What do we have in mind when we say that effects could not possibly explain their causes, but that causes do and must explain, or account for, their effects?

These questions we must defer until the next chapter. We shall consider then causation in its connections with the concepts of explaining on the one hand and theorizing on the other.

18 | Theory-Laden Language

OUR IDEAS about causation are closely linked with the activities we call "explaining" and "theorizing." How and where the link is forged are the problems to which we shall address ourselves here.

The question with which we concluded the last chapter was: Why are some inferrible precursors causes, while others are not? Why is pain *not* the cause of the scar while the wound *is,* although from both the pain and the wound one may be able to infer to the scar? Why is the whirring noise of the mainspring *not* the cause of the clock's stopping while the removal of the escapement *is,* despite the fact that the stopping of the clock can be inferred from *either* the whirring of the mainspring or the removal of the escapement?

That is our question. But before we come to grips with it, let us glance at a picture of causality that has misled all of us some of the time, and some of us all of the time. It is the causal-chain picture to which I refer.

Too often we think of the history of the natural world as if it were a kind of geological tree, each successive time slice comprising a new generation with everything prior to that time slice being its ancestry and everything subsequent its progeny.

> We say that Y is the father of Z, but the son of X.
> And X is the father of Y, but the son of W.
> And W is the father of X, but the son of V . . .
> And so on back to A, for "Adam."

Hence anyone who is a father is also a son, a grandson, a great-grandson.

Likewise with events. We are tempted to say that the course of nature is divisible into happenings, each of which has one "son," or effect, and one "father," or cause.

Thus arises the inevitable question: Y causes Z, but what caused Y? X? But what caused X? W? And what caused W? And so on back to A, the first cause.

The situation is dramatically expressed in Dryden's *Religio Laici:*[1]

> . . . Some few, whose Lamp shone brighter, have been led
> From Cause to Cause, to *Natures* secret head;
> And found that *one first principle* must be;
> But *what*, or *who*, that Universal He;
> Whether some *Soul* incompassing this Ball,
> *Unmade, unmov'd;* yet *making, moving* All . . .

And consider Laplace's dictum that were he supplied with a complete account of the state of the universe at one moment in history (plus a list of all the causal laws), then he would predict everything yet to happen and retrodict everything that has happened.[2] The state of the world two million years ago or two million years hence was, for Laplace, *written into* the state of the world as it appeared at any moment.

Consider, too, how often Bertrand Russell has appealed to the causal-chain picture. At some of the most important stages in his arguments we are suddenly brought up with the realization that we are being confronted with a variant of the old fable: For want of a nail the shoe was lost, for want of a shoe the horse was lost, for want of a horse the rider was lost, and so on to the losing of the day's battle—all for want of a nail. If a Bertrand Russell can be seduced into thinking thus of causes and effects, it is small wonder that we often find ourselves in the same straits.

All these stories have the same plot, and the same two themes: ancestry and progeny (just like a novel by one of the Brontë sisters). With this as a guide, how easy it is to construe science as being a sort of

---

[1] John Dryden, "Religio Laici or a Layman's Faith," in *The Poems of John Dryden,* ed. James Kinsley (Oxford: Clarendon Press, 1958), Vol. I, p. 311.

[2] Pierre Simon, Marquis de Laplace, *A Philosophical Essay on Probabilities,* tr. from the sixth French edition by F. W. Truscott and F. L. Emory (New York: Dover Publications, 1951), p. 4.

collective attempt to produce the Genealogical Tree of Happenings, or at least some of the more distinguished family trees. For in just the same way that the genealogist traces son through father to grandfather, so the scientist (apparently) traces happenings through their immediate causes to their more remote causes. Or perhaps the situation is that the scientist moves from one rung of the ladder of science to other rungs above it or below it, or from one link of the chain of science to the other links to which it is bound, each link being just a repetition of its neighbors.

Doubtless, in less critical moments our ideas about causality are infected with all these figures at once. But the basic story behind them all is, in a word, *bunk*. Ask anyone doing advanced research at the Cavendish, at Brookhaven, or at Oak Ridge how far up the chain he got this morning and see what treatment you receive. You would deserve it too, because the question is silly, just as silly as the fable from which it springs. This will appear more forcibly in what follows.

Back to our central question, then. Why are some inferrible precursors causes, while others are not? Why is the darkening sky with its flashes of lightning not the cause of the rainfall, while the cause *is* (or may be) the lowering of temperature and barometric pressure in a saturated region of the atmosphere? This despite the fact that we may infer to the rainfall from either.

The same question put somewhat differently: Why do causes *explain* their effects while effects do not explain their causes, despite the possibility of inferring in either direction (from cause to effect or from effect to cause, e.g., from a drop in barometric pressure to rain, or from rain to a drop in pressure)?

Again, the drop in pressure, the fall in temperature, the high humidity *do* explain the rainfall. The clouds and thunder do not.

The wound does explain the scar, the pain endured by the wounded does not.

Dark clouds and rainfall seem somehow parallel. They are related in similar ways to the pressure-temperature drop. And the pain and the scar are both explained by the wound.

"Wound" is a kind of explanatory word. "Pain" and "scar" *in this context* are not. Here they serve not as explanations but as *explicanda*. What is the difference?

What is a wound? What is it to be wounded?

Suppose we say that it is a more than superficial incision. Minor scratches and nicks incurred while slicing bread, shaving, or sewing are not wounds. But there are precautions even here. The surgeon does considerably more than scratch and nick his subjects. It is unusual, however, to speak of a surgeon as wounding his patients, though cases can be thought of in which we might say this. Does the surgeon wound his patient when the latter is fully anaesthetized and laid out in the operating theater, and when the incision is one that has been planned well in advance after consultation with other medical experts? Does the surgeon wound a patient when, on his morning ward rounds, he is just careless with a probe or a scalpel, letting it drop from his clumsy fingers into the patient's arm?

Does the plantation owner wound the rubber tree when he carves the customary V-shaped trough deep into its bark?

The answer in these cases may be that we just do not know. There may be more to find out about each of the situations mentioned that could affect our answer.

But we are quite sure that the Eskimo who is hacking great chunks of blubber off a dead whale is not wounding the whale. Throwing darts at the stuffed moose head in Aunt Agatha's parlor is not grounds for an S.P.C.A. investigation.

The carpenter who spends all day slicing, gouging, and slitting timber is not wounding the timber.

Roughly speaking, only *live* things can be wounded. And not all deep incisions in live things are wounds. An anaesthetized man undergoing an operation is not being wounded. A cut in a very calloused foot is not a wound.

At this point consider the metaphor of the wound in a ship: "She had proudly led all her sister ships since her very christening. And even here, being savagely battered from all sides, wounded and gasping she gave an account of herself that will warm the hearts of seafaring men forever."

So a wound is not just any sort of deep incision. It is one which endangers life or impairs functioning. It need not necessarily cause pain, though it very often does. Great damage can be done to the brain, for example, without our feeling the slightest pain; and yet we speak of "head wounds." And getting drunk or indulging in narcotics until one is, as the saying goes, "feeling no pain" is no guarantee that one will avoid getting wounded. If it were, the most formidable army would be the most inebriated army, for no weapon could then wound its troops.

This is why we cannot say with certainty whether the plantation owner wounds his rubber trees or whether the cork producer wounds his cork trees. This *may* impair their functioning or endanger their lives. But it is a highly technical question and we'll leave it that way.

Here is another point leading towards my thesis.

We can *see* a scar and we can photograph it. But what we photograph as a *wound* is but a deep incision. To identify it as a wound is to diagnose it as endangering life or impairing function.

We taste a liquid and pronounce that it is very sour. But that is all we taste. To say of the liquid that it is a poison is to diagnose it as a liquid capable of doing all the awful things that poisons do.

Hence a wound is a cut which endangers life or impairs function. And a poison is a substance the consumption of which endangers life or impairs function.

Words like "wound" and "poison" are, therefore, *biologically, medically, and chemically loaded words. Diagnoses* and *analyses* are built into them. So too with words like "acid," "copper," "sucrose," and "protein." In this they are remarkably similar to words like "bishop," "pawn," "trump," and "offside." The diagnostic quality of these words is reflected in the verbs with which they are often made to function, verbs which are in consequence "loaded" in much the same way, e.g., "inhale," "dissolve," "charge," "expand," etc.

The important thing to notice is that these "theory-loaded" nouns and verbs are (logically) very unlike the nouns and verbs of a completely phenomenal variety, e.g., "red patch," "smooth," "sour," "tinkle," "buzz," "from left to right," "growing larger," "disappearing," etc. Note, too, that in a completely phenomenalistic language causal connections could not even be expressed. Hence problems about causality would not arise.

The reason for this is the same as before: Events which we customarily refer to as causes are rarely thought of in neutral terms. The very mention of words like "refract," "polarize," "condense," "rarefy," "inspire," "digest," "lance," "reset," etc. brings forward comprehensive bodies of theory and experience into which these words are set, and against which they are intelligible. So talking about causes and effects is never merely talking about what is right there in front of our noses: it is, rather, talking in a relatively general way about certain ranges of experience, the elements of which are bound together by notions and theories concerning the way the world is constructed.

This point is reflected in the difference between *seeing how* and *seeing that*. I may see how the several gears, ratchets, and levers of a clock engage. I may see how they are spatially related and be able to portray this diagrammatically with complete accuracy. I may be able to do all this without having come to see that the force transmitted by the mainspring is transmitted to the driving wheel and thence through the chain of gears, finally to "escape" by measured degrees through the escapement mechanism. This is the logical difference one finds expressed in such old sayings as this: " 'I see,' said the blind man, but he did not see at all." A blind man may not *see how*, indeed, cannot see how, this particular timepiece is designed and what specifically differentiates it from other timepieces. But he still may *see that*, if it is, say, a clock, it will employ certain dynamical principles in a certain fairly regular way. He is, for example, in a position to *explain* the action of a clock to a child. The child, however, no matter how keen his vision, will only be able to *describe* the perturbations of the clock to the blind man. The child cannot say what causes this clock to behave as it does. The blind man, if he knows his horology, can say what causes this clock (and any other similarly constructed clock) to behave as it does. The blind man has what the child lacks, a knowledge of horological theory. And though the blind man lacks what the child has, normal vision, the former can say what *causes* the characteristic action of a clock while the latter cannot.

Furthermore, this difference in seeing is connected with the reason why cause-words are not parallel to effect-words; why causes explain effects, but effects do not explain causes. For cause-words are charged with theoretical implications, while effect-words, *being* those implications, are relatively much less charged with theory and hence far less able to explain their causes.

Hence the horologist explains the motion of the hands of the clock by reference (ultimately) to the coiled mainspring. (And note how "theory-loaded" is the expression "coiled mainspring," how extensive are the dynamical and horological implications of just those two words.) The horologist would say that the coiled mainspring, gear chain, and escapement *cause* the motion of the hands. He would not try to offer an explanation by way of an account proceeding in the opposite direction: He would not try to explain the action of the escapement, gear chain, and mainspring by offering only an account of the action of the hands.

Roughly speaking, effect-words are lab assistants' words. They are nurses' words. Cause-words are research scientists' words, diagnosticians' words.

"Red sky" and "rain" and "cloudy" are shepherds' words. "Depression" and "anticyclone" and "humid" are meteorologists' words. The shepherd's words are (more or less) phenomena-words. The meteorologist's words are much more than that.

When the shepherd says "lightning and thunder" he probably means "flash and rumble." When the meteorologist says "lightning and thunder" he probably means "electrical discharge and aerial disturbance."

Of course, this is a caricature. It is grossly oversimplified. There is a lot more meteorologist in every shepherd and a lot more shepherd in every meteorologist than my examples intimate. There is more diagnostician in every nurse and more nurse in every diagnostician, and more physicist in every lab assistant and more lab assistant in every physicist, than I have suggested. But the point is clear, and not less worth making because of the preponderance of borderline cases. We need not despair at twilight because we cannot say where day ends and night begins, for midnight is still very different from noon.

Now, the wider a word is, from a theoretical point of view, the more loaded it is from a causal point of view. The wider its nets of entailments, the more fertile the concept in its causal considerations.

To call a card "the three of spades" requires *some* powers of discernment and recognition. To call it "a trump" requires much more, for to call it "a trump" is already to say what it can do, e.g., it wins tricks against even the highest cards of other suits.

Compare these two sentences:

"My neighbor relieved me of $400 yesterday." You may very well ask, "By what right; what authority had he; is he a thief?"

"The income tax collector (he lives next door) relieved me of $400 yesterday." To this statement the earlier questions are inappropriate, they do not arise. For the answers to such questions are *built into* the expression "income tax collector." An income tax collector just *is* a person who has the authority and the right and the political status periodically to relieve his neighbors of considerable sums of money.

"Income tax collector" is a politically loaded expression. "My neighbor" is not. A child can understand "my neighbor," most children do. But to have even a rough idea of what "income tax collector" means, one must have some understanding of the meaning of words like "government," "citizen," "policeman," "jail," "revenue," etc. He must have some clues about the differences between "fine," "subscription," "theft," "pay," "debt," "tax," "gift," etc. The expression "income tax collector" is a member of a large battery of interlocking political, juridical, and economic expressions. It entails what it does, evokes and prevents the evocation of the questions it does, expresses the causal connections it does only against a subtly interlaced background of theoretical and experiential considerations.

Q. "Why did he take your money?" A. "*Because* he is an income tax collector."

Q. "Why does he win the trick? I have the highest cards." A. "He wins *because* he holds the three of spades, and that is trump."

Q. "What caused him to return his castle to the square on which it was originally?" A. "He would have put his king into check by moving the castle at all."

Q. "What caused that thundering noise?" A. "An electrical discharge from that cloud on the horizon; in principle, a cloud like that is just an electrostatic generator." Or, more fully, "The thundercloud produces electric charges, positive and negative, the separation of which leads to a concentration of positive charge in one region of the cloud and of negative charge in another. As charge separation proceeds, the electric field between these charge centers (or between one of them and the earth) grows until electrical breakdown of the air occurs."

Note that the cause of the thundering noise is not said to be just the flash that precedes it. The word "flash" is not roomy enough to hold within it an implicit reference to consequent aerial disturbances and partial vacuums. The words "electrical discharge" and "electrostatic generator" are, however, quite roomy enough for this.

Q. "Why is it raining?" The causal explanation of the rain would be couched in the language of the meteorologist. The shepherd's answer of "Red sky this morning" will not do.

Thus, as with "income tax collector," "trump," "check," "depression," and "discharge," whenever we are contented with the assertion that C caused E, "C" will belong to an interlocking battery of more or less cognate expressions, all of which will be *ex officio* the terms of a scientific theory, or at least of an infantile theory.

"Revoke," "trump," "finesse," "crossruff," etc. belong to the total parlance-system of bridge. You cannot learn one of these properly and remain in the dark about all or most of the rest.

Likewise "bishop," "king," "check," "gambit," and "ploy" in chess.

Likewise "pressure," "temperature," "volume," "conductor," "insulator," "charge," "discharge," "wave length," "amplitude," and "frequency" in physics; "ingestion," "digestion," "assimilation," "excretion," and "respiration" in biology. To understand thoroughly any of these terms is to understand, to some extent, all of them.

This is why science is so much more than the chronicle of what lies in front of our noses. The more "phenomenal" a word, the less theoretical it is, the more is it possible to understand it and use it *independently* of the language system in which it figures. Little children do quite well with "cold," "hot," "red," "smooth," "box," "chair," etc. And the wider and more varied their experiences become, the greater the demands put upon the language they are learning to use. By the time explanation, causation, and theorizing have become part of their daily fare, each unit of their speech will have become buttressed and supported in a thousand different ways with other units of their speech. Questions about the nature of causation are, to a surprising degree, questions about how certain descriptive expressions couple together with, complement, and entail other descriptive expressions.

In arguing this case, however, I have put all my weight (which is considerable) on the concept of "theory." I have said that "cause-words" are impregnated with theoretical considerations, while "effect-words" are relatively less impregnated with theory. The temporal priority and superior explanatory capacity of causes over effects, and the so-called "necessity" of the connection between causes and effects, have been attributed to the explanatory power and generality of theories and to the regulative control a theory exercises over inferences made about phenomena of which the theory treats.

What, then, is a theory? What does the word "theory" mean? What is it to have a good or partial grasp of physics, chemistry, biology, economics, or philology?

Clearly a theory is not just an assemblage of particular matters of fact. Having a grasp of optical theory is not just having been on the spot when there were rainbows, shadows, and half-submerged sticks to be seen, any more than understanding bridge or chess consists just in possessing records of all the games ever played and all the moves ever made.

The propositions of a theory are *general,* not *particular.* If it embodies mathematical formulae these are most surely algebraical and not merely arithmetical. Understanding Snell's law does not entail our being able to snap back with the angle of refraction the moment we are told that the angle of incidence is 34° 16′. It *does* entail our being able to *find* the angle of refraction by the application of Snell's law. Understanding the political structure of West Germany does not require knowing who is mayor of West Berlin or what the Social Democrats are to do about taxes.

A theory is many-fronted. Theories are not usually just about Venus, or about the planets, or about the planets plus the fixed stars, etc. They are, rather, about such things as these *plus* the behavior of meteorites, comets, pendulums, freely falling bodies, shell trajectories, molecular collisions, etc.

A theory is expected to be *fertile* and *applicable.* It must enable its holder to explain what he finds taking place, and to infer from what he finds happening now to other past, present, or future happenings. But what sort of fertility and applicability is this? Consider:

If I know bridge pretty well, its rules and tactics, I still do not know what cards are being dealt to me. I must look and see. But then, knowing both what cards I am holding and what are the rules of bridge, I also know straightaway what sorts of things I *cannot* do, what sorts of things I had *better* not do, and what (specifically) I *ought* to do with these particular cards.

Similarly, a cook requires not just a recipe and utensils. She requires ingredients too. She cannot produce a cake by merely contemplating the recipe. Only by juggling the ingredients in accordance with the recipe can a cook produce a cake.

Similarly again, a theory does not tell us what *has* happened, *is* happening, or *will* happen. Recipes and theories, by themselves, are neither true nor false. But with a theory and some relevant observations I can tell much more about what I am observing and what I am not observing.

The cook applies the recipe by operating conformably with these purchased ingredients. Her recipe tells her what to do with her ingredients when she has got them.

So, too, a theory tells me what to do with my observations when I have got them. A theory's applicability lies in its capacity to encompass a rather wide range of possible observations. Its fertility lies in its capacity to encompass even more types of observations than were at first thought possible.

Of course, the cook's operations with flour and currants are very different from the astronomer's operations with his telescopic observations. These latter are inferential and explanatory; the former are not.

The nurse in the clinic observes a lump on a patient's hand exuding a whitish fluid. If she is a student nurse (*very much* a student nurse), she might leave it at that, reporting "There is a lump on Mr. Smith's hand out of which a whitish fluid is oozing." But if she is a head nurse, she will report, "There is a contusion that has opened and gone septic on Mr. Smith's hand." "Septic," with its connected concept of "pus," "bacteria," and "antibody," serves as part of the theory-recipe which leads the head nurse to lance and drain the infected area and then apply, say, penicillin.

All of which leads up to the thesis these three chapters have been meant to secure:

We *explain* special matters of fact by saying what are their causes. But knowing that two events regularly occur together does *not explain* either of them. Nor need the fact that two events are mutually inferrible explain either of them. So neither "regularity" nor "mutual inferribility" will constitute all that we mean by the word "cause," for neither of these would count for very much in *explaining* phenomena. *When we use the word "cause" we bring a causal law and its associated theory to bear on a particular matter of fact so as to explain the latter by the former.*

If this thesis is tenable, then it must be clear that questions about the nature of causality are not, save in a secondary sense, questions about the world. They are, rather, questions about how certain laws (set into certain theories) guarantee inferences from impregnated propositions to their consequences.

How these laws and theories are established in their turn is another question, and in any case it is not a question about the nature of causation. More generally, questions about laws and theories need not be dependent upon questions about the nature of causation. But questions about causation are very much dependent upon other parallel questions about laws and theories.

If this thesis is tenable, the causal-chain picture goes up in smoke. For the relation of theories to laws to causal propositions is not the relation of one link in a chain to another, or one rung on a ladder to another. It is much more than that and much less than that.

It is more than that in so far as it is often a subtle and intricate task to determine whether a certain law is binding on a particular proposition describing a natural event. And it is often an expert's task, one requiring perhaps a biochemist and not an astrophysicist, or a spectroscopist and not a geneticist, and in any case not a philosopher. It is less than what the causal-chain picture suggests in so far as there is probably no natural analogy of the chain or ladder variety that will help very much with what is so largely a logical question, namely, how clusters of consequence-propositions are related within a theory to a few powerful and general propositions about the nature of things.

And if this thesis is tenable, that subversive entity "necessary connection" will have been exposed for what it is. Causes are necessarily connected to their effects, not because the world is held together by a kind of cosmic glue, but because the very notions of "cause" and "effect" are intelligible only against the background of a comprehensive theory of some sort, one which guarantees inferences from cause to effect. It is this logical guarantee that theories place upon causal inferences that explains the difference between truly causal sequences and mere coincidences. There is no necessary connection between Tommy's litmus-paper experiment and the eclipse in Southport because there are no theory and no cluster of laws to guarantee an inference from one to the other. And for the same reason there is no causal connection between my winding of the clock and then going to sleep, though no two events could be related with a more monotonous regularity. But to say that, because there is a necessary connection between the heating of sucrose in mineral acid and the resultant issue of glucose and fructose, the world must be pervaded by a kind of ineffable mucilage, like the ether in its omnipresence but unlike it, perhaps, in being even more diaphanous is to utter a colorful but palpable *non sequitur*. Causal connections are not investigable in the ways in which the insides of molecules, motor cars, and monkeys are investigable. They are not even investigable in the ways in which the presence of ether was investigable. Investigating causal connections is much more like investigating the logical character of a deductive system. I said it was *like* doing this. I did not say it was doing this. It is an investigation in which figure questions like: "Does it follow?" "Is there a rule governing this move?" "Is there a more economical way of making this step?" You will agree that such questions are not at all usual with men busy probing the insides of molecules, motor cars, and monkeys.

Some readers may be disturbed by my statement that theories guarantee inferences from cause to effect. For I have referred continually to cause-*words* and effect-*words*. Can inferences be from word to word? It is customary to say that inferences are from proposition to proposition. From the words "red" and "table" I can infer no more than I could from a grunt or a thump. But from the proposition "This is red" I can infer "This is colored." And from the proposition "That is a table" we can infer "That is an object." So from words in isolation nothing follows: It would be like turning through the blank pages of a 1964 diary and finding the word "red" in the place reserved for the engagements of October 27.

But suppose someone burst into the room shouting "Fire!" Or suppose that the quiet of our stroll through the wood is broken with the cry "Wolf!" Here are single words from which a very great deal can be inferred, indeed, *had better* be inferred. Notice, however, that the words occur in a very definite context. The propositional force of the words "Fire!" and "Wolf!" *in these contexts* is this: "There is a fire in this building; look to your own safety!" and "A wolf is prowling this wood; arm yourself or seek shelter!" The words "Fire!" and "Wolf!" are really a kind of verbal shorthand for complete propositions the exact nature of which is perfectly clear from the contexts in which they occur.

Now, many words behave in just this way, i.e., function as propositions *within a context*. The physician, after moving his stethoscope systematically over a patient's thoracic region, exclaims, "Valvular lesion," and his nurse understands him to have made a perfectly intelligible assertion. The chemist who knowingly labels a bottle of water "inflammable" is held to have told a lie, i.e., to have asserted a proposition he knew to be false. Such words as "lesion" and "inflammable" in such contexts do the service of complete propositions, and propositions are the stuff of inferences.

In a similar way, and I do not mean in the *same* way, "cause" words have an *almost* propositional force. The word "mainspring" occurring in a horological context, or "crystalline" in a physical context, or "dextro-rotatory" in a chemical context, or "oogenetic" in a biological context (clearly the list could be made infinitely long) have a logical force very like the words "Fire!" and "Wolf!" in the contexts mentioned. If a man shouts "Fire!" and then follows that up with "Run for your life!" you may feel inclined to reply, "What else?" Part of the propositional force of "Fire!" in the context mentioned is that he who hesitates is lost. Hence the added pronouncement of "Run for your life!" has a kind of compulsive obviousness to it that makes one feel that no normal person could hear and understand the shout "Fire!" and fail to run for his life.

Effect-words stand in much the same relation to cause-words. When we hear that the clock hands are moved by the mainspring, we feel this is obvious enough if we understand the word "mainspring" uttered in an appropriate context. We may even feel that the clock hands, being what they are, must *necessarily* be moved by the mainspring, it being what it is. This, I venture to say, is the whole story about necessary connection. Effect-words are compressed propositions or, better, quasi-propositions. In a specific context, and by way of a specific theory, they can be inferred from other quasi-propositions that are compressed into cause-words. It is in this way that the cry "Fire!" is bound to the word "destruction."

And in a similar way, I do not mean in the *same* way, the word "fire" in some contexts would be bound to the word "heat," in other contexts to the word "smoke," in other contexts to the word "melt," in others to the word "perish," etc.

Cause and effect talk, therefore, is always talk within a definite context. It ought not to be considered just *in general*.

Finally, if this thesis is tenable, the notion of a cause as a push from behind must be untenable. A cause is no more a push from behind than a premise is a kick from above. An effect is no more the result of a push than a theorem is the result of a kick.

Remember, the first scientific theory really to find its feet was physics, including astronomy, optics, and dynamics. The laws of this science apply to what is animate just as well as to what is inanimate. Early causal explanations in terms of impact, attraction, momentum—in short, pushes and pulls—led to the belief (one which is surprisingly persistent and long lived) that *all* causes are impacts, attractions, pushes, or pulls, i.e., that all science is, sooner or later, physics.

There is absolutely nothing in the nature of explanatory theory to justify this belief, and I urge you to take care before humming old platitudes like "But of course meteorology, economics, psychology, and sociology are not *really* sciences." Because unless you equate the term "science" with a "push-or-pull science," an equation for which I can see no grounds whatever, your remarks will be false.

For those who have been trained to think in terms of dichotomies and trichotomies, for whom nothing is intelligible unless it is sorted into logical pigeonholes, I ought to add this: I have argued that the locus of causal-talk is not in the physical world. There is nothing we can see, touch, or kick in nature that will answer to the name "causal connection." That being said, some will straightaway bark, "Oh, his is a *subjective* theory of causality." To which I would reply, "Absolute nonsense."

There is nothing in the world we can kick that goes by the name "fact" or "true statement," either. Have you ever kicked a fact? A true statement? Of course not. But this is hardly sufficient to say of facts and true statements that they are subjective, or somehow not objective.

Facts, true statements, and causal connections are all what they are because the world is what it is. Were the world different, our ideas about facts, truth, and causality would be profoundly different. But this is not to say that facts, truth, and causality are somehow built into the world like great pieces of terrestrial furniture. They are, as I have urged, to a large degree built into the structure of our language. They are not for this reason, however, subjective, chimerical notions.

What I am getting at is this: When we say that A causes B it is because A and B are the kinds of things they are. Sometimes very patient inquiry is required to decide just what kinds of things A and B are, so there is nothing grossly subjective in our ultimate decision. But beyond this, when we say that A causes B, we say *that* because "A" is a word that is, as I have said, impregnated with logical considerations. It is, as I have urged, *theory-laden, rule-governed,* and anything but neutral.

Thus when someone offers me the choice, "Is causality an objective or a subjective relation; is its locus 'out there' in the world or 'in here' in our minds?" I think it best to refuse to take the question seriously. For I doubt that I have offered any *theory* of causality that is easily discussible in this pompous language of "subjective" or "objective." I have tried only to suggest what each and every one of us means by using the word "cause" in the contexts in which we use it.

In brief, causality has to do with the nature of arguments, but there is nothing subjective about arguing validly to true conclusions: This very much depends on the way the world is.

19 | The Scientists' Toolbox

IN THE last three chapters I have been concerned to explore the concept of causality. In the next three chapters I wish to conclude this section by considering a set of related puzzles about the character of scientific laws and our methods of representing and conceiving nature.

The sciences are *conglomerations* of special languages and notations. There are many different ways of expressing truths about a particular phenomenon. These need not all be parts of one grand common language. Not every method of expressing or describing a natural phenomenon need itself be a piece of some enormously complicated linguistic jigsaw puzzle which, could we but find all the missing pieces, would give us a complete account of nature.

There are, in fact, many quite different ways of expressing, describing, or picturing phenomena, and these often stand side by side and independent in any given science. Our way of talking about some given local problem *might* one day fit together with our ways of talking about other local problems. But it is just as likely—perhaps more likely—that it will not. We are best advised, I think, to try to understand each new problem in science in its unique, particular setting, rather than just assume it to be a phase of some other problem. Should several local languages conflate to form one powerful language, e.g., Maxwell's equations, we would surely have cause to rejoice, for this is far more the exception than the rule. To have learned this much about science is to have quit the search for the philosopher's stone which, when struck, will blend all our symbolisms, descriptions, and diagrams into parts of a common linguistic structure.

In science we choose the language appropriate to our immediate needs; comprehensive, systematic elegance—though not unimportant—is but a secondary consideration. It is this feature of science that I accuse the "one-language" theorists (of whom incidentally, there is no shortage) of failing to observe.

Consider any two alternative methods of representing the same phenomenon. What is common to these methods is like what is common to two photographs of the same scene, two photographs which differ greatly in their fineness of mesh and texture. Suppose that one of them is a wire photo or radio photograph of the British atomic explosion at Monte Bello as this appeared in newspapers of the day. On a close inspection, we would have seen that the picture was composed of dots only. Some of these would be dark and large. Others would be faint and small. The composition as a whole might be compared to an impressionist painting of the turn of the century, perhaps by an exponent of *"pointillisme."*

The other photograph of the explosion (the one with which we will compare the wire photo) was brought to us by air transport. It is a very fine-grained exposure, with no dots or blobs anywhere, at least not so far as the unaided eye can detect, but just a smooth series of dark and grey and white areas. This photograph does not compare to a canvas which has been stippled with color by the repeated dabs of a fine brush point. Here we have what is more like a canvas covered with long, smooth strokes of a fairly thick and flat brush. (Of course, under high magnification this fine-grain photograph would also look like a crude wire photo, a collection of dots, but this is a complication unnecessary for the simple point I wish to make.) Thus these two pictures have meshes which differ greatly in size. But as long as the wire photo mesh is not so enlarged that we are prevented from recognizing it as a picture of the Monte Bello explosion, what we have are two pictures of the same event rendered by different methods of presentation. We might very well wonder at the motives anyone might have for denying this fact, for saying that the wire photo was *not* a picture of the explosion or, if he granted that it was such a picture, for saying that it was an incorrect picture.

There are those who, knowing something of modern physics, dismiss

Newtonian mechanics with a snap of the fingers. I suggest that such people are doing just what is done by the man who argues that the wire photo is not a photograph at all, or at best an incorrect one. They are confusing the *purposes* of what are now seen as two distinct theoretic languages or methods of representation. It is no longer a question of Newton's laws being wrong and Einstein's laws being right. Yet how many physicists, and especially students of physics, continue to pronounce their verdicts in just this manner? While there is something very true in the claim, as every scientist knows, there is also something quite misleading.

Imagine for a moment that the laboratory is an elaborate photographic studio. Our apparatus consists of cameras, film, lights, etc. Very often, with the equipment we have, our experiments will not yield pictures of grain suitable to represent what the relativist or the quantum theorist claims to be the case. The grain may be too fine or too coarse. And for these cases the Newtonian picture is quite exact.

Indeed, with respect to such medium-sized experiments, what precise meaning could be attached to the expression "more exact than Newton's laws"? The quantum theorist may run his deductions out to 15 decimal places, if he pleases. But this precise determination will make no difference whatever to our usual macro-physical experimentation. The famous uncertainty principle of quantum mechanics, when translated into a calculation involving a 1 cm., 10 gm. bullet moving at 1000 ft./sec., gives an uncertainty in position of 2×10^{-28} cm.—a totally unmeasurable quantity. Clearly quantum mechanics provides us with no method of estimating the uncertainty in the position and momentum of a bullet that is more exact than that of Newtonian mechanics. Moreover, because of its greater simplicity, the Newtonian formulation is greatly preferable to relativistic quantum-theoretic laws (even supposing that these *could* be adjusted to the requirements of the molar experiments being considered). Only when our experiments absolutely require more refined representation do we place the Newtonian formulation to one side. But then there is no question of appealing to the more *correct* theoretical language, the truer method of representation. Newtonian mechanics is simply *inappropriate* to the representation of relativistic and quantum phenomena. By exactly this same token, relativity theory and quantum mechanics are inappropriate to the representation of a good deal of the macro-physical world. (We do not distemper walls with water-color brushes, nor do we repair watches with sledge hammers.)

The fact is that in science we choose the particular notation and language we regard as adequate to the purpose in mind, just as a carpenter uses a saw on one occasion and a plane on another, each to give some desirable shape to his wood. Who expects a carpenter to use jeweler's tools when building a fence? Yet this is the absurd counsel implicit in some of the overenthusiastic accounts given of the spectacular progress of recent science. A sense of proportion must be kept.

Consider the multiplicity of methods in our study of electricity. When we use a direct current circuit, we do not worry about its inductive or capacitive properties (unless we are concerned with the effects of transients on the line). Even when concerned with transients we do not invoke the *entire* Maxwellian theory, wherein we must know the geometry of all the conductors and insulators in space. Instead, we make a schematic diagram of the circuit, inserting resistance, inductance, and capacity, and then we apply Kirchhoff's laws. Only when the whole process is a very brief one indeed do we abandon such quasi-static conceptions of the electric network and describe it with Maxwell's equations.

Analogies of this type are always instructive. They help one to learn new theories; they give one a *method* for altering one's technique of representation. This statement may be taken as another dig at those scientists and mathematicians who would banish analogy altogether. Their case is overstated. An analogy used cautiously, as the mathematical one just described *must* be used, helps one to surmount the difficulty you and I often experience in getting acquainted with a new method of representation. Once we see how a new method resembles a method with which we are already familiar, we accept the novelties provisionally. The new method is then allowed to justify itself by its successes, if it is able. In the course of time, the details of the analogy itself tend to be forgotten in our actual practice with the method. Only when we set out to teach others the new technique does the analogy strike home again.

Some of the most important analogies in science are expressed mathematically. We note similarity in the form of two sets of equations and straightaway inquire whether the operations we are accustomed to use on the familiar set will work also on the less familiar set. It was in this manner that Heaviside's operational calculus won the cautious support of science. And in biology one might almost trace progress by considering the ways in which certain analogies have led research to new discoveries. The hydraulic model of the nerve network opened up avenues of inquiry for many years until, when pressed too hard, it broke down. Then followed the electrical conduit model, then the electrochemical model. And the telephone switchboard model of the brain, while a little quaint these days, did yeoman service in its time. Consider too the camera model of the eye and the photographic plate model of the retina. The importance of these analogies cannot be overlooked, even by someone who is interested in showing where they misled people.

Again, there is no question here of right or wrong, correct or incorrect. We may characterize *certain* competitive theories about the nature of the world in this way, e.g., the planetary theories of Ptolemy and Copernicus (though even in such cases the matter is not so simple, and the choice between planetary theories was especially difficult when it had to be made before the time of the telescope). But the "rivalry" between the theories of Newton and Einstein is not a question of being right or wrong, at least not always. So, too, with the issue between classical and quantum mechanics, which, moreover, represents more than the mere choice between crudity and refinement. The carpenter who refuses to use jeweler's tools when building a fence is not showing himself to lack care and skill. He is, rather, exhibiting a keen sense of which tools are appropriate to what tasks.

So, too, the microbiologist who has learned through bitter experience the advantages and disadvantages of using high power optical and electron microscopes on delicate specimens of the order of size of 1 mu.

Nothing reveals the importance of using the right tools more dramatically than the bringing together of particle and wave ideas with regard to the electron, and it is these ideas which I will try to review now in, I hope, a fairly intelligible manner. Here is a classical instance of the piecemeal development of scientific theories. For these are two apparently incompatible methods of representation. They have jockeyed and jostled themselves so much that peace could be reinstated only by wedding the two into a single, complicated method of representing the experimental data of electron physics, which is surely the neatest trick of the century.

On the one hand is the dynamics of a particle, on the other, the theory of wave optics. With particles, the essential notion is of something at a particular place at some particular time. But what is most characteristic of wave motion is that it is not localized like a particle. Wave properties derive from a disturbance extended through space over a period of time which is long as compared with the period of vibration.

However, by an ingenious, and mathematically elegant, exploitation of the ideas of constructive and destructive interference, theoretical physicists like De Broglie, Schroedinger, Heisenberg, and Born were able to conjure up and develop the notion of a wave packet. A wave packet (or wave group or wave pulse) is a quite specific cluster of interference maxima of a great range of different yet interfering waves. By this mathematical model most of the experimental behavior of the electron (e.g., the diffraction patterns of electron beams and the deflection of such a beam by the transverse magnetic field) can be described, explained, and predicted with great accuracy.

It is this conception of a localized yet wavelike propagation of energy which gives us our logical connection between the two methods of representing the electron, as a wave and as a particle. In a picture of coarse grain, both methods give approximately the same picture. Looked at in this way, the situation is not troubling, and the mathematics will grind out our answers for us with delightful efficiency.

Should we do what our teachers will probably advise us not to do, however, viz., *stop and think of the moving electron itself,* a kind of logical and grammatical difficulty arises. In terms of one method of representation, we want to say, "The electron is a particle." In terms of the other representation, we want to say, "The electron is a wave." We may then be perplexed. And we ought to be perplexed, for the two statements do not fit together. One statement appears to postulate properties of the electron which contradict what is asserted by the other statement. It took a lot of hard work on the part of our science teachers at school to get us to appreciate this discrepancy between wave behavior and particle behavior, so why shouldn't it seem paradoxical to wrap them together as electron behavior?

You may have heard of the "principle of complementarity." (In fact the principle is a subtle and complicated affair, so we will not get too adventurous with it.) This principle of complementarity appears to grant us permission to use either expression as we see fit. It appears to license us to say either "The electron is a particle" or "The electron is a wave," as we please. What it really does, however, is to grant permission to use whichever aspect of the quantum mechanical theory which is appropriate to our immediate laboratory needs. The principle does not permit physicists to say what they please, according to preference or taste. It permits them, rather, to employ whatever mathematical features of the theory will be most helpful to the expression of their local experimental problems.

But surely we are up against the same situation in ordinary dynamics. Do we not represent the earth at one time as a particle without extension and at another time as a body extended throughout a sphere 8000 miles in diameter? In the dynamics of a gas we need no principle of complementarity to represent the gas at times as a continuous medium for acoustical problems and at other times as an assembly of discrete molecules. Why should we require an official dispensation to proceed in exactly the same way in atomic physics? That the need for such a license is felt here as it is not felt in other parts of science indicates the presence of a conceptual perplexity.

In the older parts of physics we have no difficulty whatever in changing over from one method of representation to another when it is appropriate to do so. Physicists are accustomed to this as an everyday affair. Perhaps it is that some of us have difficulty with the two expressions "e is a wave" and "e is a particle" because we naturally incline to the feeling that if one of them is correct the other must be incorrect. The principle of complementarity does not say that both expressions are correct, or that neither of them is correct. It suggests that either way of talking can be *appropriate* to certain experimental situations and inappropriate to others. It all depends on what we are trying to do with such language, what we are trying to say. The wonderful mathematics of quantum theory puts both methods of representation at our disposal by bringing them together in a formally and experimentally acceptable way. The principle of complementarity gives us freedom to adjust our mathematics to our particular experiments in the most appropriate way.

Thus the grammatical form of the sentence "The electron is a wave" can be a source of perplexity. The word "electron" is a substantive. It serves as the grammatical subject in many sentences. It is the "thing-word" in many noun clauses. In this it is like the name of many *things* in science: "star," "planet," "mountain," "plant," "cell," "molecule," etc. All of these can serve regularly as the grammatical subjects of sentences or of noun clauses.

Consider the sentences "The electron is an angular velocity" and

"The electron is an aperiodic motion." Both these statements are just *bad grammar*. We may say of some rigid body that it *has* an angular velocity. We may speak of the aperiodic motion *of* a galvanometer coil. But we never say that some rigid body *is* its angular velocity, any more than we would say that a galvanometer coil *is* its aperiodic motion.

Wave motion too is a *process*. Wave motion is not a thing, not an object. Thus an electron regarded as an object (as we very often do regard it) is not a process. Its *translation* is a process.

How about this, then: "The translation of an electron is a wave"? But a translation is not a wave. This will not do, either.

We could, however, say this: "The translation of an electron is the translation of a group of waves." We could say this without implying that the electron *is* a group of waves. All this means, however, is that *the laws we use for calculating the translation of an electron are just the laws we use for calculating the translation of a group of waves in a particular system.*

This is harmless enough but is hardly obvious from the grammatical form of the sentence, "The electron is a wave."

One must be vigilant at all times, in philosophy and in these philosophical pronouncements about and within science, to distinguish the *grammatical* form from the *logical* form of a proposition. (Remember the sentence "A coelacanth exists": It is a sentence which cries out for this distinction to be made. Grammatically, the sentence is on a par with sentences like "A coelacanth swims," "A coelacanth eats," "A coelacanth breathes." We might even be tempted to suppose that *exists* is something a coelacanth does, or that existence is a property of the coelacanth, as are the properties of being a swimmer, being an eater, or being a breather, etc. But it is clear that the sentence "A coelacanth exists" has no more logical force than "*There* is a coelacanth." Its existence is not one of the properties recorded in Professor Smith's notebooks devoted to the description of this rare fish. Likewise, the grammatical form of the expression "The electron is a wave" must be distinguished from the logical force of that expression, this logical force being conveyed, rather, by the proposition "The laws we use for calculating the translation of an electron are just the laws we use for calculating the translation of a group of waves in a particular system.")

So whenever one is clearly aware that he is using one form of language, *as opposed to another,* he is likely to be careful not to make the mistake of mixing up the grammar and the logic of the two systems. Or, should he be trapped by such logical difficulties, he might be able to extricate himself. He might, by proper consideration of the logical features of opposed types of notation, analyze his difficulties, and disentangle the grammar of one method from that of another. This is why getting clear about the logic of propositions in science, even when that logic is fairly obvious, as with measurements of weights, volumes, temperatures, and densities, is worth doing. For the logic of scientific propositions will not always be obvious, especially in the conduct of research.

In the growth of a subject, at its frontiers, some confusion is inevitable, indeed, it is highly desirable. Students of science will appeal to one of several ways of relieving themselves of their perplexities. Some will be happy to adopt the literary *tour de force* of an Eddington, a Jeans, or a Huxley. Others are at peace with themselves when they just have learned to manage the methods of symbolic calculation, usually in a wholly uncritical and unquestioning spirit; these are the tough-and-no-nonsense scientists with whom Professor Popper has such quarrels. Others will work hard to develop such a familiarity with the difficulty that it will soon just cease to bother them (much as the college senior has become immune to cafeteria food). They might come ultimately to ignore it altogether. Or perhaps they will formulate a "policy" or a "principle" with which to bridge the gap between opposed points of view.

None of these dodges, however, will appeal to the man who wishes an elucidation of the grammatical or logical situation which gave rise to his difficulty in the first place. He will refuse to be bullied by the thought that someone must already have supplied the analysis of that situation, or that questions of the sort we are raising are too simple for the attention of really accomplished scientists. If the questions are so simple, then the answers ought to be simple, and they ought to be given early in Physics I or Biology I, so as to avoid later confusions. And the reply to this is that the answers are not given.

Logical and philosophical questions are never simple. They must be pursued tenaciously and in a wholly critical spirit.

The necessity for tenacious, critical study of logical and philosophical questions is brought about by the fact that the sciences are collections, often untidy collections, of distinct notations, languages, and ways of representing. This is only the natural effect of our choosing just those methods of representation which are appropriate to our immediate scientific needs. The difficulties which arise are very like those of a nearsighted carpenter who grasps for a hammer and picks up a wrench instead. Doubtless, he will have problems when he tries to make a wrench do a hammer's work, problems that are not overcome by his being told that, after all, wrenches and hammers are both tools and hence not *really* different. The language of science is a toolbox.

Some students, recognizing that there is this multiplicity of techniques of representation in science, proceed to invent a scale of methods as follows: They consider methods having a coarse grain (in the sense discussed) to be but *approximations* to those having finer grain. Methods which represent nature in *greater* detail are taken to be *better,* scientifically more valuable, than those which are less detailed. These students arrange methods in a series according to the degree of fineness with which they can present pictures of phenomena. From this point of view, of course, the last member is considered the best. And should this last method in the series have enjoyed a general success in its use, e.g., quantum mechanics or electron microscopy, then the temptation arises to regard it as *the best possible of all methods.* All reference to the series of methods is thus forgotten.

The expression "laws of nature" might have been based originally on an analogy with the laws a ruler imposes upon his subjects. The only difference is that in science, when we encounter what ought to correspond to disobedience in a subject, namely a phenomenon which does not obey the law, we suppose that we did not really know the *correct* law of nature in the first place. Ignorance is our excuse, whereas when it is a matter of disobedience to a ruler we say ignorance is no excuse. Or, we might regard the aberrant phenomenon as a privileged exception to the law, governed by some special rule, as if the ruler had granted an indulgence or special dispensation.

By this token, so long as no experimental fact breaks the law, we are sure that we know the law of nature and are tempted to regard an infringement of the law as impossible. It is impossible, we may say, that sucrose (when heated with dilute mineral acids) should *not* take up water and be converted into equal parts of glucose and fructose. It is impossible that hydrogen should fail to produce water vapor on combustion.

There is no need to labor the point. Every scientist knows of instances where the confident theorist has gone wrong in his predictions.

Still, again and again theorists will claim that a particular experimental result must be wrong on *theoretical grounds only*. Only experimental scientists are competent to judge whether a result is experimentally correct, and this they do *after* having investigated the matter in the laboratory. If the experimental evidence is shown to be sound *by experimental standards,* then we may have to say that the law has failed, and we will change the law, irrespective of the ingenuity and mathematical elegance of the theory that supports it.

However, if a law of nature can be *altered* in this manner, what is the source of its authority? We have been very seriously misled by our analogy if we think that the correct law for which we are groping in our experiments is the law imposed on the world. *Must* phenomena be governed by such a law? Just what is the meaning of "the correct law" in this connection? Can we ever be sure that one hundred years hence some given law will be as it is today in every detail, unaltered in any respect?

The expression "the correct law of nature" is an improper one from a logical point of view. How can we establish the truth (or the falsity) of a statement employing an expression of this form? What would count as evidence one way or the other? If these questions cannot be answered, then we have given no meaning to the statement in the first place. And this is precisely what I am advocating.

Impressed by the fact that the "laws of nature" *are* subject to change, some writers have gone too far the other way. They would dismiss such laws as those of the conservation of energy and of momentum as "mere conventions." They are but a convenient yet quite arbitrary shorthand for expressing certain self-imposed regulations upon our study of nature. This view, it seems to me, overlooks entirely the very difficult processes by which physical scientists came to these ideas. So, while I would not have laws of nature represented as immutable, neither do I embrace the view that they are mere conventions.

Before offering you a suggestion about how these laws may be represented, let me pull together the threads of the foregoing discussion:

(a) There are many methods of representing a given natural phenomenon. There is as little good sense in denying that statement as there is in denying that we have many different sorts of tools for doing some single job. It is startling that one should even *have* to deny that the language of science is a wholly elegant and logically homogeneous mixture.

(b) But, once recognizing this multiplicity, we are sometimes tempted to set our representational techniques into an order. At the pinnacle we place those methods that depict a scientific happening in *great detail,* and to the bottom we relegate those that are only *approximations* to such highly detailed accounts. We might even go so far as to say of the method that had achieved the greatest detail that it was the *best* method, perhaps even *"the* correct law of nature." This, however, is like a ruling that if we really want to build our fences and garages properly, we must all fit ourselves with jeweler's eyepieces and watchmaker's tools. The recommendation (and the language in which it is expressed) should be scrutinized carefully, and tortured logically.

Thus I would summarize, roughly, what this chapter has tried to state so far. I turn now to a suggestion I would like you to consider. Suppose a man first stated baldly that: *The laws of mechanics are the laws of the notation by which we represent mechanical phenomena.* Since we consciously choose a method of representation when we describe the world, it cannot be that the laws of our notation say anything about the world.

Permit an analogy: When we choose to represent the surface of the earth upon a flat sheet of paper we must project the globe into a place.

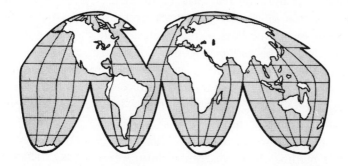

We may use a sinusoidal projection, an orthogonal projection, a homolographic projection, a Mercator projection, a Lambert azimuthal projection, etc. Each one of these is governed by entirely different formal rules for projection, interpretation, and correction. There are the "laws" of cartographic projection. The world is a sphere, or an oblate spheroid at least. We never see it sinusoidally, orthogonally, or homolographically, we just represent that world this way. Likewise we never see the mechanical world all differentiated and equated and labeled with mathematical names. Nor is that world full of point particles, geometrically straight lines, and space-time axes, but we surely do represent it in that way. The rules for our techniques of representation are *built into* all the formulae and symbolism of mechanics. That the world can be represented in this way is surely an important fact about the world, but are our laws of mechanics facts about the world? Do not just answer, "Obviously yes" or "Obviously no." *Think* about it. For perhaps the laws of mechanics stand to the specific formulae and sets of formulae in mechanics as the law of a cartographic projection (Mercator's or Lambert's) stands to the specific delineations and contours we make on our two-dimensional map. Our rules of projection control what lines it is permissible to draw on the plane. Our rules of mechanics control what formulae it is permissible to construct as representing phenomena and, in a way, what phenomena are representable by our formulae.

The authority which we have tended to attribute to our laws of nature (in one way) ought perhaps to be attributed (in another way) to the logic of our notation.

In short, if we wish to make pictures of the world according to a particular scheme, then we *must* follow the rules of that scheme.

This is not to say, as no one would dream of saying, that the scheme determines what must be the form of the actual pictures we do draw. But it does decide what sorts of pictures *are possible*. In a very important and analogous way, the laws of mechanics, while they do not determine what values are to be assigned to the variables in a specific dynamical problem, *do* determine what form the problem is to take. They determine what forms are permissible for propositions within the problem. And, of course, they determine which phenomena shall be noted as appropriate to mechanical representation.

Perhaps what we have called "the laws of nature" are only the laws of our methods of representing nature. Perhaps laws show nothing about the natural world. But it does show something about the world that we have found by experience how accurate pictures of the world (of a certain degree of fineness and/or simplicity) can be made with the methods we have learned to use.

We *can* describe the action on a cinema screen as though it were continuous. But to do this we must issue a rule to the effect that observations are not to be allowed to get so refined as to show up the actually discrete pictures. And could not this rule be expressed as a *law*, a *regulation?*

We *can* describe the motion of sub-atomic particles by classical dynamics provided that the dynamically significant fields of force in which they move are not given structure on too small a spatio-temporal scale. And do we not express this as a rule or a law about how far our classical theory can be pressed?

We *can* describe the motion of the planets around the sun over a few years without introducing relativity theory, provided that . . . and so on.

In all these cases, the structure of the physical situations being represented is shown by virtue of the appropriateness or success of a particular notation or technique of representation. This way is, so to speak, a reflection of the rules of that method of representing the world. The rules of the language of science cast shadows on the natural world. And it is the regularities in these shadows that we call "laws of nature." The real force of this oracular pronouncement can be appreciated when one considers how the facts of science are always expressed, and are always expressible, *in language* (whether mathematical or otherwise). The expression "an unexpressible fact of nature" is wholly vacuous for science. It is provocative to consider how we hunt first for those things in nature about which our language allows us to form questions. Medieval science floundered largely because it consisted of a search for those things in the world that corresponded to the subjects and predicates of medieval Latin grammar. *Our* science does *not* flounder, apparently. But this does not mean that it has slipped the linguistic yoke; this it quite clearly has *not* done. Our language, our techniques of representation, screen the world in advance: The very conception of "a meaningful question" is fundamental to our attack on the mysteries of nature. And it is a logical or a linguistic conception. It is all this that is meant by saying that language throws shadows on the world, and that laws of nature are regularities in those shadows.

Next I will try to get a little light into the umbrous corners of this thesis.

20 | Laws, Truths, and Hypotheses

OUR LAST chapter concluded with the suggestion that at least some of our laws of nature are the laws of our method of representing nature. The laws of mechanics, thus, are just the laws of the methods by the use of which we represent mechanical phenomena. Or, to put it another way, the laws of mechanics are the laws which relate our mechanical *concepts*. To this was appended the even more alarming thesis that, all this being so, the laws of mechanics are not facts about the world at all. Laws of nature are not facts about nature. Clearly, this needs support.

Are not laws of nature sometimes true and sometimes false? What makes them true or false, if not nature itself?

I will try to persuade you that epithets like "true" or "false," and even "probable" or "improbable," are not appropriate to laws of nature. The reason? These laws are not just generalizations like "All cows are herbivorous." Yet, on some accounts that are given, we are invited to believe that there is no logical difference at all between a *law of* nature and a *generalization about* nature.

First, then, concerning the "truth" of natural laws: Take Snell's law. It is stated thus: A ray of light incident at the surface separating two media is bent so that the ratio of the sine of the angle of incidence to the sine of the angle of refraction is constant for those two media.

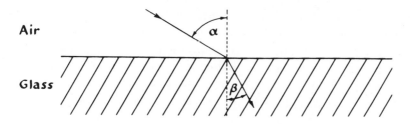

In other words, there is a simple relation between $\angle \alpha$ and $\angle \beta$ which is fixed for any two substances.

Now what marks this statement off from the statement that cows are herbivorous?

Notice that Snell's law is stated by use of the expression "ray of light." Hence the law can be given a physical meaning only where the expression "ray of light" is intelligible, that is, within geometrical optics. Where optical phenomena elude the powers of geometrical optics to explain them, Snell's law does not even have an interpretation.

Again, as Stephen Toulmin has pointed out, a statement of Snell's law and a statement about the range of circumstances in which Snell's law holds are two different types of statement.[1] Consider: "Most transparent substances have been found to reflect light in such and such a way." This is not what we call "Snell's law." Such a statement just reports the facts. It tells us when and in what circumstances Snell's law holds or fails to hold. It is likely to mention also that certain crystalline materials (like Iceland spar) do not behave as the law suggests. There are circumstances in which, when Iceland spar is the refractive medium, Snell's law does not hold.

To every law of nature there corresponds some such statement or set of statements. These indicate what Toulmin calls the "scope" of the law.[2] A great deal of research is required to discover just what is the *scope* of a law. Assertions about what is that scope, therefore, are *true* or *false*, or *probable* or *improbable*.

Laws of nature, however, are very unlike statements of scope in precisely this respect. Suppose our refracting medium is *glass* in the

---

[1] Stephen Toulmin, *The Philosophy of Science, op. cit.*, p. 78ff. (Hanson's argument here is a close paraphrase of Toulmin's on pp. 78–80.—Ed.)
[2] *Ibid.*

first instance. In terms of experiments with just this one type of medium, we may consider in but a tentative way, as an hypothesis, the formula offered by Snell. For Snell's formulation is not restricted with respect to the transparent substances it asserts will refract light in the specified way. As an hypothesis the formula will guide us in future experiments, and we may very well be heard to ask, "Is Snell's hypothesis true or false?" When we ask that question, however, we mean, "Have any limitations been found for Snell's formula?"

Once the hypothesis becomes established (when we have come to treat it as a law), we no longer ask, "Is it true?" We ask now, "When does it hold?" This "shift" indicates that what was once a tentative hypothesis has now become a standard part of, say, optical theory. Exceptions to the law, viz., double refraction and anisotropic refraction, are thought of as anomalies which require an explanation in a way in which ordinary cases of refraction do not.

Earlier we considered how laws of nature differ from the laws a ruler imposes on his subjects. A similarity may now be noticed. Regulations enforced by a ruler are not themselves true or false. Statements about the scope of such regulations are true or false. If it is ruled that alcoholic beverages will not be consumed on Sundays, we do not inquire whether that ruling is true or false. Consider this statement, however: "The ruling about the consumption of spirits on Sundays applies everywhere except in hospitals and certain churches and college rooms." This statement is true or false depending on whether or not the rule has the application cited. Thus we do not ask, "Is the rule true?" but only, "Is this a case where the rule applies?" Only of the statement of scope or application do we ask, "Is it true?"

This is not to deny that scientists seek the truth. It merely shows that the term "truth" is wider than the term "true." Not every sort of statement with which science deals is such that we can speak of it as "true," "false," "probable."

To say that a law holds universally is not to say that it is true always (instead of just under conditions). As Toulmin observes, the contrast "holds-does not hold" is as fundamental as the opposition "true-untrue." [3] The former cannot be reduced to the latter.

Suppose we learn that our optical ray diagrams must be modified

---
[3] *Ibid.*, p. 80.

when temperature is brought into experiments on Snell's law as an operative factor. The concept of *refractive index* thus augments our optical theory. But one thing is clear straightaway. Questions about refractive index are meaningful *only in so far as Snell's law holds;* they have a point only on the assumption that the law is applicable. Snell's law is an indispensable feature of the theoretical background against which the notion of refractive index must be discussed.

Statements offered at one level of scientific discourse draw their meaning only from other statements already offered as established. Scientific disciplines are *stratified* in so far as this is the case.

We must distinguish between the hypothetical and the established parts of science. And we must be alert to the fact that it is this aspect of scientific thinking which is most likely to suffer distortion in philosophical accounts.

Consider the pronouncement of so many philosophers that *all* empirical statements are hypotheses, hypotheses which can (strictly) never be more than *highly probable*. They support this contention by observing that it is a mark of any genuine empirical proposition that we can imagine evidence the discovery of which would tend to disconfirm the proposition. Now I do not deny the logical utility in making such a move whenever we wish to assess the precise meaning of statements made within scientific discourse. But to put the matter in just this way is to pulverize words like "hypothesis" and "hypothetical."

A word which is made applicable to everything rapidly becomes good for nothing. E.g., there is a good, homely distinction that you and I make every day between things that are large and things that are small. Ocean liners, oak trees, and elephants are large. Fleas, grains of sand, and pin holes are small. Suppose that a pedantic microscopist insisted that all these things were *really* very large; the electron microscope makes elephants of fleas, boulders of grains of sand, and craters of pin holes. Suppose, too, that American and English grammarians, impressed by the microscopist's great learning in these matters, jointly issue an edict forbidding the use of the word "small" to describe enormous things like fleas, grains of sand, and pinpoints. "Henceforth," they decree, "these things will all be characterized as 'large,' as in all accuracy they should be."

We must still distinguish between elephants and fleas. But we are no longer allowed to do so in terms of the dichotomy "large-small." Now *both* elephants and fleas are large. Let us call the elephant "Large," with a capital "L," and the flea "large," with a tiny "l."

The grammarians persist, however. Besides microscopists they number certain philosophers among their learned acquaintances. The philosophers assure the grammarians that the only propositions which can be said to be "certain" are the *logically necessary* ones of logic and mathematics. All empirical propositions must remain logically dubitable; their truth can never be established with certainty. Some future discovery may show up what we now take to be true as grossly false.

With this in mind the grammarians rule, "Henceforth the word 'certain' will not be used to characterize empirical propositions; anyone who does so speak will show himself to be unrefined, uneducated, base, common, and popular." Another good contrast, that between "certain" and "uncertain," is lost to us. In truth all empirical propositions are in a sense uncertain, but how now to distinguish what we say when the motor of our car coughs and sputters, namely, "The gas tank is *probably* dry," from what we say when we draw the dry dip-stick out of the tank, namely, "The tank *certainly* is dry"?

Of course we are losing our patience with these linguistic dictators. When they go on to insist that all empirical statements are (in so far) mere hypotheses, we are likely, no, we are *certain*, to urge that they be divested of their professional credentials. Their recommendations have quite lost touch with the actual speech habits of experimental scientists and ordinary men.

In any science it is easy to distinguish between the problems currently being discussed and earlier problems the solutions to which must be taken for granted if we are even to state our current problems. One does not, indeed *cannot,* question the adequacy of Snell's law and continue to talk about refractive index.

This is not an argument to prove that natural science is really less empirical than is usually supposed. It is unquestionably important to see that every statement in a science *could* be called into question. We could always consider what the world might be like were some empirical statement false instead of true. To deny this could be to have failed to learn any important lesson of modern science, and to have opened the door for the dogmatic rationalism of the Cartesians.

Still, it is equally important to see that in any actual experiment most empirical propositions would *not* be called into question; indeed, they *could not* be called into question and leave us with the same experiment, for by questioning some empirical statements we deprive others of their very meaning. The hypothesis of an inquiry is designed to single out just what is being called into question. This indicates, too, what will not be questioned. Every experiment rests on facts. But the facts of an experiment are mentioned in just those empirical propositions which, for the purposes of that inquiry, are assumed without question.

Hence the contention that all empirical propositions are just hypotheses must be scrutinized very carefully. If it means that one could always consider circumstances in which the propositions might be doubted, then it is harmless enough. (Indeed, as such it is a clear tautology, for it asserts only what we *mean* by the expression "empirical proposition." An empirical proposition is a proposition for which we could always imagine circumstances in which it might be disconfirmed.) If, on the other hand, the contention is that *all* the propositions of science are actively being doubted in any and every experiment, i.e., that every proposition of science is now and always an hypothesis, it is a grotesque falsehood. It constitutes a serious invitation for us all to develop ulcers over the question, "Is science possible?" Clearly, as here presented, the possibility of our ever coming to *know* anything through science is the possibility Achilles had of catching the tortoise.

Granted, there are often second thoughts about matters which we thought had been settled before. But putting these right is not like replacing a faulty brick in the top of a wall. It is more like knocking the wall flat and rebuilding again from the bottom. Men who uncritically insist that *all* empirical statements are corrigible make it sound as if the putting right of any faulty stone in the great wall of science were a simple matter of replacement. This is notoriously not the case. The discovery of aberrations in the perihelion of Mercury and of the shift in the position of the fixed stars during an eclipse, the discovery of electron diffraction and radioactivity, required a lot more than a simple bit of taxonomic repair work.

Very few scientists, however, make a full-time job of knocking down the wall of science and repairing it from the ground up. Copernicus, Galileo, Mendel, Darwin, Einstein, and Heisenberg spent their time messing about with foundations. But most experimenters are perched atop the wall coping with the problem of setting in some individual stone. *That is the only way the wall gets built.*

The generalizations of natural history—for example, that no swans are green, that cows are herbivorous, that pigs do not fly—are fairly high up in the structure of the wall. A green swan, a meat-eating cow, or a flying pig will cause changes, and a great deal of astonishment, no doubt. But the whole wall of science will not have to be rebuilt; no more than was this necessary when the duck-billed platypus turned up. And of these generalizations in natural history we may reasonably ask, *"True* or *false?"*

Of full-fledged laws of nature, however, we do not say *true, false,* or *probable*. For this language suggests that the wall of science is quaking and near to collapse beneath the weight of experiment going on at the top. This is surely not the attitude of practicing scientists, and philosophers do themselves discredit by using language which suggests that it is.

Hence the words "established" and "hypothetical" mark a fundamental distinction in science. It is almost wicked to suggest that nothing is ever really established and that all of science is merely hypothetical. While there is a sense in which that is true, there is also this important sense in which it is false.

The statements at the foundation of the wall of science are well and truly established in the best and most respectable sense of that word modern science can supply. It is worth noting that the logicians' standard of *established* is, by definition, out of reach of experimental science. This being so, it is quite an inapplicable standard for any scientific use of the word. The meaning of the word "established" as it occurs in natural science is very different, of course, from the meaning of the word "established" as it occurs in deductive science. But this is only to say that physics, chemistry, and biology differ from mathematics and logic. *We knew that already*. How odd to insist on the point with the mysterious language "All empirical statements are hypotheses." This verdict is, apparently, the result of observing that no empirical propositions are established *in the mathematical sense of "established."* This is nothing if not true. Thence to the conclusion "If empirical propositions are not established in this sense, then they are mere hypotheses"; and this is nothing if not false. For there is an important non-mathematical sense of "established" which is applicable to some but not all empirical propositions. It should be clear that it is *this* use that is standard in science and not the mathematical use, which has no application there.

Statements that are established in empirical science, then, are established in an empirically scientific way. (What hard work it must require to fail to see that!) It is these statements which support and give meaning to the particular and local sort of statements "above" them. In this way Snell's law supports and gives meaning to statements about refractive index.

These foundations-statements are, roughly, of two types: laws of nature and statements about the circumstances in which these laws have been found to hold. Neither type of statement need be treated as "only highly probable" or fairly well-established hypotheses. The laws of nature evoke from us, not the question, "True or false?", but, rather, "When and where?" The statements about the scope of these laws, being simple reports about past fact, are logically dubitable, but about as open to practical doubt as the statement "Japan was at war from 1941 to 1946." Yet both laws and statements of their scope are legitimately called "empirical."

So, while the generalizations of natural history might be termed "highly probable," for after all a pig *might* fly, a swan *might* sprout the plumage of a parrot, a cow *might* turn predatory, it is to carry a point of logic beyond the bounds of common sense to insist on the same treatment for laws of nature.

This, I think, answers our initial question, which was, "How do laws of nature differ from generalizations like 'All cows are herbivorous'?" It answers another question too, if only in the negative, for we see that the distinction between laws and hypotheses is a *logical* matter. It involves far more than the simple considerations of probability theorists, such as the differing degrees of confidence with which we advance laws (as opposed to theories) or the number of confirming instances we have observed.

What now of the distinction between laws and principles? We have seen the difference between hypotheses and laws, hypotheses such as that about the refractive index of some transparent or near-transparent substance and laws like Snell's law. But what of the difference between Snell's law and the principle of the rectilinear propagation of light? And what of the difference between Newton's second law and the principle of uniformity in nature? And what of the difference between Mendel's laws of segregation and independent assortment and the principle of natural selection?

Consider the principle of the rectilinear propagation of light. The principle is at the very heart of the science of geometrical optics. A geometrical optics which had a different law of refraction may be hard enough to imagine. But it is not inconceivable. Considerable changes would be entailed in our ways of thinking about optics, our ways of grinding and assembling lenses, and our ways of coping with elementary problems in astronomy. Yet geometrical optics would still *exist as a subject* despite this quite serious alteration at the foundations. Our thinking, our techniques of lens grinding, and the design of optical instruments would eventually accommodate to the change.

However, the situation in which the principle that light travels in straight lines is challenged is a far different one from the situation in which Snell's law is challenged. For to abandon the principle of the rectilinear propagation of light is to abandon geometrical optics *as a whole*. It is less like tinkering with the foundations of a wall and more like having all its mortar turn to dust. To question the principle of the rectilinear propagation of light is to put the entire discipline called geometrical optics at stake. This principle, therefore, is not *open to falsification in any simple, straightforward way*. If it is unrealistic to insist that laws of nature are just hypotheses (always and forever entertained with suspicion and doubt), it is absurd to say the same things of principles.

Yet these principles are surely empirical. They have not just become empty tautologies or truths by convention or stipulation. In circumstances very unlike those at present, we might decide to give up the principle of rectilinear propagation. And by this I mean not merely that we would *modify* the principle even more than Einstein has invited us to do. I mean circumstances wherein we would question the principle even as applying to our middle-sized world, the world of dimensions from (roughly) 10^{-2} centimenters to 10^{-2} light years. But in these circumstances, understand, it is *all* of geometrical optics which must go overboard. For the discipline draws its strength, its structure, indeed, its very meaning, from the unqualified acceptance of this principle. What these radically different circumstances might be is another interesting (and difficult) question. But the magnitude of the revision required in this case is such as to make the lumping together of hypotheses, laws, and principles with the one label "hypothetical, dubitable propositions" no more than a stroke of noble incompetence on the part of certain commentators on science.

The place of Snell's law is *within* geometrical optics. It holds a fairly fundamental place there, but nonetheless it is still inside the discipline. The principle that light travels in straight lines, however, is *not* within geometrical optics, save in a very special sense. The principle is, rather, *built into* the very geometrical mode of representing optical phenomena. Geometrical optics would not be possible without this principle cementing together, as it does, all of our various pieces of optical knowledge—big foundation pieces like Snell's law and smaller, higher-up pieces having to do with refractive index and the like.

This picture of a stratified language in science is sometimes presented in a very misleading way, however. It is suggested that the relations between, for example, principles, laws, and hypotheses is a *deductive* relation. The result is that any given scientific discipline comes to be spoken of as a "deductive system." There are several objections to this view.

The role of deduction in science is *not* to transport us from the more abstract to the more concrete. The relationship among principles, laws, and hypotheses is not a deductive one. Consider a strict, rule-guided inference in science—in astronomy, for example. Suppose we wish to work out where a planet was last month from a knowledge of its present position, velocity, etc. This information is *not deduced* from the laws of motion. It is manifestly not the case that in proceeding with our inference about the planet's position we begin:

> Premise number I: Nature is uniform; what happens once will, under sufficiently similar circumstances, happen again.
> Premise number II: A particle under the action of no external force remains at rest or moves with constant velocity.
> Premise number III: The resultant force on a particle is the product of its mass and its acceleration.
> Premise number IV: And so on . . .

This is clearly not how we set about making our inference. We do not deduce our astronomical conclusion *from* such principles and laws. Rather, we make our deduction *in accordance with* them. Laws and principles do not stand at the "head" of scientific inferences. They stand, rather, if one may say so, "alongside" of them. For our astronomer, the premises are:

    I. At noon today Venus was at such and such a place.
   II. The velocity of Venus is such and such.

From these premises he concludes where Venus was one month ago *in accordance with* such general principles and laws as those set out a moment ago. He reasons correctly when he obeys those laws and principles. But this is not to deduce his conclusions from the laws and principles he obeys. We deduce nothing from the proposition, "Keep off the grass," either, though we behave properly or improperly in terms of that proposition. It is not seriously different in science.

Statements in terms of "refractive index" are not deduced from Snell's law. There is a strong logical connection between the law and the statements, naturally enough. But this is simply because the term "refractive index" is introduced by reference to Snell's law, and not because the two classes of sentences are mutually deducible. *It is the terms, the words which appear in statements at one level of scientific discourse, not the statements themselves, that are linked logically to statements in the levels below.*

Moreover, on the "deductive system" account of a scientific theory, the most abstract statements of all present us with a very strange picture. For on this account each abstract statement would just assert a coincidence. Consider the following apparently deductive argument.

    1. All ungulates are herbivorous,
                therefore
      Horses eat grass
                and
    2. Cows eat leaves,
                therefore
      Dobbin eats grass
                and
    3. Bessie eats leaves.

If it were a coincidence not only that Bessie eats leaves and Dobbin eats grass but also that all other cows and horses eat leaves and grass, then we might daringly generalize from these collections of coincidences and say, "All ungulates are herbivorous." But then how much more of a coincidence that $\lambda = h/mv$, to cite one of De Broglie's abstract principles! Like all such coincidences, moreover, the latter must be prone to sudden upset, especially so since they are even more abstract than the assertion "All ungulates are herbivorous." For while we must always be prepared to hear of some wild Tasmanian pig that is predatory, or some pony in Madagascar that is carnivorous, how much more must we be prepared to hear of exception to $\lambda = h/mv$? De Broglie is not concerned with making wide generalizations from experiment.

It is no wonder that the deductive system model of a scientific theory gets its primary appeal by way of habit statements about horses, cows, swine, and the like. But however well such an interpretation might represent certain departments of natural history, it wholly misrepresents the logical structure of a good deal of natural science, including most of physics and chemistry and a most important part of biology.

We have been using the expression "law of nature," however, as though it were the single name of a single collection of things. So too with the expressions "hypothesis" and "principle."

There is a wide range of conceptual entities in science that we call "laws of nature." On the one hand, there are fairly straightforward propositions like Boyle's law which involve no theoretical terms at all. Boyle's law just asserts that the pressure and volume of a gas vary inversely at a given temperature. No terms like "force" or "mass" or even "light ray" intrude here. On the other hand, there are the laws of motion and Maxwell's laws of electromagnetics. Unlike Boyle's law these do not just express the form of a *regularity found in phenomena*. They are much more like the axioms of a calculus (though that does not mean that they *are* such axioms). The test for abstract laws like these is not that they account directly for the phenomena that we observe. It is, rather, that they must provide a framework into which can be placed laws like Boyle's law, Charles' law, and Avogadro's law.

Snell's law is an intermediary. Since it includes the theoretical term "light ray," it differs from Boyle's law and Charles' law (in which no such terms occur). Still, it is much more like these than like the laws of motion, or of electromagnetic attraction, or of thermodynamics, which, after a time, come to be related to their several disciplines as the principle of the rectilinear propagation of light is related to geometrical optics. Indeed, the word "principle" is often used interchangeably with "law" in the latter cases.

But, despite these logically very important differences, all laws of nature still have one feature in common. *Laws of nature do not, by themselves, tell us anything about nature.* Rather, they express the *form* of a regularity, the scope of which is stated elsewhere, as in, e.g., "Most transparent substances refract light so that the ratio of the sine of the angle of incidence to the sine of the angle of refraction is constant; a notable exception to this is Iceland spar." This statement expresses the scope of Snell's law. It does not state Snell's law itself. Of this statement of scope we may very well ask, "True or false?" After all, it may be false that most substances do so behave, or that Iceland spar is an exception. Of Snell's law itself, however, we would not ask, "True or false?" We ask, rather, "To what systems can the law be applied?" or "Under what circumstances does the law hold?"

There are certain scientific laws which we do describe as true or false, however. Surely Kepler's three laws of planetary motion *are true*. They correctly represent the orbits of our planets, and if they did not they would be said to be untrue. We do not say merely that they *hold* in our solar system; we go much further and say that they truly characterize planetary behavior.

But consider, Kepler's laws do not tell us about planets in general. They tell us about *the* planets, viz. Mercury, Venus, Earth, Mars, Jupiter, Saturn, Uranus, Neptune, and Pluto (of these, the last three were unknown to Kepler). That is, Kepler's laws summarize the observed behavior of all members of the class of the sun's planets, a very limited class. They are not explanations in terms of the nature of things. They are even more like the generalizations of natural history than are Boyle's law, Charles' law, and Avogadro's law. Hence, no physicist would ever speak of Kepler's laws of planetary motion as laws *of nature*. Summary generalizations of behavior in a restricted class are not laws of nature.

If you are unwilling to be bound by such a ruling, call them all "laws of nature" if you wish. But then, at least, mark the important logical differences between *generalizations* about cows and planets, *general* remarks about the behavior of actual and ideal gases, *statements in general form* which include theoretical terms like "light ray" or "geodesic," and highly abstract sentences ranging from Newton's laws of motion and Maxwell's laws of electromagnetic attraction to Einstein's $e = mc^2$ and De Broglie's $\lambda = h/mv$. To collect all these under the same label and to refuse to distinguish further is just dim.

To return briefly to Kepler: there are now three laws of nature which correspond to Kepler's laws of planetary motion. However, these are undoubtedly expressed, not in terms of "the nine planets," but in terms of "bodies moving under the influence of gravitation alone." These laws would explain Kepler's ingenious observational summaries. To identify such general laws with those Kepler formulated, however, would be a mistake.

Newton's three laws have been a perennial puzzle. Sometimes they have been assigned a status alongside Kepler's laws. Sometimes they are thought to resemble laws like Snell's law; at other times they are made highly abstract. Elsewhere they are treated like aristocratic definitions of "mass," "force," and "momentum." The whole problem of the status of the laws of motion is treated with a few embarrassed remarks in Elementary Physics (remarks not always consistent with themselves).

None of these accounts squares very well, however, with all the actual uses to which the laws are put. Newton's laws of motion are neither generalizations like "Cows are herbivorous" nor tautologies like "Cows are animals." They are different from both "The line from the earth to the sun sweeps over equal areas in equal intervals of time" and "The earth is a planet."

Newton's laws do not set out to tell us anything about the actual motions of particular bodies. They provide, rather, a form of description to use in accounting for these motions. The point is put almost paradoxically in Ludwig Wittgenstein's *Tractatus Logico-Philosophicus*: ". . . the possibility of describing the world by Newtonian mechanics tells us nothing about the world; but what does tell us something about it is the precise *way* in which it is possible to describe it by these means." [4]

Notice that it is no slur on Newton's laws, or any laws of nature, that they tell us nothing about the world. For this is not a failing. Laws of nature were never set out to do any such thing. Laws of nature express only the *forms* of regularities, or, as Professor Herbert Dingle puts it,

---

[4] Ludwig Wittgenstein, *Tractatus Logico-Philosophicus, op. cit.*, p. 139.

formal relations between our concepts. They do not bear the burden of our experimental observations. It is the statements of scope that do that. It is these statements that indicate how the laws of motion are to be used to represent the actual motions of planets, projectiles, leaves, ships, and waves.

There is in science a clear division of labor between laws themselves and statements about when, where, and how those laws are to be applied. To return to a metaphor, the laws of nature are the tools in the toolbox called "scientific explanation." By themselves, in the toolbox, they do not do anything; there they are certainly not true or false. But statements about when, where, and how they can be used (statements not unlike this one: "Hammers are used to drive nails, tacks, and staples into wood and plaster but not into metals; hammers cannot be used to tighten bolts or to cut wood or paper.") are true or false. They are, moreover, clearly based upon our experience. And they tell us something about the world (in this example, about hammers specifically). But *hammers* tell us nothing about the world, and not just because they refuse to talk. They were not made for such a job. Thus it is not slanderous to say of tools that they do not build fences. Nor do they even tell us how to build a fence. Neither is it a slur to say of natural laws that they tell us nothing about the world, for they were not made for such a job.

What job were they made for? We are a little uncertain; shall we say that they define terms like "force," "mass," and "momentum"? Or shall we say that they tell us about modes of measuring force and rest?

There are good reasons for this uncertainty.

The laws themselves do nothing. It is *we* who do things with them. And there are several kinds of things that we *can* do with them, just as there are several kinds of things we can do with a hammer—we can drive nails, pound sheet metal flat, tap a chisel or a gauge, or force home a wedge. There is no more need for us to be puzzled by the question of whether the laws of motion are descriptions, definitions, or assertions about methods of measurement than there is for puzzlement over the question of whether hammers are for driving nails, flattening sheet metal, or splitting ice.

Rather, it is up to us to distinguish certain applications of the laws from other applications. We must not confuse the case where the scientist describes by way of these laws the motion of a freely falling body with the case where he uses them in defining, say, "electromotive force," or with the case in which he invokes them in devising a mode of measuring the mass of a new type of particle.

The laws are not ambiguous or vague. It is, rather, that scientists are versatile in applying these laws.

There is a general philosophical point here towards which I will only gesture. A great many of the words and concepts that are used in science seem to be slippery in a peculiarly logical or philosophical way. Here are some examples: "fact," "true," "hypothesis," "theory," "definition," "measure," "entails," "infer," "principle," "premise," "necessary," "probable," "cause," "explain." There are hundreds. Like "law" and "law of nature" their complexities inhere, not in the words and concepts themselves, but in the enormous versatility that we display in using those words for different linguistic jobs at different times and places. Scientists often steer clear of looking too closely at such words because they think that the word itself is a kind of Pandora's box which, once opened, will release confusion and chaos into the otherwise untroubled scientific world. But it is not the *word* that is confusing. It is the hasty assimilation (on the part of scientists and philosophers) of a variety of uses which the word may have into one or two standard uses. Thus philosophers wear out everyone's patience with their hunt for *the* meaning of "true," or "fact," or "cause," or "infer," often failing to see that the words mean what we make them mean in certain clearly specifiable contexts. And scientists shower abuse upon such philosophers for the manifest failure of their efforts. Yet they dare not sally forth themselves to engage the same problems because they imagine that it is the words that are making the trouble when really the trouble is made by the people who use and analyze the words. Or, if scientists do try to wrestle with these philosophically perplexing notions, they use definitions as an all-purpose weapon, reminding one of the time Baby fixed Dad's wrist watch with a hammer.

So long as we remain short of the perfect science in which every word is clearly defined in advance, we must be prepared to see important words used in a multiplicity of ways. Our job as philosophers of science is, then, to say not "this word *must* mean such and such" but "*in this context*, scientist X is likely to have meant such and such."

In any case, the moral is that we must philosophize about science a little at a time, not all at once. We must learn what is meant by a word as it is used here and there and there before we can begin to say how it is used everywhere, if ever we can say that at all. For words have lots of uses in science. That is partly because science is a conglomeration of languages.

Laws of nature, particularly, show up the mosaic character of the language of science. But the pattern into which all the several uses of the expression "law of nature" fit is one that marks it as inappropriate to be questioned as "true or false." Laws of nature tell us nothing about the world, but only about what procedures to follow in representing the world.

21 | Principles and Platitudes

ONE THING we noted in the last chapter was how the label "law of nature" is applied to several quite different types of statement. Boyle's law is a straightforward law of nature. So is Snell's law. Snell's law, however, involves a theoretical term—"light ray." Boyle's law requires no such term. The laws of motion and of electromagnetics and thermodynamics are laws of nature, too, and they are riddled with theoretical terms, being less like formal summaries of what we observe and more like the axioms of a calculus. Thus, we are versatile in our use of the expression "law of nature."

We will consider in this chapter one of the uses of the word "principle." It is a use very different from those found in expressions like "principle of inference," or "principle of the rectilinear propagation of light," or "principle of least action," or "principle of equivalence," or "principle of natural selection." The principle to which we will attend is the "principle of uniformity in nature."

The principle has been expressed in many forms. Often it hides disguised. But it can always be forced into the open in statements like these: "The laws of nature are immutable; they are now as they have always been, everywhere in the universe." "Repeat the cause of X and X will occur repeatedly." ". . . there are such things in nature as parallel cases, what happens once, will under a sufficient degree of similarity of circumstances happen again."

This last is a quotation from John Stuart Mill's *System of Logic*.[1] In that book Mill sets out his so-called "experimental methods." These were, for Mill, methods of discovering causal connections in nature. They served as the foundation of his experimental logic. A short survey of these methods will show why the principle of uniformity in nature assumed the importance that it did for Mill.

To find the cause of some effect Mill reckoned that all the factors relevant to the experimental situation ought to be varied one at a time. Mill's methods are offered as a systematic way of varying these factors.

(a) The first method is called the "method of agreement." It instructs as follows: "If two or more instances of the phenomenon under investigation have only one circumstance in common, the circumstance in which alone all the instances agree is the cause (or effect) of the given phenomenon."[2]

Suppose that last spring the buds were found to be withering in several gardens. We note that the gardens differ in many ways—the quality of the soil, the kinds of flowers grown, their size, their location—and also that the temperature fell sharply just before the withering was observed. This factor would affect all the gardens in the area. Hence the drop in temperature, it being the only invariable circumstance, must be the cause of the withering of the buds.

(b) A second method invoked by Mill was his "method of difference": "If an instance in which the phenomenon under investigation occurs, and an instance in which it does not occur, have every circumstance in common save one, that one occurring only in the former; the circumstances in which alone the two instances differ is . . . the cause, or an indispensable part of the cause, of the phenomenon."[3]

Thus we note two buds on the same shrub. Both of these appeared at the same time and have grown to the same dimensions. But one of them has withered, while the other has not. A real difference between them is that one of them, the one that did not wither, is high up on a trellis set in against a chimney. The other bud is close to the ground and at some distance from the wall. Perhaps the heat from the chimney bricks prevented the first bud from withering when the temperature fell.

---

[1] John Stuart Mill, *A System of Logic Ratiocinative and Inductive* (London: Longmans, Green & Co., 1965 (reprint)), p. 201.
[2] *Ibid.*, p. 255.
[3] *Ibid.*, p. 256.

(c) Next, Mill combines these two methods to form the "joint method of agreement and difference." He expresses it this way: "If two or more instances in which the phenomenon occurs have only one circumstance in common, while two or more instances in which it does not occur have nothing in common save the absence of that circumstance, the circumstance in which alone the two sets of instances differ, is the . . . cause, or an indispensable part of the cause, of the phenomenon." [4]

Consider two occasions when we note a withering of buds (in entirely different local conditions). The only factor common to both occasions is the sharp drop in temperature. Now, if every occasion where buds in these differing conditions do not wither is also an occasion when there has been no drop in temperature, then the drop in temperature was the cause of the withering of the buds.

(d) Mill's next method of experimental inquiry bears the grand title, the "method of concomitant variation." Where his other methods are highly qualitative in nature, there is a quantitative twist to this one. Consider: "Whatever phenomenon varies in any manner whenever another phenomenon varies in some particular manner, is either a cause or an effect of that phenomenon, or is connected with it through some fact of causation." [5]

Clearly this method is effective only where degrees or magnitudes of causes and effects can be distinguished. Here Mill advises us to try to correlate the magnitude of the fall in temperature with, say, the extent to which withering took place. Measurement and statistical techniques are indispensable in this connection. They were absent in the earlier methods.

(e) Finally, the "method of residues." It is stated thus: "Subduct from any phenomenon such part as is known by previous inductions to be the effect of certain antecedents, and the residue of the phenomenon is the effect of the remaining antecedents." [6]

Thus, suppose that the temperature drops *and* the wind rises. Next morning we note that the buds have not only withered but are also strewn all over the ground. We know from past experience that it was the fall in temperature that withered the buds. Hence that they were parted from their axes must have been caused by the wind.

---

[4] *Ibid.*, p. 259.
[5] *Ibid.*, p. 263.
[6] *Ibid.*, p. 260.

Now it is not for us to appraise these "methods." That is done very well in Cohen and Nagel's *Introduction to Logic and Scientific Method*.[7] There the actual utility of Mill's canons of inquiry is seriously questioned; in fact, they all come out of it very seriously scathed.

What does concern us is that these methods of induction forced Mill to formulate explicitly the principle of uniformity in nature. Why?

The claim that the experimental methods can demonstrate universal and invariable connections, and do so with certainty, rests on the belief that "nature is uniform." Mill says that induction consists in inferring from a finite number of observed instances of a phenomenon that it occurs in *all* instances of the class of phenomena which resemble the observed instance. However, this very statement of what induction is requires an assumption concerning the order of the universe.

Mill puts the matter this way:

> The Induction, "John, Peter, etc. are mortal, therefore all mankind are mortal," may be thrown into a syllogism by prefixing as a major premise . . . that what is true of John, Peter, etc., is true of all mankind. But how came we by this major premise? It is not self-evident; nay, in all cases of *unwarranted* generalizations, it is not even true. How then is it arrived at? Necessarily, either by induction or ratiocination [deduction]; and if by induction, the process, like all other inductive arguments, may be thrown into the form of a syllogism. This previous syllogism it is, therefore, necessary to construct. There is in the long run, only one possible construction. The real proof that what is true of John, Peter, etc., is true of all mankind, can only be that a different supposition would be inconsistent with the uniformity which we know to exist in the course of nature. Whether there would be this inconsistency or not may be matter of long and delicate inquiry; but *unless* there would, we would have no sufficient ground for the major [premise] of the inductive syllogisms. It hence appears, that if we throw the whole course of any inductive argument into a series of syllogisms, we shall arrive by more or fewer steps at an ultimate syllogism, which will have for its major premise the principle, or axiom, of the uniformity of the course of nature.[8]

The logical situation in which Mill finds himself is schematized as follows:

---

[7] M. R. Cohen and Ernest Nagel, *Introduction to Logic and Scientific Method* (New York: Harcourt, Brace, 1934), pp. 251–67.

[8] Mill, *op. cit.,* pp. 202–3 (Hanson's italics).

A typical *deductive* syllogism would run

    All S is P
    $\alpha, \beta, \nu, \delta, \ldots$ are S
    $\therefore \alpha, \beta, \nu, \delta, \ldots$ are P

Or in words

    All men are mortal
    Socrates, Plato, John, Peter, . . . are men
    Socrates, Plato, John, Peter, . . . are mortal

An *inductive* syllogism, however, is different. Mill asks, "How do we get from

    Socrates, Plato, John, Peter, . . . are mortal

to

    All mankind are mortal?"

Schematically, how do we get from

    $\alpha, \beta, \nu, \delta, \ldots$ are P

to

    All S are P?

The only way to do this, says Mill, is to insert the major premise that what is true of $\alpha, \beta, \nu, \delta$ is true of all S; that what is true of Socrates, Plato, John, and Peter is true of all mankind.

How did we get that? If we appeal to experience, then we only appeal to another inductive syllogism. Ultimately, our proof that what is true of Socrates, Plato, John, Peter, . . . is true of all mankind rests on the contention that to suppose anything else would be inconsistent with the uniformity which we *know* to exist in the course of nature.

Comment about this argument is unnecessary. Its back is broken. Still, it articulates very clearly the thesis that if induction is to be accepted as a rational policy in science, it must be built upon the presupposition that the course of nature is regular and uniform, that the creator of the world is not merely a dice-playing god, that whatever happens anywhere and any time in the universe happens in accordance with laws that govern happenings everywhere else and at every other time in the universe. Mill's argument also manifests the typical logician's prejudice that inductive argument at its best is only second-rate deduction. The philosopher's game thus becomes one of patching up induction until it can pass as a species of deduction, of trying to transform the reasoning of natural science into something of which Euclid might have approved.

We need not go far to find scientists explicitly paying homage to the principle, or even adducing evidence in its support. Studies of the pleochroic haloes in mica (dear to the hearts of students of geology) have supported the hypothesis that they were produced by the various groups of particles emitted during the radioactive transformation of uranium and the products of its disintegration. This has made men want to say that if the radioactive theory of their origin is correct, these haloes (which must have been in existence some hundreds of millions of years) provide evidence that the laws of radioactive transformation have not changed in all that time.

A variation on this theme is that of the paleontologist who, on analyzing certain early pre-Cambrian rocks, discovers in them concentrations of the carbon isotopes, C^{12} and C^{13}, in ratios *now* characteristic of organic concentrations. If the laws of these isotopic ratios were then as they are now, muses the paleontologist, we have evidence of the existence of pre-Cambrian organisms.

In this way the issue is harmless and unexceptional. But do we stop there? We do not! Somehow all this gets twisted into a proof that nature is uniform, that the truths of science are universal in application, that the pattern of cause and effect now is the reflection of the structure of all things always.

Before we get completely swept away by this cosmic cross-examination, let us scrutinize the principle of uniformity in nature. For some men, like John Stuart Mill, the principle solved the problem of induction (a problem, incidentally, which can be acquired as one's very own only with a good deal of industry). It provided the logical bridge between our observation of things past and our predictions of things to come. Other men, different from Mill in this respect, read the history of the world and of science itself as evidence of uniformity already revealed. Still others treat the principle as an article of scientific faith, an expression of the scientist's confidence in the possibility of solving his problems.

Principles as Platitudes | 351

Thus, just as the word "principle" is made to do many jobs, so too the expression "principle of uniformity in nature" is kept busy. Here we can only hint at a line of attack which would require special adjustments if it were to fit the special demands of each of the uses this troublesome expression enjoys.

Consider an extreme case: Suppose it were urged (as thinkers from Mill to Russell have urged) that by themselves the arguments of science are logically unsound. The holes in them must be plugged up with some major premise about the uniformity of things in general. One cannot move from a finite number of observations of X to assertions about *all* X without this logical insurance policy.

And of course the actual behavior of scientists at work does suggest that they are busily making assumptions. Notice how they calculate the motions of falling apples, of our moon, and of Jupiter and double stars. In all these cases they appeal (usually) to the same law of gravitation. Scientists never seem to consider whether the forms of the equations they use should be modified when passing from one system to another.

Surely they *ought* to consider that, though. Surely the law of gravitation might be different on Mars, or in the nebula in Andromeda. Surely the law might have been different 1000 million years ago in the pre-Cambrian era. This being so, is it not the case that scientists *do* assume that the laws of nature take, will take, and have always taken one form everywhere, galaxies away and eons ago no less than here and now?

If we treat laws of nature as if they were generalizations *about* nature this conclusion is irresistible. If the law of gravitation is made the bedfellow of statements like "Cows are herbivorous" or "Pigs are not arboreal," then the Mill-Russell tradition has given us the appropriate clue.

For suppose that when Dan Dare *does* land on Mars he is greeted by an honest-to-goodness cow. Daring as he is, will Dan just reject out of hand the hypothesis that Martian cows may have diets quite unlike our Jersey cows' diet? Not likely! Martian cows might turn out to relish the taste of chicken. And Martian pigs might spend their lives in the branches of Martian trees (which latter are doubtless also unusual in their own peculiar way). Thus, generalizations like "Cows are herbivorous" and "Pigs are not arboreal" will need scrutiny when such Martian species are reported.

Natural historians will soon become even more cautious about jumping to general conclusions after Dan Dare returns. They will hardly ever say things like "Since there are cows on Mars there must also be grasses for them to eat."

What about gravity? Might not *it* be different on Mars too? Do not physicists now run the same risk as would the natural historian who infers from the existence of Martian cows to the existence of Martian grass? They surely assume as much about gravity on Mars as our incautious naturalist assumes about cows on Mars.

We have noted again and again how useful it is to ask what a statement is meant to *deny* when we are out to determine what it is meant to assert. Consider in this way our two assumptions (1) that cows eat the same sorts of food everywhere in the universe and (2) that the law of gravitation takes the same form everywhere in the universe. What the first denies is this: that cows on Mars eat foods they do not eat on earth. The second denies this: that the law of gravitation takes a different form on Mars than it takes on earth.

The first is clear enough and meaningful enough. We all know quite well what it would be like to discover cows eating chicken rather than grass. But what about the second?

Let us deflate an ambiguity before it gets too puffed up.

We must distinguish a law of nature from a restricted application of that law. We have seen how Kepler's laws of planetary motion are not really laws *of nature* at all. They are in fact restricted applications of three proper laws of nature which are expressed, not in terms of "the sun's planets," as are Kepler's, but, rather, in terms of "bodies moving under the influence of gravitation alone."

Likewise, distinguish Newton's inverse-square law from the statement that freely falling bodies accelerate at 32.2 ft./sec.^{2} This latter we might call "the law of gravitational acceleration on the earth." We are not likely to call it a "law of nature," however.

Now clearly, it would be risky to assume that freely falling bodies accelerate at 32.2 ft./sec.^{2} on Mars. This would be on a par with the assumption that Martian cows are herbivorous. The rate at which freely falling bodies accelerate on Mars is undoubtedly different from the rate at which they accelerate on earth. In this sense, then, the law of gravitational acceleration will be different on Mars. And in this sense the assumption that physical laws operate in the same way everywhere (though it makes perfectly good sense) is entirely unfounded.

Of course, scientists never make such an assumption at all.

However, the fact that gravitational acceleration varies in no way establishes the need for a principle of uniformity of nature. *This* "law" of acceleration is not a law of nature at all. It is an empirical discovery made by *applying* the law of gravitation to the special circumstances that obtain on earth.

With a real law of nature the situation is very different. For it is not now possible to say what it is that is being denied or ruled out.

If the study of gravitational phenomena on Mars obliged us to amend the law we could not simply let things rest with the qualification ". . . except on Mars." We could not blithely say, "The law of gravitation takes a different form on Mars," as we might say, "Freely falling bodies accelerate at a different rate on Mars," or "Cows eat different foods on Mars." The very conception of a law of nature is such that we are required to show how (if a law must be amended) the modified law continues to explain all the phenomena it previously explained, as well as the new phenomenon being considered. A discovery that forced us to amend the law of gravitation represents not a mere non-uniformity but an inadequacy in our present ideas about gravity. Such a discovery would reveal a defect in our current theory, a defect with implications as well for the earth and the extra-galactic nebulae as for Mars.

Hence, the suggestion that the law of gravitation might be different on Mars is a very queer one. The suggestion that scientists make a risky assumption in taking the law of gravity to be no different on Mars than it is on earth is also very queer. It is not apparent to me, at any rate, that scientists are assuming anything very remarkable when they apply the same law of gravitation to phenomena in different places or at different times.

Thus, that a scientist expresses his law of gravitation in an identical form on all occasions proves nothing at all about the uniformity of nature. What it may be said to prove is that we accord the title "law of nature" to nothing which is not expressed in an identical form on all occasions.

Does this much entitle one to say, however, that *no* assumptions are being made in any given application of a law of nature? It may be a little misleading to put it in just this way.

If we begin by taking a phenomenon to be purely gravitational, but it turns out that our current theories cannot explain it, then either of two things may happen. Either we will say of the phenomenon that it is not after all purely gravitational, looking elsewhere for our explanation, or we will call the whole current theory of gravitation into question.

But if it is the latter course we take, questioning the whole current theory, the revised law will be just as free from restrictions of time and place as was the former. Einstein's theory of gravitation was as universal in its application as was Newton's. In neither case will we talk of the law of gravitation as having a different form at different places or times. Nor will we say it has the *same* form in different times and places, except in a misleading way. For whenever we speak of *the* law of gravitation we have but *one* universal expression in mind.

Generalizations are different. The pronouncement "All cows are herbivorous" is liable to sudden upset. We would be more prudent, perhaps, to say, "All known species of cow are herbivorous." This is like whispering, "On Mars, who knows?" Such a guarding clause is out of place with a law of nature.

All this is tied in with the fact that, while the natural historian is far from free in the use of his language, the natural scientist may take great liberties. What is or is not a cow is (to a large extent) for the public to decide. It is very different with electrons and mesons, chlorine and copper sulphate, stomate and chitin. Every word in the sentence "Cows eat grasses" had a comprehensive meaning before the generalization could even be formulated.

A petulant naturalist might have refused to call those black Australian birds "swans" because of his conviction that all swans must be white. Suppose he takes a similar line on Martian cows. Are we to let him dictate thus: "These cowlike animals on Mars are to be called 'scows' to distinguish them from 'cows,' for cows are herbivorous while scows are not"? I doubt that we would. He may mess about with the taxonomic implications of the existence of carnivorous cows all he pleases. But we know a cow when we see one, even one that eats chicken. And we know a swan when we see one, even if its plumage is black.

The naturalist's move is something like this. He says, "No Harvard man would cheat at cards." When the *Crimson* suggests otherwise, he saves his claim from falsification by the amendment, "No true Harvard man would cheat at cards." This, though it be in the best interests of the university, is logically untidy and scientifically unacceptable. Cows are cows, swans are swans, Harvardians are Harvardians, however many the varieties.

In physics, chemistry, and biology, however, the terminology is not fixed beforehand, least of all by the non-scientific public. Theories, notations, and terminologies are often introduced all at once. Thereafter it is a technical question (a very technical question) whether or not some given metallic specimen will be called "cadmium," or whether some parabola on the scope of a mass spectograph is characteristic of Neon 22.

Thus the phenomena which form the scientist's field of study are classified in a systematic way. This is reflected in the terms with which, and the methods by which, he sets about explaining them. It is the systematic nature of reclassification within science that distinguishes it from *ad hoc* distortions of an existing classification, which was all that the "no *true* Harvard man" move consisted in.

Hence, that the law of gravitation is spoken of as one and the same thing regardless of time and place involves the making of no particular assumptions. It would not be a law of nature were it spoken of otherwise —if, for example, it were restricted to place, as are Kepler's laws, or time, as are pronouncements about the extinction of ammonites.

Scientists *would* be making assumptions only if, for example, they supposed that all the systems they studied were purely gravitational, ignoring the possibility that other theories besides gravitation theory might be required to account for their behavior.

This assumption is never made, however.

The most that scientists do is to *presume* (not assume) a) that existing theories will suffice to explain the behavior of each new system brought forward for study, and b) that each new system will behave pretty much as do the familiar systems which it resembles in structure. Thus the motions of a double star are explained by the same law which explains the motion of falling apples and the moons of Jupiter. The fact that the same law is used to explain these different phenomena represents a certain uniformity in our techniques of dealing with these three systems. And this uniformity reflects the presumption that the three phenomena are similar in type, that is, all gravitational in nature. The presumption is not present where scientists are at work in different fields, viz., gravitation and magnetism.

Moreover, these are but *initial* presumptions. Our student of gravitational attraction is on the alert for deviations. If gravitation theory fails to account for the behavior he observes, he will ask why this is so.

Now, if the scientist were not on the lookout for such deviations, if he did not even bother to ask whether the theory of gravitation could by itself explain the motions of a double star or whether other forces were also involved, *then it might very justly be said that he was presupposing or taking something for granted which required independent justification.* As we all know, however, scientists are always ready to reconsider their initial presumptions. They are always ready to question, when there is any reason to do so, whether double stars and solar systems are strictly comparable.

Further, when there *is* such reason for questioning the presumption b) that new systems will behave pretty much as do the old ones they resemble in structure, the deeper presumption a) that existing theories will account for the new phenomenon will also be re-examined. Existing theories may have to be modified, or supplemented, or wholly scrapped in the process.

What has looked like an assumption, then, to scientists and philosophers since (at least) the time of Hume turns out to be just a bit of common sense. The use of the same form of law in widely differing situations marks, not a daring presupposition about the uniformity of nature, but a decently methodical procedure. And when we try to say just what this presumption is, it turns out to be, not some grandiose philosophical principle about things in general, but only a platitude, namely: Unless there is some reason to suppose that a phenomenon cannot be explained by the theory it is natural to turn to first, why then there is every reason to turn first to that theory.

This is not a very dangerous assumption, is it? Does it require any sustained defense? Yet, *there* is your principle of uniformity in nature, the core of so many disputes about the nature of science and the so-called "problem of justifying induction."

If a scientist makes a mistake, if, for example, he takes for a purely magnetic phenomenon what further investigation reveals to be partly electrical, this will soon enough show up in the work itself. Once his error becomes apparent the scientist is warned: He knows what to expect the next time he encounters a similar system. *So it is not nature that is* uniform, but scientific procedure. And scientific procedure is uniform only in this, that it is methodical and self-correcting.

Let me conclude with an attack on the principle from another quarter. Suppose that, despite all my arguments, someone just digs in and flatly asserts, "It's true!"

What's true?

"The principle that nature is uniform, that what happens once will, under a sufficient degree of similarity of circumstances, happen again."

I have been arguing, remember, that the principle is not so much an assumption of science, a logical finger to plug up the holes in the dikes of scientific reasoning, as it is just good common sense. Neither is it a faith expressing a fervent hope of scientists that everything will come out all right. It is not logically risky to adopt the principle; it is just stupid to refuse to follow the innocuous counsel it suggests, namely, turn first to the theory which seems naturally to explain a certain phenomenon, unless there is some reason not to do so.

Now *you* come along and say, "It's true!" This is an even stronger claim than is made by the scientists and philosophers who prate about the principle in the manner I have been abusing.

What's true?

Look, the principle is nothing if not vague; ". . . what happens once will under a sufficient degree of similarity of circumstances happen again." But *what is* a sufficient degree of similarity? The principle, as Mill and others put it, does not tell us. In any particular inquiry we must rely on entirely different criteria, if such there are, to determine what are the circumstances material to the occurrence of a phenomenon. Does the principle instruct us to ignore the disposition of Jupiter's satellites when comparing the masses of sub-atomic particles? It does not. Does the principle warn us not to overlook variations in temperature in our inquiries about the refractive properties of the earth's atmosphere? It does not.

Again, the principle does not assert that *every* pair of phenomena are invariably related. Only *some* pairs are. But which pairs? The principle does not say. It is therefore useless in particular investigations. Buds may wither due to a fall in temperature, or some type of virus, or some genetic aberration, or drought, or an impoverished soil, or any of a number of other factors taken singly or in combination. Mill's methods of experimental inquiry, and the principle of uniformity in nature which is supposed to be their logical guarantee, do not specify which of the innumerable causal connections are invariable. They only assert that some are. But do we need a principle with a capital "P" to be persuaded of that? The purpose of any particular inquiry is to determine whether some *already designated* pair of factors is in causal connection. To determine this requires science, not Mill's slippery logic.

So, what is it that is true? To say of some string of words that it is true is to say that it can be used to assert what is the case. What does the principle of uniformity of nature assert that can be shown to be the case? Indeed, what does it assert?

No, the principle of the uniformity of nature is not an hypothesis about the world at all. *We are not prepared to substitute an alternative hypothesis.* It is, as I have suggested, a misleading way of characterizing what scientists do. They do just as we do. They try the obvious avenue of explanation first. Remember all this when, in your reading, you find yourself hip-deep in justifications of induction and the foundations of probability theory.

Here in conclusion are some puzzles with which to delight your friends and confound your enemies:

A. Recall our earlier reference to the pleochroic haloes in mica. Recall also the enthusiastic conclusions that some scientists have drawn from the existence of this phenomenon, namely, that it provides evidence that the laws of radioactive transformation have not changed in hundreds of millions of years. Is this a proper conclusion about the world? What if the haloes had, in fact, a different structure than they do have? Would we still hold that their origin was radioactive in nature? Obviously we could not do this without some additional hypothesis to explain the structure of the haloes we would then actually observe and the structure predicted by the unqualified theory of their radioactive origin. Suppose that the additional hypothesis is of this form: "The laws of radioactive transformation were different in the past." We would then look for a physical cause of the changed behavior of radioactive substances. But then how could one be certain that this physical cause operated in the distant past according to laws established by experiments made in the present era? And so on.

That is one for your enemies. Here is another.

B. Suppose that a representative assortment of our present laws of nature just *had* to give way to an entirely new assortment due to certain fundamental changes in the gravitational, electromagnetic, and thermodynamic properties of the universe. Suppose that terrestrial gravitational attraction was suddenly oriented towards the North and South Poles instead of towards the earth's center as it now is. All of us, and all unbuttressed objects on this side of 0° latitude, would "fall" to the North Pole.

Suppose that electromagnetic waves began propagating themselves like bubbles, and that light, magnetic attraction, and radiowaves were revealed in pulses and bursts not unlike the popping of soap bubbles one by one. Suppose that hot bodies diminished in size relative to the degree of their heat such that before a specimen of tungsten melted it disappeared, and before a sample of helium liquefied, its molecular agitation became so great as to make it fill all space.

A queer world, you will admit. But suppose that science somehow continued unbroken, absorbing these cataclysmic changes in the behavior of the physical world. Would this new science have a principle of uniformity in nature? If it did, would the principle be one whit different from what it is now?

Here are two questions for your friends.

C. Once upon a time there was a little girl named Emma. She had never eaten a banana in all her life, nor had she ever taken a journey on a train. On one occasion circumstances made it necessary for her to journey from New York to Pittsburgh alone. To relieve Emma's anxiety her mother gave her a large bag of bananas to eat on her railway journey west. At Emma's first bite of her banana, the train plunged into a tunnel. At the second bite, the train broke into the daylight again. Emma, being a bright little girl, takes a third bite. Lo! Into a tunnel. A fourth bite and into the daylight again. And so on all the way to Pittsburgh (and all the way to the bottom of the bag of bananas). Is Emma justified in saying to the people who meet her at the station, "Every odd bite of a banana makes you blind; every even bite puts things right again"?

D. Finally, examine evidence for the conclusion drawn in the following argument, one which might very well have appeared in the columns of the *Daily Sketch*: "Last week I got into trouble after having had too much brandy and water. Yesterday it was whisky and water. Two months ago I had an awful day following an evening with beer and water. Clearly water is causing all my difficulties." How do Mill's methods and the principle of uniformity in nature conspire to secure this conclusion, which is, to say the least, questionable?

Part IV | PROBABILITY AND PROBABLE REASONING IN SCIENCE

22. Frequencies and the Mathematics of Probability
23. Using and Interpreting the Probability Calculus
24. Elements of Statistical Technique
25. The Principle of Uniformity Revisited

BIBLIOGRAPHY
PART IV

Barker, Stephen F. *Induction and Hypothesis*. Ithaca, N. Y.: Cornell University Press, 1957.

Braithwaite, R. B. *Scientific Explanation*. New York: Harper & Row, 1960.

Carnap, Rudolf. *Logical Foundations of Probability*. Chicago: University of Chicago Press, 1950, 1962.

Fisher, Sir Ronald A. *Statistical Methods and Scientific Inference*. 2nd ed. New York: Hafner, 1956.

──────. *Statistical Methods for Research Workers*. London: Oliver and Boyd, 1954.

Hacking, Ian. *Logic of Statistical Inference*. Cambridge: Cambridge University Press, 1965.

Jeffrey, R. C. *The Logic of Decision*. New York: McGraw-Hill, 1965.

Keynes, John Maynard. *A Treatise on Probability*. Introduction by Norwood Russell Hanson. New York: Harper & Row, 1962. (First edition, 1921.)

Kneale, William. *Probability and Induction*. Oxford: Oxford University Press, 1949.

Kyburg, H. E. *Probability and the Logic of Rational Belief*. Middletown, Conn.: Wesleyan University Press, 1961.

Nagel, Ernest. *Principles of the Theory of Probability*. Vol. I, Number 6, of *International Encyclopedia of Unified Science*. Chicago: University of Chicago Press, 1955.

Popper, Sir Karl. *The Logic of Scientific Discovery*. New York: Basic Books, 1959.

Reichenbach, Hans. *The Theory of Probability*. Berkeley: University of California Press, 1949.

Salmon, Wesley. *Foundations of Scientific Inference*. Pittsburgh: University of Pittsburgh Press, 1966.

Savage, Leonard I. *The Foundations of Statistics*. New York: John Wiley and Sons, 1954.

Von Mises, Richard. *Probability, Statistics and Truth*. 2nd revised English ed. New York: Macmillan Co., 1957.

22 | Frequencies and the Mathematics of Probability

OF THE many sorts of questions that can be asked of a specific piece of reasoning in natural science the following three may be distinguished:

1. *Does it follow?* That is, are the conclusions so related to the evidence (or premises) that they may be said necessarily to follow from them? This first question is independent of any question of truth or falsity. It is designed strictly in terms of what might be called *consistency*.

2. *Is it true?* This usually comes to the question *Are the premises true?*, or, *Are the data correct?*, though it must be remembered that true conclusions can follow from false premises. For that reason we must distinguish questions about the truth of an argument from questions about the consistency of the same argument. Ptolemaic astronomy is consistent but false, and there are arguments in modern physics which, though they turn up answers that are true, are nevertheless formally inconsistent.

We have had occasion to discuss all this before. But not much has been said about the following question:

3. *How is the truth of the premises established?*

When discussing the nature of formal reasoning we observed why it was that such rigor attached to the proofs of mathematics and logic. The premises in these disciplines are, so to speak, closed: They are certain because they are *accepted* as certain. The mathematician's task is only to trace the implications of the premises, not to establish their truth. And in certain departments of applied mathematics, even where the criteria for the acceptance of premises requires not just acquiescence but evidence as well, this *certain* character of the premises, though always logically distinguishable from mathematical certainty, is not very different practically. Thus evidence for the premises of problems in classical mechanics or thermodynamics or optics is so complete and so strong that one who insists on their fallibility is thereby virtually making a point concerning inductive logic.

But such complete and conclusive evidence is rarely available. Indeed, one might say that in any living, thoroughly experimental science such evidence is never available, which is my roundabout way of saying that classical mechanics and optics are neither living sciences nor experimental sciences. Today they are more like catalogues or algebra textbooks, and hardly the kinds of discipline from which to draw morals about the nature of scientific thought and inquiry. This was not true of these sciences three hundred years ago, of course.

Suppose we are faced with the following issue: We conjecture that a certain philosophy professor likes wine with his meals, and our evidence for this is the fact that most philosophy professors like wine with their meals. As a rigorous proof this is hopelessly inadequate. It is obviously possible for the proposition "Jones is a philosophy professor and therefore likes wine" to be *false*, even though the proposition "Most philosophy professors like wine" is true. And yet it would be absurd to assert that the fact that most philosophy professors imbibe wine is altogether irrelevant as evidence for this particular professor's probably doing likewise. In fact, a person who regularly makes inferences of this type, i.e., "Most α's are β's, γ is an α, therefore γ is a β," will in the long run be right more often than he will be wrong.

An inference of this type, which from true premises gives us conclusions which are true in most cases, is called a *probable inference*. It is meant to be contrasted with the inferences of a formal system, which, beginning as they do from certain uninterpreted, tautologically true premises, always render up uninterpreted, true conclusions. Even so, as the Latin etymology of the word "probable" indicates, such ordinary, practical kinds of evidence are popularly felt to be a kind of proof, even though not conclusive.

Naturally, where the evidence in favor of a proposition is partial or incomplete, the probability of the inference *may* be increased by additional evidence. Later we shall consider when and how we can measure the degree of probability, and what precautions must be taken to insure that our inferences shall have the maximum probability attainable. Now we must note that deductive inference enters as an element in every such determination. Let us consider, first, the case where a probable argument leads to a *generalization* or *induction* and, second, the case where such an argument leads to what we may call a *presumption of fact*.

Suppose, for example, we wish to learn whether a certain substance, say, benzoate of soda, is generally deleterious. Clearly we cannot test the reactions to this substance of every living human being (a phrase which itself requires examination). We must select a number of persons who volunteer to take benzoate of soda with their food. These persons, moreover, must be such that we would regard them as typical, or representative, human beings. Then we should set ourselves to observe whether the ingestion of this substance produces any noticeable ill effect. If in fact all of our volunteers show some positive ill effect, we should regard that as evidence in favor of the general proposition "Benzoate of soda is deleterious."

It is notorious, however, how frequently such generalizations turn out to be false. For the volunteers may not be at all typical or representative. They may all be science undergraduates in the Easter term, or they may all be high-strung, half-starved, near-neurotic artists, or they may be used to certain specific sorts of diets or subject to some other unnoticed condition not prevailing in all cases; e.g., if the food (plus the benzoate of soda) were brought to the volunteers by a seven-foot gorilla we should hardly wish to ascribe any unusual digestive effects strictly to the chemical. But lots of other factors less obvious than seven-foot gorillas might be equally responsible for any effects we might too hastily ascribe to the benzoate of soda. We try to minimize the force of such doubts by employing the inferred rule "benzoate of soda is deleterious" as a premise. We then deduce its consequences as to other individuals living under conditions markedly different from those under which our volunteers lived. Should our new observations agree with the deduction from our assumed rule, the probability of this rule would be increased, though, of course, we cannot thereby eliminate all doubt, no matter how often deductions from the assumed rule turn up conclusions which are supported in experiment and observation.

If on the other hand there should be considerable and regular disagreement between the several deductions from our general rule and what we find in our new observations, the rule would itself require modification in terms of the general principles of deductive reasoning. Thus while generalizations from what we take to be typical cases sometimes lead to false conclusions, such generalizations frequently enable us to arrive at conclusions which are true in proportion to the care with which our generalization is formulated and tested.

The second form of probable inference of which I have spoken is the reciprocal of the first. This second form does not consist in inferring from a lot of observed α's to what is probably the case with all or most α's; rather, it consists in inferring from our conviction about what is the case with all, or most, α's to what is *probably* the case with this particular α.

If the office of the Philosophy Department should be burgled, and the only item stolen turns out to be the final examination paper in philosophy of science, our first inference would be that some undergraduate studying philosophy of science was responsible for the dastardly deed. But the inference is obviously not a necessary one. Our evidence does not rule out the possibility of someone unconnected with the university having played the thief. Nonetheless the inference is of the type which would quite often lead to a correct conclusion; the probability of the inference is increased when we consider that had someone other than an interested party committed the theft other valuables would have disappeared as well, something that did not occur.

Let's try another slightly less morbid example.

You notice one morning that your laboratory instructor is highly irritable and his coloring is slightly on the green side. You know also that severe headache can, as it were, put one off color in just this way. And it is well known that the graduate students of the science division had a gathering the night before. You conclude, therefore, that your instructor is enjoying the usual hangover that follows collegiate celebrations of this sort. But what does the inference involve? We have first a proposition asserting the existence of a particular observed state of affairs, viz. *that the instructor is irritable.* Another proposition, i.e., *that there was a graduate students' party the night before,* takes its place beside the first proposition. These are *the data.* Along with these data we entertain a brace of more complex propositions which assert as a kind of rule, or principle, both *that all (or most) guests at graduate school parties end up with hangovers* and *that all (or most) hangovers are accompanied by irritability and a slightly greenish tint to the cheeks*. But that *the instructor has a hangover* does not follow necessarily from these premises. His irritability might derive from many other possible causes. Nonetheless the inferences will lead to a true conclusion in a large number of cases. And we test the truth of our generalization (or induction) by deducing its consequences and seeing whether they hold in new situations.

Another example: We all know that various substances, like oxygen, copper, chlorine, sulphur, hydrogen, etc., when they combine chemically do so according to fixed ratios of their respective weights. And when the same amount of one substance, say chlorine, combines with different amounts of another substance, say oxygen, to form different compounds, it does so in ratios that are small integral multiples of the number *one*. (This is the observed event, albeit a highly complex one.) We know further that when different clusters of ordinary solid bodies are composed of units that are similar in weight and mechanically indivisible, the clusters could be combined only in respect to integral ratios of their weights, just as any note sounded on a piano (where the distances between adjacent notes is fixed and indivisible) stands to all other notes in a determinate numerical relation (something that is unfortunately not true of instruments like the violin). From this consideration we may put it that if each of the substances mentioned—oxygen, copper, chlorine, sulphur, hydrogen, etc.—were composed of similar atoms, or mechanically indivisible particles, the substances would combine in integral ratios of their weights (this is a kind of general rule). We conclude that such substances *are* composed of atoms.

Now from the point of view of logical implication this inference is hopelessly invalid, just as invalid as the argument which runs, "All persons named John are male, this person is a male, therefore this person is named John." This is, of course, a ludicrous example; but the scheme we have been considering is identical in form, for it runs, "All substances composed of atoms combine in integral ratios of their weights; oxygen, copper, chlorine, sulphur, and hydrogen are composed of atoms." It is, of course, possible that the cause of these substances combining in ratio to their weights is one altogether different from the assumed atomic constitution of matter. Nonetheless our evidence in favor of this assumption is highly probable because the general proposition that *mat-*

ter has an atomic structure generates all sorts of other consequences that have been found to be true in observation and experiment.

Our daily lives are, of course, riddled with situations like these, where we must reason, not along lines of pure logic or mathematics, but in terms of what will *probably* be the case. If a philosophy undergraduate refused to walk on wooden or concrete floors because it has not been proved absolutely that they will not disintegrate or explode, we should think him to be taking his studies much too seriously. We might even question his sanity.

Yet undoubtedly, as long as we are short of omniscience and do not know all of the future, our generalizations are logically fallible, i.e., probable. It is not logically impossible that the floor disintegrate one minute from now.

But what exactly is the nature of this concept of *probability?* It is a very loosely used word, full of snares and ambiguities as we ordinarily employ it. And, indeed, logical theorists are forever disputing as to what is the correct analysis of the term. Despite all this conceptual confusion, however, we continue to plan for births, deaths, marriages, holidays, experiments, etc. on the basis of rational evidence which, although not conclusive, is, we are sure, *probable*.

As we begin to thread our way through the logical cluster called *probability* it may help to keep one rather rough and provisional definition before us. That definition is this:

An inference is probable if it is one of a class of arguments such that the conclusions are true with a certain relative frequency when the premises are true.

Consider first the familiar observation that magnetized bars of iron, needles, nails, etc. have two unlike poles. Unlike poles attract one another and like poles repel each other. Moreover not all metals are equally magnetizable. Even the best metal magnets lose their magnetic properties in time. How can all this be explained?

One possible explanation is this: We assume that all metallic substances are composed of small particles, fundamental magnetic units. Each of these is a permanent magnet with two poles. Each is capable of rotation around a fixed center. In one arrangement these particles will neutralize one another completely. The entire bar then will display no magnetic properties. In another arrangement, only *some* of these particles will neutralize each other. The entire bar will then show magnetic properties. Changes in the relative positions of these particle-magnets, and the ease of their rotation about fixed centers, will then explain why a metal bar acquires or loses magnetic properties.

Again, this hypothesis as to the nature of magnetism is not *demonstrated* logically when it is shown that its consequences agree with observation. To think it is is to commit the fallacy known by logicians as the affirmation of the consequent, the self-same fallacy that vitiated our earlier arguments about John and all males, and about certain chemical elements and the atomic theory. Other assumptions might equally well explain magnetism. Indeed, we can never be certain that our own hypothesis will not one day turn up logical consequences that disagree with actual observations. Nonetheless, our hypothesis about magnetism is probable in the sense indicated before.

This example should make it clear that arguments within natural science are not a kind of second best to arguments typical of mathematics and logic (which are necessary and complete demonstrations). One must come to see the reasonings of natural science in a wholly different way, for there are no necessary demonstrations at all in natural science, so there is no point in characterizing in such terms the reasoning that does typify natural science.

Let's take our magnetism argument and recast it in the form of what logicians call an hypothetical syllogism. This is done as follows:

1. If each metal is composed of imperceptibly small particle-magnets free to rotate around a fixed axis, then these metals will exhibit phenomena of magnetism under specified conditions (this is our general rule in this piece of reasoning).

2. But this, that, and other metal bars exhibit phenomena of magnetism under the specified conditions (this is a true statement of what is observed).

3. Conclusion: Therefore each metal is composed of imperceptibly small particle-magnets, each free to rotate around a fixed axis (a probable conclusion though formally invalid as we saw before).

In a series of such inferences made when the premises are true the conclusion would be true with a considerable relative frequency.

"But what does this mean?" you may ask. "What is it for the conclusion to be true a certain fraction of the time that the premises are true? How can we ever *know* that this particular conclusion is true when the particle-magnets are, almost by definition, imperceptibly small? Does not this make it impossible ever to establish the truth of the conclusion?"

The answer to this objection is as follows: We must distinguish the *meaning* of a proposition from the *evidence* in favor of its truth. So far as I know we have no evidence for believing there to be a twin-peaked mountain on Uranus. And the proposition "There is a twin-peaked mountain on Uranus" is wholly meaningful though there is at the moment no evidence for or against it. So, too, the meaning of probability in general may be quite definite, though in specific cases we have not enough evidence to determine its particular numerical value. It is possible and desirable to speak of the *status* of conclusions in probability arguments independently of the question of what evidence there is for or against those conclusions, just as it is possible and desirable to speak of the formal properties of spheres and cubes independently of the question whether in a given case the object examined is in fact a sphere.

The complete analysis of the probability of propositions which cannot be *directly* verified, e.g., that metals are composed of imperceptibly small particle-magnets, is rather more complicated than anything so far indicated. For in such cases the argument depends on the hypothesis involved having as its logical consequences propositions that *may* be directly verified, and which may lead to the observation of phenomena *other* than those for which the theory was originally proposed.[1] As we shall see, the evidence upon which a set of hypotheses is probable consists of *samplings* made from all the necessary consequences of the hypotheses. With respect to our magnetism example this comes to no more than the following:

The hypothesis that all metals are composed of imperceptible, freely rotating particle-magnets has as one of its logical consequences that hammering or heating a magnet should make it lose its magnetic properties. Either of these procedures would upset the rather delicate alignment of the axes of electronic spin in the metal and destroy the gross magnetism of the entire metal body, which (according to our premises) depends on such atomic alignment. This phenomenon does actually occur. Beating or heating a magnet destroys its magnetic powers. A more complete statement of our argument, therefore, should run this way:

1. If this, that, and other metals are composed of imperceptibly

---

[1] Cf. Braithwaite, *Scientific Explanation, op. cit.*, Chapters I–III.

small particle-magnets, then they *should* exhibit phenomena of magnetism under specified conditions (this again is our general rule).

2. But these metals exhibit magnetic properties under the specified conditions (this is a true statement of what is observed).

3. Conclusion: Therefore, each metal is composed of imperceptibly small particle-magnets, each free to rotate around a fixed axis. This conclusion, however, is an inferred fact, or perhaps it is a theory; it is incapable of direct verification.

4. If, however, metal bars *are* composed of permanent magnets, then hammering or heating a magnetized needle, e.g., will make it lose its magnetic properties (this is an inferred fact which is directly verifiable).

The purpose of such a deductive elaboration of the premises is just to supply us with as many verifiable consequences as possible. The argument *for* the hypothesis proceeds *from* a proposition known to be true, *to* other propositions directly verifiable, *via* the hypothesis (which is not itself directly verifiable). The argument is thus rather more complicated than our earlier bits of reasoning. This does not, however, alter the inference employed. The hypothesis is probable on the evidence because the argument for it belongs to a *class* of arguments in which the relative frequency of the truth of the conclusion does not necessarily amount to the number *one*.

Let me now state in order some of the essential characteristics of probable inferences. We have already noted some of these; others are in anticipation of discussions to follow.

1. Probable inference, like all inference, is based upon certain relations between propositions. No proposition is probable *in itself*, but only in relation to the other propositions serving as evidence for it.

2. But whether or not a proposition has a degree of probability does not depend on the state of mind of the person who entertains the proposition. Questions of probability, like questions of logical validity, are to be decided entirely on objective considerations, and not on the basis of whether or not we feel an impulse to accept a conclusion.

3. An inference is probable only in so far as it belongs to a definite class of inferences in which the frequency with which the conclusions are true is some determinate ratio of the frequency with which the premises are true. This is to say that the very *meaning* of probability involves

relative frequencies. Other philosophers might disagree with this, and for their arguments I advise you to consult the book entitled *Probability and Induction*, by William Kneale.

4. Since the probability of a proposition is not an intrinsic character of it, the *same* proposition may have different degrees of probability, depending on the evidence marshaled in its support. (I will not raise the subtle and difficult question of whether we could in two situations encounter the same proposition, i.e., a sentence with the same meaning, where the evidence differs in each case.)

5. The evidence in support of a proposition may have different degrees of relevance; e.g., the frequency with which Jupiter's inner satellite eclipses is not thought to be relevant to determinations of the refractive index of glycerine, even though indirectly and by way of a complicated theory the two phenomena can be seen as not altogether unrelated. In general, that evidence is chosen for a proposition which will make its degree of probability as great as possible. But the relevance of evidence cannot be determined on formal grounds alone; nor can it be effectively assessed without some at least provisional theories as to the nature of the subject matter with which one is dealing.

6. While the *meaning* of the *measure* of the probability of an inference is the relative frequency with which that type of inference leads to true conclusions from true premises, it is nonetheless true that in most cases the definite numerical value of the probability is not known. This will become clearer as we proceed. Compared to the number of cases wherein we judge a proposition to be probable on the basis of certain evidence, we are very infrequently able to determine the exact degree of such probability. This in no way qualifies what I have so far said about the *concept* of probability, however. We can quite well know what probability is *in general* without being able to supply in a given case evidence that will be sufficient to assign numbers to the value of the probability.

The remainder of this chapter will be given over to a discussion of the elements of the calculus of probability. Actually I wish only to begin this discussion here; I shall conclude it in the next chapter.

Most ancient and modern discussions of probability are addressed to questions to which numerical answers can be given. The prospect of success at gambling and games provided a strong initial impetus to the study of probability. "What is the probability of getting three heads in four throws with a coin?" "What are the chances of getting seven with a pair of dice?" "What are the chances of black turning up on the next spin?" Such questions as these were the prime concern of 18th Century probability theorists and mathematicians. Today almost every branch of research science finds that it too must employ a mathematical calculus of probability in one form or another.

Immediately the matter is put in this way, however, a potential difficulty crops up. For mathematics studies only the *necessary* consequences of a given set of assumptions. It is not in the least concerned with the objective truth or falsity of its premises. Only the business of tracing the implications of quite arbitrary premises absorbs the mathematician; the fact that the Bishop of Rochester does not walk diagonally fails to make the chess player give up his game as false.

Hence, no *purely* mathematical theory can determine the probability of propositions about matters of fact. True, a mathematical theory can determine a factual proposition's probability when assumptions are made concerning it: The theory can, for example, tell us exactly what are the consequences of these assumptions. Hence the theory of probability can be a purely mathematical theory only when it is restricted to questions of necessary inference, that is, only when it is a formal game built on largely arbitrary definitions and rules. Viewed in this way, the theory of probability is just a branch of pure mathematics, so the elements of the calculus of probability I am about to bring forward now are to be seen for what they are: purely mathematical devices for developing and unpacking other, previous assumptions about matters of fact.

Consider first this simple problem: What is the mathematical probability of obtaining a head when a penny is tossed? But let us not now discuss the probability of the proposition "This penny will fall head uppermost" (as we have been doing). Let us adopt the older, more usual classical terminology of *events;* let us discuss the probability of the *event* "getting a head." *Heads* and *tails* are, then, the possible *events*, or the possible *alternatives*. If we should be hoping for *tails*, then *tails* is said to be the *favorable* event, all other possibilities being unfavorable.

The mathematical probability is then defined as the fraction whose numerator is the number of possible favorable events, and whose denominator is the number of possible events (i.e., the sum of the number of favorable and unfavorable events), all provided that each of the possible events is equiprobable, that is, equally probable.

If, therefore, a penny has two faces and the coin can fall in no other way than to yield heads or tails, the probability of getting a head is ½. In general, if f is the number of favorable events, and u is the number of unfavorable, and if the events are equiprobable, the probability of the favorable event is defined as

$$\frac{f}{f + u}$$

Clearly, this is always a proper fraction; its values will always be between 0 and 1. A probability of 0 indicates that the event in question is impossible (e.g., the probability of someone's building a *perpetuum mobile* of the second type is 0). A probability of 1 indicates that the event in question must necessarily take place (e.g., the probability of there being an uncertainty in our present estimations of the position of an atomic particle with known momentum is 1).

Now, the condition that the possible events must be equiprobable is of fundamental importance. The older mathematical theory cannot get under way without this. And yet *equiprobability* is remarkably difficult to define. In fact, some of the most serious errors in the history of probability theory derive from the attempt to provide such a definition.

Roughly, what is meant is that one possibility should occur as frequently as any other. And it has been often maintained that two possibilities are equally probable when we know no reason why one should occur rather than another, an argument which is dangerously like observing that, since a donkey standing midway between two haystacks has no apparent reason for choosing one haystack over the other, the animal will remain stationary and starve to death on the spot. It is this same *principle of the equal distribution of our ignorance* (or *principle of indifference,* as it is sometimes euphemistically called) that vitiates Archimedes' classical demonstration of the principle of moments.

Whatever the nature of these difficulties, however, it is not the mathematician's business to discover criteria for equiprobability. He is concerned only to find the necessary implications of such an assumption (irrespective of questions of its truth or falsity). But the importance of the equiprobability condition is clear from the following example:

Suppose we wish to learn the probability of getting a six with a die. We argue as follows: There are two possibilities, either getting a six or not getting a six; one of these is favorable, consequently the probability is one-half (½).

This anwer, however, may turn out false unless we make the factual assumption that the two alternatives, six and other than six, are equiprobable. But this factual assumption would not generally be made, because, as experience has taught us, the possibility of getting other than six is made up of five subsidiary alternatives—getting a one, or a two, or a three, or a four, or a five, each one of which is equiprobable with getting a six (if, that is, the die is fair and not loaded). Hence, if the six sides are assumed to be equiprobable, the probability of getting a six with a die is 1/6.

For the same reason it would be unwise to translate what we have learned about the probability of getting a head with a tossed penny as if this information were applicable to all coins. The English threepenny bit, for example, presents besides a chance of a head or a tail a distinct third possibility, which would qualify by a small but theoretically important amount our probability estimate of ½ for heads or tails.

We will proceed in the next chapter with this survey of the mathematical foundations of the theory of probability. Then we will discuss the probability of a joint occurrence of two events and the probability of a disjoint occurrence of one but not both of two events. We will also examine several rival interpretations of what the theory really is from a philosophical point of view.

But before leaving the question of equiprobability I must bequeath the following puzzle to you:

The probability of life existing on a remote planet as yet unobserved must be ½, since we have no more reason to assume the existence of life there than to deny it. But then the probability that there are elephants on such a planet is also ½, our ignorance of the existence of elephant life being equally deep. So too the probability of the existence of cats, dogs, coelacanths, and Siamese twins is also ½ in each case. But the probability that any combination of these last events will not take place is the *product* of the corresponding probabilities, as you will see in the next chapter; that is, the probability of no elephants, no cats, no dogs, no coelacanths, no Siamese twins, . . . is ½ times ½ times ½ times ½, . . . etc. This amount can be made as small as we please by finding more and more (independent) living organisms and tacking their names onto the ever growing proposition about the probability of such organisms not existing on the planet. Hence the probability that no life of any kind exists on the planct can be made as small as we please, making it ever more probable that life does exist there. This flies in the face of our original 50/50 view of the likelihood of life.

There had better be something wrong with this argument, because if there isn't the universe is likely to get pretty cluttered up with elephants, cats, and dogs.

23 | Using and Interpreting the Probability Calculus

WE HAVE been discussing some of the fundamental features of the classical calculus of probability. The equiprobability of rival events was seen to be a major assumption of the calculus. Moreover, it is an assumption which the pure mathematician need not bother to justify. He need only present his formal system as follows:

If all the alternatives are equiprobable, then my system provides the complete machinery for calculating the probability of alternative events occurring.

But *whether* actual alternatives, say in a laboratory experiment, *are* equiprobable is not for the pure mathematician to say. He is concerned only to work out the consequences of a system based upon that assumption.

Let us look further at the system itself, therefore.

It may be said that the main burden of the probability calculus is to determine the chances of a *complex* event occurring, from a knowledge of the probability of its component events. We require a few further definitions:

1. Earlier an argument was said to be a *probable* argument if it is one of a *class* of arguments such that when the premises are true the conclusions are true with a certain relative frequency.

2. The mathematical probability of an event is, then, defined as

the fraction whose numerator is the number of possible favorable events and whose denominator is the number of possible events (i.e., the sum of the number of favorable and unfavorable events), provided that each of the possible events is equiprobable with the others. Thus:

$$\frac{f}{f+u}$$ where "f" = favorable events,
"u" = unfavorable events,
and where f and u are equiprobable.

3. *Equiprobability*, as we saw, is itself very difficult to define. But what is meant is that one possibility should occur as frequently as any other. It has been maintained that two possibilities are equally probable when we know no reason why one rather than the other should occur.

These definitions were tendered in the last chapter. (I should state that in offering them I was not going back on my general *reluctance* to appeal to definitions in philosophical discussion. Be quite clear that these definitions and those of the present chapter are offered, not in a spirit of pseudo-exactitude or precision, but only in the spirit of brevity. Further discussion on each of these points is not ruled out by the fact that I have called them *definitions*.)

A further definition has to do with the *independence* of events:

4. Two events are *independent* if the occurrence of one is not affected by the occurrence or non-occurrence of the other.

Of course, we may note again that the assertion that two events are in fact independent is a *material,* or *factual,* assumption which must be explicitly stated. Serious errors can arise when the calculus is applied in physical and biological contexts where there are inadequate grounds for assuming the independence of events; and *ipso facto* errors arise when the whole principle of independence is neglected.

This puts us in a position where we can discuss the probability of a joint occurrence of two independent, equiprobable events, to which we shall turn now.

For example, what is the probability of getting two heads in tossing a coin twice or tossing two coins simultaneously? This is a complex event whose component events are (a) getting a head on the first throw and (b) getting a head on the second throw. If these two events are independent, and if the probability of getting a head on each throw is 1/2, then (according to the calculus) the probability of the joint occurrence of heads on each throw is the *product* of the probabilities of heads on each individual throw. Thus the probability of getting two heads is $1/2 \times 1/2$, or $1/4$.

We can see by inspection why this result follows necessarily from our assumptions; all we need to do is enumerate all the possible events when we throw the coin twice. These are:

1st throw	*2nd throw*
H	H
H	T
T	H
T	T

Hence, on the assumptions made, there are just four equiprobable possibilities. Of these, only one, HH, is favorable. Thus the probability of getting two heads is, again, 1/4, just as we saw when we multiplied together our independent probabilities of 1/2.

In general, if a and b are two independent events, and P(a) is the probability of the first event, and P(b) is the probability of the second, the probability of their joint occurrence is P(ab) or P(a) × P(b).

Great care must be taken, when calculating the probability of such complex events, to enumerate *all* the possible alternatives. If, for example, we require the probability of getting *at least* one head in two throws with a coin, our enumeration above shows three favorable events, namely *HH*, *HT*, and *TH*. The probability of getting at least one head is therefore 3/4. But even quite eminent scientists have sometimes failed to note all the alternatives. The great D'Alembert, for example, estimated that selfsame probability of getting at least one head in two throws of a coin, not as 3/4, but as 2/3. He argued that the possibilities were just *H*, *TH*, *TT*. If a head comes uppermost on the first throw it is not necessary to throw again in order to get at least one head. This reasoning led D'Alembert to collapse the alternatives *HH* and *HT* into the single possibility *H*, giving him his ultimate calculation of 2/3 probability of turning up at least one head in two throws. The analysis is as faulty as the puzzle about elephants on an unexplored planet with which the previous chapter concluded: The enumerated possibilities are not equiprobable. The first alternative *H* must be regarded as including *two* component alternatives each of which is equiprobable with the others, while *H* is twice as probable as the others.

We can sometimes calculate the joint occurrence of two events even when the events are not wholly independent. Thus, imagine a bag containing 3 white buttons and 2 black buttons. It is taken to be equi-

probable that any one of the five buttons may turn up as we remove them from the bag one at a time. What is the probability of withdrawing 2 white buttons in succession in the first two drawings (when the buttons are not replaced after having been withdrawn)?

Obviously, the probability of drawing a white button out of the bag is 3/5. Should a white button in fact be withdrawn on the first trial (and not be replaced), there remain 2 white and 2 black buttons in the bag.

Hence the probability to get a second white button *if the first button drawn has been white* is 2/4. It follows that the probability of getting 2 white buttons in two withdrawals from the bag is $3/5 \times 2/4$, or $3/10$. (This again conforms to our earlier definition of mathematical probability. For the total number of ways of drawing any 2 buttons (irrespective of color) from a collection of 5 is $\frac{5 \times 4}{1 \times 2}$, or 10. And the number of ways of drawing 2 white buttons from a collection of three whites is $\frac{3 \times 2}{1 \times 2}$, or 3. This is the number of favorable events. Therefore, the probability of getting 2 white buttons under the conditions set out before is 3/10, just as by our other derivation.)

In general, if P(a) is the probability of an event (a), and P(b/a) the probability of an event (b) when (a) has already occurred, the probability of the joint occurrence of (a) and (b) is $P(ab) = P(a) \times P(b/a)$.

Having discussed the probability of a joint occurrence, we may turn now to the problem of estimating the probability of *disjunctive* events. This is the probability that *either one* of two equiprobable and independent events will occur. First, however, we must make clear what is meant by *strictly alternative* events, i.e., *disjunctive events*. Two events are disjunctive if both cannot simultaneously occur (i.e., if one does occur, the other cannot do so). In tossing a coin, the possibilities (heads and tails) are assumed to be disjunctive. It can be shown that the probability that either one of two disjunctive events will occur is the *sum* of the probabilities of each.

What, then, is the probability of getting either 2 heads or 2 tails in tossing a coin twice—assuming that the probability of getting a head on a single toss is 1/2 and assuming also that any two tosses are independent of each other? As we saw, the probability of getting 2 heads is the product of the probabilities of getting 1 head on the first throw and 1 head on the second, or 1/4. Similarly, by the same reasoning, the probability of getting two tails is 1/4. Hence, the probability of getting *either* 2 heads *or* 2 tails is 1/4 + 1/4, or 1/2.

The same result comes directly from a consideration of our four possible events *HH, HT, TH, TT*. Two of these are favorable to either 2 heads or 2 tails. Thus the required probability is 2 out of 4, or 1/2.

In general, if P(a) and P(b) are the respective probabilities of two exclusive events (a) and (b), the probability of obtaining either is $P(a + b) = P(a) + P(b)$.

These two theorems, the product theorem for independent events and the addition theorem for exclusive events, are fundamental in the calculus of probability. These theorems, and extensions from them, provide the means whereby quite complicated problems can be easily solved.

Suppose, for example, we withdraw one button from each of two bags. The first bag contains 8 white buttons and 2 black buttons. The second bag contains 6 white buttons and 4 black buttons.

382 | *Chapter 23*

Each withdrawal is assumed to be equiprobable to every other withdrawal.

What, then, is the probability that, when we withdraw a button from each bag, *at least one* of the buttons will be white?

Well, the probability of withdrawing a white button from the first bag is 8/10. The probability of getting a white button from the second bag is 6/10. Shall we *add* these fractions, then, to obtain the probability of withdrawing a white button from either bag? *No,* it would be a serious mistake to do so. The result of so doing would be an answer of 14/10 or 7/5, a number greater than 1. This would make it impossible to withdraw two black buttons from the bags, and this is absurd (as you can see).

We simply cannot add in this case. Why? Because the events are not *exclusive:* The expression *at least one* conflates distinct withdrawals in the same manner we criticized earlier in the D'Alembert example. However, we may calculate as follows:

The probability of *not* getting a white button from the first bag (i.e., the probability of getting a black button) is 2/10. And the probability of not getting a white button from the second bag is 4/10. Therefore, assuming these withdrawals to be independent, the probability of getting a white button from *neither* the first *nor* the second bag is 2/10 × 4/10, or 8/100. Consequently, since we must either get no white button from either bag or get at least one white button from either, the probability of getting at least one white button is 1 − 8/100, or 92/100.

Here is another, even more complex example: Assuming that in the simultaneous tossing of several coins each throw is equiprobable and independent, what is the probability of getting *three heads in tossing five coins?* This problem will bring us face to face with a most important formula in the calculus of probabilities.

Perhaps we ought to reason as follows: Since five coins are thrown, and the probability of getting a head on each one is 1/2, the probability

Using and Interpreting the Probability Calculus | 383

of getting three heads is $1/2 \times 1/2 \times 1/2$, i.e., $1/8$. But we want *exactly* three heads, no more and no less, so the other two coins must fall tails, the probability of which happening is $1/2 \times 1/2 = 1/4$. May we conclude, then, that the probability of getting *first* three heads (i.e., three heads and two tails) is $1/8 \times 1/4 = 1/32$?

This result, however, is far from correct, which will be seen readily in the writing out of all the possible ways in which the five coins can fall.

The possible (and equiprobable) alternatives are:

1 possible way of getting 5 heads, 0 tails   H H H H H

5 possible ways of getting 4 heads, 1 tail
$\begin{cases} \text{H H H H T} \\ \text{H H H T H} \\ \text{H H T H H} \\ \text{H T H H H} \\ \text{T H H H H} \end{cases}$

10 possible ways of getting 3 heads, 2 tails
$\begin{cases} \text{H H H T T} \\ \text{H H T H T} \\ \text{H T H H T} \\ \text{T H H H T} \\ \text{H H T T H} \\ \text{H T T H H} \\ \text{T T H H H} \\ \text{H T H T H} \\ \text{T H T H H} \\ \text{T H H T H} \end{cases}$

10 possible ways of getting 2 heads, 3 tails
$\begin{cases} \text{H H T T T} \\ \text{H T H T T} \\ \text{H T T H T} \\ \text{H T T T H} \\ \text{T H T T H} \\ \text{T T H T H} \\ \text{T T T H H} \\ \text{T T H H T} \\ \text{T H H T T} \\ \text{T H T H T} \end{cases}$

5 possible ways of getting 1 head, 4 tails
$$\begin{Bmatrix} H & T & T & T & T \\ T & H & T & T & T \\ T & T & H & T & T \\ T & T & T & H & T \\ T & T & T & T & H \end{Bmatrix}$$

1 possible way of getting 0 heads, 5 tails   T T T T T

There are, then, 32 equiprobable possibilities. Ten of these are favorable to the event. Thus the probability of getting three heads and two tails is 10/32—a result, then, ten times as great as the 1/32 we obtained by the incorrect method.

Why the earlier method was incorrect should now be obvious. That method failed to take into account the different orders, or arrangements, in which three heads and two tails can occur. We require, therefore, a method of evaluating the numbers of different arrangements which can be made of five symbols three of which are of one kind and two of which are of another kind. There is a very simple formula which will yield the required probabilities quite easily.

The number of possibilities favorable to each category of this complex event (i.e., 1 possibility for 5 heads, 0 tails; 5 possibilities for 4 heads, 1 tail; etc.) is nothing other than the appropriate coefficient in the expansion of the binomial $(a + b)^5$, which is $a^5 + 5a^4b + 10a^3b^2 + 10a^2b^3 + 5ab^4 + b^5$.

In general, it is rigorously demonstrable that if p is the probability of an event, and g is the probability of its sole exclusive alternative, then the probability of a complex event with n components is obtained by selecting the appropriate term in the expansion of the binomial $(p + g)^n$. This expansion is as follows:

$$(p + g)^n = p^n + \frac{n}{1} p^{n-1}g + \frac{n(n-1)}{1 \times 2} p^{n-2}g^2 + \frac{n(n-1)(n-2)}{1 \times 2 \times 3} p^{n-3}g^3 + \ldots + g^n$$

Let's have a final example of this binomial formula. Into a bottle of ink have been dropped two white buttons and one red button; we are to make four drawings from the bottle, replacing into the bottle after each drawing the button we fished up. We assume that our withdrawals are equally probable, and that they are independent, having been thoroughly mixed up after each replacement. What, then, is the probability of withdrawing three white buttons and one red button in four tries? Well, the probability of fishing out a white button in the first try is $p = 2/3$ and the probability of getting the red button is $g = 1/3$. To obtain the required answer we need only expand $(p + g)^4 = p^4 + 4p^3g + 6p^2g^2 + 4pg^3 + g^4$, and substitute the indicated numerical values in the term which represents the probability of getting three white and one red button. This term, clearly, is $4p^3g$ and the required probability is $4 \times \left(\frac{2}{3}\right)^3 \times \left(\frac{1}{3}\right)$, or in other words 32/81.

Naturally we have only begun tallying the numbers of possible theorems of the calculus of probability. It should be amply clear that from here on the calculus develops like the purely mathematical system it is. But before proceeding to discuss certain interpretations of the calculus I must reiterate a warning to which I earlier called attention. The warning is this:

The mathematical theory of probability studies the necessary consequences of our assumptions about a set of alternating possibilities; it cannot inform us as to the probability of any actual given specific event or set of events.

How, then, can we determine the probabilities of actual, given, specific events? Under what material circumstances may the theorems of the mathematical calculus of probability be applied to the real world?

Our first interpretation of the probability calculus will be as a kind of measure of the *strength of our beliefs* about natural events.

According to the logician De Morgan, "probability" means "the state of mind with respect to an assertion, a coming event, or any other matter on which absolute knowledge does not exist."[1] De Morgan holds that the expression "It is more probable than improbable" means "I believe that it will happen more than I believe that it will not happen."

God, apparently, would have no need of probable inference. He would know every possible proposition as wholly true or as wholly false. But since we are none of us gods, and hence lack omniscience, we must rely on probabilities. Probability, then, is a kind of measure of our

---

[1] Augustus De Morgan, *Formal Logic* (1847), ed. A. E. Taylor (London: Open Court, 1926), p. 199.

ignorance. When we *feel* altogether sure that an event will take place, its probability is 1. When, on the other hand, we are possessed with the overwhelming belief that the event in question is impossible, then its probability is 0. When our belief regarding a possible event is intermediate between the certainty of its occurrence and the certainty of its non-occurrence (as is the normal state of affairs in natural science), the probability is some fraction intermediate between 1 and 0.

On this interpretation the calculus of probability can be employed only when our ignorance is distributed equally between several alternatives.

As we have so often seen, the mathematical calculus of probability *can* answer the question "What is the probability of three heads turning up when a coin is tossed three times?" but it can do so *only* when outside information is supplied concerning

1) the number of alternative ways in which the coin can fall,
2) the equiprobability of these alternatives, and
3) the independence of the different throws.

When probability is interpreted as a measure of belief, this information is easily obtained. For, as we had occasion to note earlier, the theory employs the famous (or *infamous*) criterion called the *principle of indifference,* or the *principle of insufficient reason,* according to which the following obtains:

If there is no known reason for describing a subject of discourse in one way rather than in any of a number of alternative ways, then (relative to this knowledge) the assertions of each of these alternative descriptions have an equal probability.

And similarly for independence:

If there is no known reason for believing that two events are independent, it is just as probable that they are independent as it is that they are not.

Two alternatives are equally probable if there is "perfect indecision," if "belief inclines one neither way," if "our ignorance is evenly distributed," etc. Thus when we are completely ignorant about two alternatives, the probability of one of them occurring must be (on this view) 1/2. If, then, we are confronted with a strange coin, since we have no reason for believing *heads* a more likely result than *tails,* we must say that the probability of tossing heads is the same as the probability of tails.

But there is something very suspicious about this interpretation of the foundations of probability.

First of all, our ability to predict and control events by means of probable inferences—as, for example, in thermodynamics and statistical mechanics—is altogether inexplicable if such inferences rest on nothing but our ignorance, or the evenly dissipated strength of our beliefs. How hopelessly disastrous it would be for an insurance company to conduct its business on the basis of estimating the feeling of expectation which its officers and shareholders may have. Conversely, consider how much in evidence such strong feelings actually are in the field of politics, government, and public administration, and compare the presence of these feelings with the relative absence of any effective use of probability techniques in the civic sphere.

A second mark against the rendering of probability theory in terms of strength and distribution of beliefs is this:

Whose beliefs are to be measured? That beliefs about the same event vary with different people, both in content and in intensity, is notorious. No lawyer can proceed very far in his career without learning this. We are all sufficiently mercurial and temperamentally uneven to have felt our beliefs with respect to some event (at some time) range all the way from certainty to utter doubt, sometimes for quite trivial reasons. Which of all our possible states of expectation regarding, say, next year's Indiana-Purdue game is to be taken as a measure of probability?

Again, probabilities can be added and multiplied, but can beliefs be dealt with in a corresponding manner? There are no known operations by which beliefs can be added; indeed, it is difficult to find any sense in the expression "adding beliefs." All the morals of our discussion of additive and non-additive properties apply here.

Finally, this interpretation of probability can lead us straight into absurdity (unless, that is, we severely restrict the range of its application). Suppose, for example, we know that the volume of a unit mass of some substance lies between 2 and 4. On the interpretation of probability in terms of strength of beliefs, it is just as probable that the specific volume lies between 2 and 3 as it is probable that it lies between 3 and 4. But if the volume of a unit mass is v, the density is $1/v$. Hence the density of this substance must lie between $1/2$ and $1/4$ (i.e., between $4/8$ and $2/8$) and therefore it is just as likely that it will lie between $1/2$ and

3/8 as between 3/8 and 1/4. This, however, is equivalent to saying that it is just as likely that the *specific volume* will lie between 2 and 8/3 as between 8/3 and 4. This contradicts our first result, where we found it to be just as probable that the volume would lie between 2 and 3 as between 3 and 4.

This sort of difficulty made the application of the classical calculus of probability, with all its apparatus of equiprobability, independence, and distribution of beliefs, most difficult to effect. It forced theorists to look in a new direction for a workable interpretation of the calculus. It forced them to interpret probability as the *relative frequency* with which an event will occur in a class of events.

When we say that the probability of a given coin falling heads is 1/2, we are not necessarily saying that we have no strong reason for believing a tail to be more likely than a head—i.e., we are not making a public confession of our ignorance. What we mean by saying that the probability of heads is 1/2 is that, as the number of tosses increases indefinitely, the ratio between the numbers of heads thrown and the total number of throws will be about (i.e., will not materially differ from) 1/2.

This, of course, constitutes an assumption or an hypothesis about the actual course of nature—as such it requires factual evidence. Such evidence either may be deduced from previous knowledge as to the nature of things or may be statistical. We may, for example, infer from our knowledge that pennies are symmetrical and (from our knowledge of mechanics) that the forces making a penny fall tails are bound to balance those which make it fall heads. Or we may rely on more purely empirical observation as evidence that *in the long run* neither heads nor tails predominate. In science and in our everyday affairs we rely on both types of evidence.

Statistical evidence, however, is apt to be more prominent, so we must be on our guards not to identify the meaning of a proposition with the actual amount of statistical evidence available for it at any given time. For, as we ought to remember quite clearly, though hypotheses about nature often make assertions about all possible members of a certain class (that's what they mean), we can never prove such hypotheses by any finite number of observations. But of several hypotheses the one that agrees best with statistically formulated observations is naturally preferable.

Perhaps now we can understand more clearly the practical function and application of the mathematical theory of probability.

Consider the hypothesis that the probability of the birth of a human male is 1/2. The calculus of probability can then be used to deduce, and predict, the frequency with which human families with two male children, or with two children of opposite sexes, should (in the long run) occur.

Suppose, however, that all the children born in Texas during one whole year are female. Does this disprove the assumption that the probability of a male birth is 1/2? Not in the slightest! The calculus cannot show that such an event is impossible, however improbable it may be. The calculus *can* show, however, that such an exceptional occurrence as, for example, 10,000 female births in sequence may be in conformity with (or would be less improbable than) some other assumption. A very large number of repetitions of "exceptional" occurrences may diminish the probability of our original hypothesis and increase the probability of another one. This has actually happened in *vital statistics* and the probability of a male child in Europe is now set at 105/205 rather than 1/2.

The calculus of probabilities thus systematizes our experience on the simplest available assumptions. Again, no given hypothesis can be conclusively refuted by any finite number of observations. Nonetheless, statistical results can show certain hypotheses to be less probable than others.

Hence, on the relative frequency interpretation, probability is not concerned with the strength of subjective feelings. It is grounded rather in the nature of classes of events. Objective data are needed to determine their probability.

On this view, of course, it is meaningless to speak of the probability of a unique event. Even when we appear most obviously to be discussing the probability of specific individual events, it must be that we are speaking elliptically: We are really considering only that phase of the event in question in which it resembles other events of a certain kind. When it is said that the probability of getting a head with a *given*, particular coin on a *definite*, specific toss is 1/2, what is meant is that in a long series of such tosses heads will turn up about one-half of the time. And when we say that in tossing a coin twice the probability of two heads is

1/4, what we mean is just (again) that in a long series of sets of double throws double heads will turn up approximately 1/4 of the time.

Hence the relative frequency view of probability provides us with a telling caution against what might be called alternatively "the Monte Carlo fallacy" or "the gambler's fallacy" or "the expectant father's fallacy." For consider:

We join in a game of chance. The coin with which the game is played is assumed to be *fair:* We assume, that is, that the probability of heads is 1/2 and that the tosses made with it are independent. Suppose we have just encountered an amazing run of 50 tails and now we are obliged to bet on the next throw. Many players cannot help thinking that the probability that a head will turn up on the 51st toss is much greater than 1/2. For, after all, if the coin is fair, heads and tails must "even up," so to speak. This conclusion is, however, tragically erroneous and it indicates how fatal is the principle on which are based gamblers' systems devised to make winnings secure when playing with *fair, unbiased* gaming instruments. "Fatal" is just the word, too, as the suicide rate at Monte Carlo will indicate. If a coin is indeed fair, the 50 tails already tossed do not in the slightest affect the 51st toss, nor would the fact that the last 500 babies delivered at the Mill Road Maternity Hospital were female affect in the slightest Mr. Throckmorton's chance of having a son.

When we say that the probability of heads on the 51st throw is 1/2, or that the probability of a boy on the hospital's 501st delivery is 1/2, we are speaking elliptically of a very large series of similar events. If, on the other hand, the coin is not fair but loaded to favor heads, then clearly the probability of heads on the 51st throw is greater than 1/2— but this was exactly the situation on the first throw as well. Similarly, if Mr. and Mrs. Throckmorton are genetically so constituted as to favor the conception of a male baby rather than a female, then obviously Mr. Throckmorton's chances of becoming the father of a boy are better than 105/205, and the recent past record of deliveries at the maternity hospital is wholly irrelevant.

Let us note finally that no event is probable *intrinsically.* The probability of an event is assessable only in terms of its membership in certain classes of events or series of events. E.g., the probability of heads turning up when a coin is tossed by hand may be altogether different from the probability of heads when the coin is thrown from a mechanical cup. The name "heads" when used to apply in both cases is really being made to cover *two* different classes of events. In general, the class of events to which a given event belongs must be carefully described and explicitly noted when we are evaluating the probability of the event. Geneticists have learned this lesson the hard way.

Of course, there are some serious objections that can be raised against the frequency theory of probability as I have so far outlined it. The theory certainly is not capable (in any obvious way) of interpreting what we mean when we speak of a theory being *probable,* or probably true. Often we may be heard to say such things as that the heliocentric theory is more probable than the geocentric theory. But what can this possibly mean according to the frequency theory of probability? In this connection I urge you to read P. F. Strawson's chapter on *Probability* in his book *Introduction to Logical Theory*.[2] You might also think about how (i.e., in terms of what considerations) we should in fact be satisfied that one of two rival hypotheses in a science was more probable than the other (and on this consult Braithwaite, *Scientific Explanation*).

Moreover, we frequently assert singular and particular propositions of the form, "It is probable it will rain tonight," "It was improbable that Molotov and Dulles would ever have agreed on China," "It is probable that had Hitler invaded Britain in 1940 he would have won the war." Such statements find no very easy or natural interpretation within a frequency theory. Many thinkers, however, do not regard these objections as being fatal. They have devoted great effort and ingenuity to the altering of the technical expression of the frequency theory, correcting and strengthening every weak logical link until now there are several frequency-type theories of probability available to scientists, the major differences among them consisting largely in the varying stress they lay on the importance of *events, samples, statements,* and *limits* in their individual formulations. For further information about these several systems consult Nagel's monograph on *Principles of the Theory of Probability,* or Braithwaite, or Kneale.[3]

---

[2] P. F. Strawson, *Introduction to Logical Theory* (London: Methuen and Co., 1952), Chapter IX: "Inductive Reasoning and Probability."

[3] See the bibliography of the beginning of Part IV for exact references.

24 | Elements of Statistical Technique

WE HAVE surveyed the logical foundations and some of the principles theorems of the mathematical calculus of probability. We have considered, also, some of the important interpretations of that calculus for certain experimental contexts.

Why should we be so confident, however, that the mathematical calculus of probability will have an interpretation, or be applicable to the empirical world? What is it about our experimental results which makes assessing them in statistical, or probable, terms a desirable procedure?

During the previous chapters we have discussed several ways in which scientists seek to make their ideas clear, to themselves and to others. Precise observations, exact measurements, and careful numerical analysis have all yielded large collections of reliable data, often embarrassingly large collections. And so, very early in the history of recent science, the need was felt for methods of handling the great multitude of numerical data. Statistical analysis is the result.

For example: Suppose we are interested in comparing the lengths of the forelegs of cattle, our suspicion being that this leg length is markedly affected by certain environmental factors. We may, accordingly, measure (or have measured for us) the legs of several million head of cattle. However, several million figures having to do with the leg length could not possibly be compared with an equally unwieldy collection of data from studies of the environment *unless we found some way of compressing each set of data into something conceptually more manageable.* We are all limited psychologically; we can attend only to a relatively small number of things at any given time.

The physicist is no better off. If he measures again and again the wave lengths of a certain line in the solar spectrum, it is supremely unlikely that he will get exactly the same value each time. Hence he must find some way of *summarizing* his results, if ever he is to compare the wave lengths of different spectrum lines—or, for that matter, if he is ever to compare any large class of qualitatively similar data with any other class.

Even where measurements can be made with some degree of uniformity the numbers of independently varying factors may still be very large indeed, as for example in meteorology. Nonetheless, although we cannot predict the weather very accurately, the comparison of large collections of meteorological data does help us to find *some* correlations. It is important, therefore, for us to survey the techniques employed in compressing and comparing the data obtained from enumeration and measurement. In short, we will review here the techniques which comprise the science of statistics.

Of course, the obvious first step in simplifying any kind of numerical data consists in classifying the information under suitable headings. *Frequency tables* are often helpful in assessing such material. Thus if we measure the foreleg lengths of cattle and find that they vary from 2 feet 3 inches to 4 feet, we may then find how many cattle have a foreleg length of between 2 feet 6 inches and 2 feet 7 inches, and between 2 feet 7 inches and 2 feet 8 inches, and so on. Naturally, no general rules can be set out in advance regarding how large the intervals on such frequency tables should be. And yet the actual selection of these intervals is an important and delicate matter.

The distribution of the frequencies among the different intervals must be expressed in a very much more summary way, however. We cannot just divide up a million numbers into columns indicating foreleg lengths and let it go at that. One such summary is the *statistical average;* this indicates the position of distributions, the value around which different items center. Another type of summary is called *dispersion* or *deviation;* this is an indication of the extent of variation of the data with respect to one of the averages. Two sets of data may have the same central tendency although the amount of deviation of the two sets be very different. The sets may be 3, 4, 5, 6, 7, and 1, 2, 5, 6, 11; the *statistical average* is the same in both cases, but the amount of dispersion around a common center of distribution is markedly different.

Let us consider in more detail these two types of statistical numbers, then—the *statistical averages,* or centers of distribution, and the *statistical deviations,* or dispersion around such centers.

Statistical Averages

How shall we represent the central tendency of a set of data? By a number? How do we choose the numbers? Under what conditions?

As you probably know, there are several kinds of averages, each one of which is good, not for every single purpose, but only for some particular purpose. But our reasons for using averages at all usually include the following: (1) Averages provide a *synoptic representation* of a range of data, (2) they permit us to compare different groups of data, and (3) they characterize a group on the basis of *samplings* made from it. If they are to do all this, then obviously averages must have certain quite definite properties. These properties are:

a) Averages should be defined so that their numerical value does not depend on any whim of the individual doing the calculating.

b) Averages should be a function of *all the data* in the group; otherwise they could not represent the whole distribution.

c) Averages ought to be *simple,* mathematically speaking.

d) Averages should be amenable to *algebraic development;* i.e., if we know the averages of two series of length measurements, we should like to be able to compute from this the average of a larger series gained by combinations of the original series.

e) Averages should be *stable*. Different samplings from the same group may yield different averages. We aim always to achieve an average wherein such differences due to sampling will be as small as possible.

We will consider four sorts of statistical averages only: the *arithmetic mean,* the *weighted mean,* the *mode,* and the *median.*

A. The *arithmetic mean* is obtained by adding all the numerical data and dividing this sum by the number of terms.

If the times of the successive laps in Roger Bannister's mile were as follows: 58.3 sec., 61.5 sec., 60.5 sec. and 59 sec., the arithmetic mean time would be about 239/4, or 59-3/4 seconds per lap. This does not, of course, correspond to the actual time Roger set up for any given lap. And this indicates what must never be forgotten: Averages represent characteristics of *groups* of data, and not of any individual list of data in the group.

An appearance of accuracy can too easily be attained by the uncritical and sometimes unfair use of decimals. Thus we can express Bannister's mean lap as 59.76379 . . . seconds, an unmeasurable precision so far as today's stop-watches go. This is like saying of a man that he works, on the average, 7.8697372 . . . hours a day. *Arithmetically* this is all right. But we must always realize that precision which results from purely numerical computation is fictitious unless the individual observations have themselves been made with the same degree of precision.

Another feature of the arithmetic mean which must be mentioned comes out in the following example:

Suppose that in a college discussion class 12 undergraduates are found to have coinage in their pockets as follows: *6 have $1.25, 2 have $2.40, 3 have $3.10, and 1 has $39.00*—doubtless he is the treasurer of some university club. The average wealth of the class is thus $5.05 per head. Suppose further that in a second discussion class 13 undergraduates are wealthy as follows: 5 have $3.25, 3 have $4.00, and 5 have $5.00, the average pocket money coming to about $4.10 per person. The first class is thus, on the average, wealthier than the second class, *despite the fact that 11 out of 12 in the former group have less than $3.25,* a sum less than which not one of the second group's pockets contain. We should have been inclined to say that the average second-group undergraduate had more in pocket than the average first-group chap; but the arithmetic mean tells us otherwise. And this shows up a potential danger in the arithmetic mean. For its value is much influenced by isolated extreme variations. This could lead to considerable distortion in one's appraisal of a set of data were he not careful. The mean supplies no information about the homogeneity of a group. This is why measures of dispersion are always required in all statistical work.

Despite this drawback, however, the arithmetic mean is important both for the ease with which it can be calculated and for the obvious facility it affords for algebraic manipulation, for we need only add the arithmetic means of two separate sets of data and divide by two to get the arithmetic mean of the two sets combined (if the two sets are the same size).

This arithmetic mean is furthermore connected with the mathematical theory of probability. Thus a chemist making several thousand determinations of the weight of oxygen will discover, if he had not already anticipated it, that no two of his estimations are likely to be identical. Nonetheless, if we are prepared to assume that all our measurements were made with equal care, the *most probable value* of the weight of oxygen is given by the arithmetic mean.

B. The *weighted mean*. Simple arithmetic is not always enough. Suppose that in 1950 prices per unit of *wheat, beef, iron, rubber,* and *glass* were as follows: wheat 120, beef 110, iron 105, rubber gloves 50, pipecleaners 40. Now, if we take the 1940 price per unit as par, i.e. = 100, we see that by comparison the arithmetic mean of these 1950 prices is 85. Shall we conclude that the cost of living fell between 1940 and 1950? Not likely! For the articles listed are not generally regarded as, we should say, *equally important*. We may then choose to *weight* the items according to their relative importance in some such way as follows: The weights 10, 9, 7, 2, and 1 will be worked into our calculation thus:

$$\frac{10 \times 120 + 9 \times 110 + 7 \times 105 + 2 \times 50 + 1 \times 40}{10 + 9 + 7 + 2 + 1}$$

The result of this is the figure *105.7*. This indicates that the cost of living rose between 1940 and 1950 (as if we didn't know!).

How to determine the weights assigned, of course, is another matter, and a very, very, very delicate one at that. Into this we shall not venture now.

C. The *mode* is nothing other than that element in a set of numerical data which occurs most often. Popular references to the *average American* are really references to the *modal American*. So, too, the *mode* of the pocket-money sums of our first discussion class is $1.25.

Like all averages, the mode represents a kind of distribution within a group. But it will usually indicate the nature of a group of data better than the arithmetic mean. The reason is that the mode indicates the largest subclass of a collection of data and thus locates what will be found most frequently. The modal value is not at all influenced by extreme fluctuations in a set of data. It is, therefore, often a fair basis for comparing different groups.

However, most of our earlier conditions are not satisfied by the *mode*. E.g., the determination of that item in our data which turns up with maximum frequency may differ according to our principles of classifying the group. If 47 candidates in an examination are distributed as follows:

$$\frac{0-20}{4} \quad \frac{20-40}{7} \quad \frac{40-60}{11} \quad \frac{60-80}{15} \quad \frac{80-100}{10}$$

the modal grade obviously lies between *60 and 80*. But consider how the candidates might have been grouped according to a different classification, e.g.,

$$\frac{10-30}{8} \quad \frac{30-50}{8} \quad \frac{50-70}{13} \quad \frac{70-90}{14} \quad \frac{90-100}{4}$$

Here the modal grade would be located between *70 and 90*. If a passing grade had been determined by the lower boundary of the modal interval, fewer candidates would have been passed by the second method of determining the mode than by the first, though no other feature of the situation had changed.

Moreover, it is usually quite rare for there to be a single well-defined type in a group. Suppose, for example, we find two or more wage scales which occur with relatively high frequencies. Here we could not speak of *the mode*. The existence of such "peaks" in the distribution indicates that the group under consideration lacks homogeneity, a not infrequent occurrence, particularly in natural science.

Indeed, the mode might not even be typical of its group. If among our examinees just referred to our classification was made, not according to intervals, but according to the specific grades obtained, we might find 6 candidates with a mark of 25 exactly while no other two candidates of the 47 had a grade in common. Then, no matter how atypical 25 is, it would be the modal mark for that examination.

Nor is the mode a function of *all* the items in a group. Clearly from this same example we could eliminate the performance of many candidates, reducing the total number of 47 quite drastically without in the least affecting the modal mark of 25, established because more examinees drew that particular grade than any other.

Finally, no simple arithmetic is in general sufficient for calculating the mode. In practice this determination is difficult and usually inaccurate.

D. As the name suggests, the *median* is the middle term of a series of numerical data (when these latter have been ordered in terms of increasing magnitude). Hence a group containing an odd number of data will always have a median. When the collection has an even number of members, the median is usually set at the mean between the two middle terms. In the series (1, 3, 5, 7, 9) 5 is the median. In the series (2, 4, 6, 8) 5 is the median again.

The median is but slightly influenced by extreme fluctuations within a group of data. It is a most suitable comparison of sets of data with respect to their middle terms. But, alas, the median has no algebraic properties such that its value for a complex group can be determined from some combination of the medians of the component groups. So, naturally, the median finds a place in those disciplines where *fundamental, additive* measurement is out of place, disciplines, that is, like psychology and sociology, where too often only series or scales of data can be arranged—as, for instance, with I.Q. testing and many standard-of-living appraisals. Logically these are very like the testing of the hardness of metals.

Statistical averages are, as we have seen, ways of determining the central tendencies of sets of data. Now we shall consider the techniques of estimating dispersion, that is, the *spread* of the data throughout a set.

Statistical Deviations

A. Consideration of *range* of variation is often of statistical importance. This is just the numerical distance between the smallest and largest items in a group of data. But in itself this estimation is not very satisfactory. Extreme values of a variation may remain unknown (as when I give the range of annual incomes in the United States as $9,999,-900 since the lowest is near $100 and the highest near $10,000,000). Nor is the distribution clear from knowing only the range. The number groups 1, 5, 5, 6, 6, 7, 7, 10 and 1, 2, 2, 3, 10 have the same range but, as you can see, the distribution is vastly different.

B. Next we have the *mean deviation,* or *average deviation,* as it is sometimes called. Suppose the foreleg lengths of a group of cattle run 21, 23, 24, 25, 26, 25, 27, 28, 29, and 32 inches. The mean length is 26 inches. We calculate the deviation of each measure from the mean by subtracting the mean from each measure: thus $-5, -3, -2, -1, -1, 0, 1, 2, 3, 6$. Now we can neglect the signs of these deviation numbers and calculate their arithmetic mean, which is called the *average deviation.* In this case the average deviation is 24/10, or 2.4.

Of course, all the considerations raised earlier regarding the arithmetic mean apply here.

We must note, however, that a "large" average deviation need not necessarily mirror a "large" fluctuation in the values of the group. Repeated measurements of Mt. Everest might show a mean value of 29,200 feet with an average deviation of five feet. Compared with the mean, the average deviation is a very small number. If, however, we were remeasuring the Iffley Road Track after Roger Bannister ran his great four-minute mile, an average deviation of five feet would be considerable —and altogether perplexing. For this reason the average deviation is sometimes divided by the average from which the deviations are taken; this is then called the *coefficient of dispersion.* In our earlier example of the cattle and the length of their forelegs, the coefficient of dispersion would be 2.4/26 or .92+.

C. Another measure of dispersion is the *standard deviation,* particularly useful where applications of the theory of probability are involved.

We obtain the standard deviation by dividing the sum of the square of the deviation from the mean by the number of entries made, and from this quotient we then extract the square root. Thus:

$$\frac{\text{squares of deviations}}{\text{number of entries}} \quad \frac{25+9+4+2+0+1+4+9+36}{10} = 9$$

and $\sqrt{9} = 3$.

In general, if the deviations from the arithmetic mean of *n* items are $x_1, x_2, \ldots x_n$, then the standard deviation δ_x is determined by the formula $\delta^2 = \Sigma \frac{x^2}{n}$.

The standard deviation, of course, emphasizes the extreme values of the deviations. By squaring each deviation, the larger errors are given greater weight in the sum than the smaller ones. Standard deviations are influenced less by fluctuations in sampling than are other measures.

D. The *quartile deviation* is still another simple measure of deviation. We obtain it by arranging numerical data in order of magnitude and then by finding three entries which divide the series into four equal parts. Now, if Q_1 is the first quartile, i.e., the number at the upper limit of the first equal part, and Q_3 the third, the quartile deviation is defined as $\frac{Q_3 - Q_1}{2}$. Clearly, half the items of this set of data must lie between the first and third quartiles; hence the extent of dispersion is roughly measured by the quartile deviation. Thus with our cattle data:

$$21,\ 23,\ \underset{Q_1}{24},\ 25,\ 25,\ \uparrow\ 26,\ 27,\ \underset{Q_3}{28},\ 29,\ 32$$

Hence

$$\frac{28 - 24}{2} = 2 = \text{quartile deviation}$$

Regarding the term halfway between the first and third quartiles (i.e., 25.5), 1/2 of all the data will lie between this term and the quartiles themselves. Hence the quartile deviation is sometimes not altogether properly called the *probable error*. For if we write 25.5 ± 2.0, there will be as many measures *within* the indicated limits (i.e., 23.5 and 27.5) as outside these limits (you can see this from the data). In other language, we assume that when we select at random it is just as likely that we will pick a measure lying between these limits as that we will choose one lying outside them. More should be said about the use of the word "probable" in this context, for it is not an unmixed blessing—but why this is the case lies beyond the boundaries of this chapter.

After the foregoing discussion of statistical averages and measures of dispersion, we may now note a little more generally that, just as it is the business of all inquiry to discover significant relations within a subject matter, so it is the business of statistical inquiry to facilitate the discovery and expression of relationships between different sets of data.

Having discussed some methods of comparing statistical data, we may now note that there are occasions when estimates of averages or of deviations cannot be used to answer our questions. Particularly is this so when we are seeking to determine the correlation between two or several sets of data. It should be clear that comparing sets of data by determining their averages and deviations is different from correlating the sets in terms of certain interdependent characters of the two.

Suppose a botanist undertakes to examine several thousand leaves of trees with the idea of recording their lengths and widths. Is there a connection between these two characteristics of leaves? Shall we say that the longer the leaf the wider it will be? In such a large class of data our observations cannot be made hastily, for we can neither remember all the features of each individual leaf examined nor detect significant relations between these features.

Suppose we arrange all the leaves in order of increasing length. We will then be in a position to determine whether or not this order is also the order of increasing width. If the two orders are the same we should infer (probably) that there is some connection between the length and the width of a tree leaf. If the coincidence is only partial, we might still suspect a relation. Here, however, we would require some *numerical measure* of the degree of correlation between the length and the width of tree leaves. Two variables are thus said to be correlated if in a series of corresponding instances an increase or decrease in one of the variables is accompanied by an increase or decrease in the other. When the changes are in the same direction, e.g., increase length—increase width or decrease length—decrease width, the correlation is said to be *positive*. When the changes are in opposite directions, e.g., increase length—decrease width or decrease length—increase width, the correlation is said to be negative.

There are many, many types of measures of correlation. A very clear example of one of these, *Pearson's coefficient*, is set out on pages 314–315 of Cohen and Nagel's *Introduction to Logic and Scientific Method*. Another clear statement of Pearson's and other measures of correlation is found in F. C. Mills' book *The Elements of Statistics*.[1]

It should be fairly evident now, from our survey of statistical procedures, how all this ties in with our earlier study of probability theory. Until a really large and complicated body of data has been compressed, simplified, and made generally manageable by applying some of these standard statistical techniques, the estimation of which of two rival hypotheses about the same set of data is the more probable cannot even

---

[1] F. C. Mills, *The Elements of Statistics* (New York: H. H. Winston and Rinehart, 1956).

be undertaken. The calculus of probability is the sausage machine that turns out an invaluable product once the right sort of entity is fed into it. But you cannot put cows or pigs—or would it be horses?—into sausage machines just as they are. They must be chopped up first. So, too, one million numbers *neat* do not provide a plausible opportunity for a probability calculus to show what it can do. These million figures must be classified, compressed, averaged, correlated, etc., before hypotheses with respect to them will even make sense.

So statistical methods are indispensable to the science of measurement, explanation, experiment, and prediction. Without these methods we could not even begin to choose between hypotheses.

Despite the intrinsic worth of these methods, however, they can be, and often are, terribly abused. The remainder of this chapter will be given over to the raising of warning flags about certain nuisances to which statistical methods have been put. These are hard lessons to learn. Geneticists, animal behaviorists, biochemists, and physical chemists have taken a long time to learn them and are continually relearning them each day for special cases. In summarizing them I am only calling attention to one set of painful features of the history of scientific inference. I am not asserting but perhaps I am at least timidly suggesting that that history might have been different in important ways had the possible dangers and fallacies of the uses of statistics been more widely appreciated.

One fairly obvious danger in the use of statistics would be to regard what is true of a group as being true of the members of that group. This is a statistical variation of the *Monte Carlo fallacy* or the *expectant father's fallacy* we encountered in our review of certain interpretations of probability theory. From the statistical knowledge that male births comprise about one-half of all births, an expectant father (even if he be the most gifted of statisticians) can infer nothing about the sex of his forthcoming child.

Statistical averages cannot by themselves, and independently of further study, be interpreted as describing *strictly invariable* relations within a group. The number of suicides in a country may remain constant over a period of years. But it certainly does not follow from this that a certain number of suicides *must* be committed. Initially we do not know the precise factors (if, indeed, there are any) which make for the occurrence of a suicide. Nor do we know that these factors will be operative every year. The contrary supposition is logically indistinguishable from the frustrated student's remark, "Well, someone had to get the lowest mark in organic chemistry—that's why I did, it was a statistical necessity."

A third danger is this: Certain coefficients of correlation are so generally defined that often any two groups whatever may be seen as in some way correlated. Hence, we may, by statistical means alone, relate the deviations of the ages of women listed in *Who's Who* with the deviations of the number of pages in the books listed in the university library. The coefficient may have a high numerical value too. But we would not take this to indicate any significant connection between the two groups. Indeed, it has been shown that expenditure for the British Navy is highly correlated with the sale of bananas. Many correlations of this nature are altogether fortuitous and lacking in causal significance. The tobacco companies tried to persuade the cigarette-buying public that the surgeon general's report of a correlation between smoking and lung cancer was of this fortuitous, coincidental variety. And had the evidence in favor of the correlation been purely statistical such a retort would have drawn blood. Fortunately for the future of medical science, however, a very great deal of non-statistical experimental evidence also buttressed the report. You can be sure that if in this case the statistical story were as easy to puncture as in the other cases I cited about naval expenditure and the consumption of bananas the tobacco companies should not have proved nearly so philanthropic with their grants for research into the subject.

Suppose further that the number of police arrests in New Haven increases over a period of years. Can we say that Yale undergraduates are becoming ever more naughty as the years pass? Or shall we say that the police are becoming grouchy and ill-tempered? Or shall we say that the number of bars (frequented by townspeople) has increased? Or perhaps it is the policy of the new police chief to increase arrests and warnings and censures, in the hope that this will lead to a decrease in sentences of fine and imprisonment and, indeed, a decrease also in the occurrence of crime. Which correlation shall take precedence? Again, is it safe to find a causal connection between climate and the character of a civilization just because civilizations of a certain type are found in regions with a certain climate? After this discussion I hope you will agree that it is not; but then you may enjoy even more reading a book by Markham called *Climate and the Energy of Nations* in which this very thesis is skillfully advanced, perhaps just a little *too* skillfully.[2]

Another potential pitfall in the use of statistics is this: We may fancy there to be a significant connection between types of events just because we observe them to be associated. This should remind you of my hackneyed example of how I always wind my alarm clock before going to sleep, but I will put the same point in a new way. If 90 out of 100 inorganic chemists are seen to be exceptionally dull, lethargic, untroubled individuals, may we infer that the study of inorganic chemistry dulls one's intellectual sensitivities? Despite an inclination to the contrary I must say that this would *not* be a legitimate inference, not unless we also had further information about the relative numbers of people who are not inorganic chemists but who are nonetheless dull, lethargic, and untroubled. Again, the full moon and fine weather are frequently found together. But if village magicians and old wives (and some meteorologists) were only to notice the frequency with which fine weather also occurs in the absence of the full moon they might read rather less significance into the joint occurrence of these two events.

A still further snare is encountered when high correlations are obtained just by mixing together two sets of data in each of which no correlation whatever was discovered before mixing. If the ages of husbands and wives are uncorrelated in each of two communities, *some* correlation will be found when the records of the two communities are mingled. This fact has been demonstrated with maximum mathematical rigor. The correlation will arise from the purely formal (i.e., mathematical) properties of the two sets of data, and cannot on any reckoning be taken as evidence of an invariable connection between the ages of husbands and those of their wives.

It is in some such way as this that Mr. Spencer Brown of Oxford University has challenged the alleged discoveries of modern statistical investigations of psychical phenomena and the extra-sensory powers of mediums and mystics. Spencer Brown has achieved correlations equally as good as those of psychical researchers like Dr. J. B. Rhine simply by mingling and then correlating Sir Ronald Fisher's random

---

[2] Sydney Frank Markham, *Climate and the Energy of Nations* (London: Oxford University Press, 1942).

number tables, tables which are relied upon by geneticists to be paradigms of unrelated and unconnected clusters of numbers.[3]

This points up difficulties arising from the activity called *sampling*. Correlations cannot be calculated for groups containing an infinite number of members. Obviously only a finite number of terms can receive statistical consideration. And yet we rely on the items we actually do consider as constituting fair samples of, or being representative of, groups of data that are almost sure to be larger and more inclusive than the one with respect to which we established our coefficient of correlation. But, clearly, the coefficient (and all other statistical numbers as well) is subject to fluctuations of sampling. Indeed, there may even be cases where the highest correlation is only casual and fortuitous. On the basis of throwing two dice 1,000 times, a coefficient of correlation may be calculated between the dice which is considerably greater than zero, although if the two dice are really independent and fair, the coefficient of correlation should be exactly zero. Whether this is to be interpreted as indicating some actual interdependence between the dice cannot be determined simply by considering and reconsidering the coefficient of correlation itself.

Of course arguments from samples are often fallacious just because the samples have themselves been selected in favor of the correlation sought. As such they may not be representative of the entire collection. Thus the discovery that married Italian males require less medical attention than unmarried Italian males does not prove that marriage has hygienic values. It may all be due to the unwillingness of chronically ailing Italian men to marry at all. The better health of the married group is then seen to be the effect of a prior selection in the data composed.

Statistics prove that the death rate in the British Navy during World War II was lower than the death rate in London or New York. Are we to conclude that it is safer to be in Her Majesty's armed forces during time of war than to remain a civilian? Of course not: The death rates are not strictly comparable. The death rates of London and New York during the war included infant mortality as well as deaths resulting from old age, sickness, and acute boredom, whereas Her Majesty's armed

---

[3] Brown's work was not published. For a discussion and explanation see *CIBA Foundation Symposium on Extrasensory Perception*, eds. G. E. W. Wolstenholme and Elaine C. P. Millar (Boston: Little, Brown and Co., 1956), pp. 43–44.

forces were composed of men who were neither infants nor markedly senile nor physically ailing. Nor was there much opportunity in the war for servicemen to undergo acute and prolonged boredom. An adequate comparison of two death rates requires that the two groups concerned be homogeneous with respect to age, sex, health, and nature of occupation; in none of these respects were Londoners or New Yorkers comparable with British servicemen of World War II.

Illustrations of this type of fallacy abound, not only in everyday affairs but in natural science particularly. Keep your eyes peeled for selections operating rather like these. They may bias and determine one's conclusions even before all the evidence is in. Remember what Sir Ronald Fisher has shown us—even the results of the great Gregor Mendel are from a statistical point of view just *too good to be true!* [4]

These and many other sources of potential errors in the use of statistics should warn us to use great caution even with such an indispensable tool of research. For like all sharp, powerful tools it can cut both ways: It can clarify, but unless employed with caution it can seriously distort.

Probability and statistics are thus seen to be complementary and mutually interdependent. Statistics by itself supplies us with manageable but theoretically insignificant data. It provides us with no hint about how we should predict on the basis of such data. The calculus of probability does supply us with this predictive power—but only after manageable bits of information have been fed into it. Probability and statistics are thus dovetailed foundation stones in the disciplines we call inductive science. But in a way probability and statistics are conceptually weak just as induction itself is conceptually weak. Or perhaps it is better put in this way: If induction is logically weak, then probability and statistics are logically weak too. If the inductive inference from some data of a certain sort to all data of that sort is weak, then probable inference and statistical inference share in this weakness.

---

[4] Sir R. A. Fisher, *Annals of Science, op. cit.*

25 | The Principle of Uniformity Revisited

I SHOULD like finally to consider a problem which, because it has so long exercised philosophers, has come to give scientists cause to worry about their own procedure and their own methods of reasoning. The problem has to do with the justification of induction.

We have dealt in detail with the techniques and assumptions of probability and statistics. The details, however, should not obscure the principal points. It should be clear, for instance, that when we make statistical predictions on the basis of observed relative frequencies in the occurrences of events, we tacitly assume that such frequencies have a certain *constancy,* we might even say a certain uniformity. This assumption is not unjustified, for if I toss a fair penny 1,000 times the relative frequency of the occurrence of heads *will in fact* be approximately 1/2. But, of course, it is *logically* possible that entirely different frequencies should turn up. Hence that they *do* turn up is a factual matter, not a logical one.

Some philosophers past and present regard this *uniformity in nature* as a wonderfully fortunate thing, and something without which inductive inference could not even get under way. However, the *principle of uniformity* in nature and its cousin the *inductive principle* are equally difficult to establish logically. The one says that the course of nature is, in a way, regular, uniform, and constant (i.e., same cause, same effect) while the other puts it that the policy of reasoning from *samples* of a class to statements about the *whole of class* is a sound policy. But how came we by this remarkable knowledge? Clearly, the only ways in which science can avail itself of knowledge are *formally* (as in logic or mathematics) or *empirically* (as in experiment and observation). If these *principles of uniformity and induction* are established empirically, then we are faced with a vicious circularity, for if the principles are true, then they must be operative and their truth presupposed in every empirical procedure—even the one by which we hope to prove the truth of the principles themselves. Hence, to gain knowledge of the truth of these principles by experiment and observation is to presuppose in the search the very existence of that for which we are in search. If the principles are true, we cannot learn of it empirically, for the essence of the principles is that their truth is presupposed in every empirical enquiry.

Neither can the principles be established as true by appeal to the purely formal reasoning of logic and/or mathematics. For if everything we have said about the purely formal, uninterpreted character of a strictly deductive system applies here, the truth status of the principles must be left untouched by pure reasoning. Formal deductive systems are not *about* anything, and *ipso facto* not about the *principles of uniformity and induction*.

There is the difficulty, then: the only knowledge that is acceptable in science is *either* observational or formal. If we are to have knowledge of the validity of the principle of induction, therefore, we must get it either observationally or formally. But if we try to gain this knowledge observationally, we presuppose the validity of the principle in our very observations. This is circular. If on the other hand we try to come by this knowledge through recourse to logic and/or mathematics, we fail again, for these disciplines can supply no limit whatever about any factual matters, and induction is a supremely factual matter.

Many writers on the subject, including both philosophers and scientists, claim this difficulty about the logical status of induction not to be a genuine one at all. They would say that it is an insoluble problem only in the sense that the construction of a round quadrangle presents an insoluble problem. Sometimes I incline this way myself. At the moment I am so inclined and in Chapter 21 I so expressed my views. But I shall suppress my inclinations and views in order to resurvey the supposed problem as objectively as I can.

We shall sail into these troubled waters by way of that already-mentioned backwater, the *principle of uniformity in nature*.

It is not at all easy to make clear what is asserted by the principle of uniformity (which scientists are always supposed to be supposing). At least two different forms of the principle might be distinguished. Suppose we distinguish at least two types of law of nature first; this will help us with the parallel we require for discussing the principle of uniformity.

All acids have the power to turn blue litmus papers red. That might be a kind of law of nature; but notice how it differs from a law like *If equal masses collide with equal and opposite velocities, they rebound with equal and reversed velocities.* The law about acid and litmus asserts an invariable coexistence of properties. The law about colliding masses asserts an invariable succession of kinds of events.

Suppose we call the antecedent of our second law (of invariable succession) p (equal masses collide with equal and opposite velocities). And we shall call the consequent q (i.e., they rebound with equal and reversed velocities). Then when we use the law *if p then q* as a means of predicting what will happen next time the conditions expressed by p are realized, we are really arguing *from particulars to particulars:* In particular observed cases of colliding bodies such and such has happened, therefore such and such will happen in the next particular case of colliding bodies.

And it is just here that the ugly question raises its awful head: How can we justify this inductive inference from past to future, from particular observed instances to particular unobserved instances?

If the question means, "How can we show that the conclusion necessarily follows from the evidence?" then it is just the question, "How can we transform this inductive inference into a deductive inference?" Apparently we need another strong, logical premise to bind our inference from n observations to the $(n+1)^{th}$ observation; and apparently the principle of uniformity will do the job. For, once the principle is included in the premises, the arguments become purely deductive, as follows:

1. Similar conditions are always and everywhere attended by similar effects (this is either the principle of uniformity or a statement of a specific law);

2. The conditions here and now (head-on collision of equal masses) are similar to those which in the past were attended by equal and opposite rebound of the masses;

3. Hence (I predict that) the conditions now prevailing (head-on collision) will be attended by equal and opposite rebound of the bodies.

Apparently we have justified this specific inductive inference by transforming it into a deductive inference.

The same sort of thing will work also with our law of co-existence of properties—this was where we spoke of *all acids having the power to turn blue litmus paper red*. But here, obviously, we are inferring, not from particular events to other particular events by way of a law, but from a cluster of particular events *to* a law.

So two types of inductive (or probable) inference may be distinguished here, inference from particulars to particulars, as in the collision examples, and inference from particulars to laws, as in the acid example. If we were to call the first sort of induction *projection* and the second sort *generalization,* we should see that an argument like the one about colliding masses can be made deductive without the principle of uniformity—we need only supply some general law about the behavior of colliding bodies. But then, apparently, this general law will have been the result of inductive generalization from particulars, and in this the principle of uniformity is required as "an ultimate major premise" (to use John Stuart Mill's expression). But, as you probably detect, this talk of generalizing from particulars to laws as if laws were just summary shorthand ways of referring to particular observations is *very* misleading.

Even if we do not require the principle of uniformity in inferring from particular to particular (for here we can make the inference deductive by inserting a law of the colliding bodies variety), we do require the principle (apparently) in the very establishment of that law itself. For how could an inference from observations of a few α's that are β's to the proposition that *all* α's are β's be deductively valid were it not covered by some super-general principle which lays it down that there is a regular constancy in nature such that observation of a few α's that are β's is just an *indication* that in nature all α's are β's?

What exactly is gained, however, by invoking the principle of uniformity in this transformation of an induction to a deduction? Has this move saved the rationality of our predictions? Not in any obvious way.

It may even be doubted that the principle constitutes a definite assertion of any sort. It is perhaps more like a sentential function than a definite sentence: It does not say specifically that a few observations of colliding masses (and the effects of such collisions) indicate that in nature all collisions of masses will show these same effects. Nothing nearly so specific is asserted. The principle is, rather, full of blank spaces, or variables, where for the α's and β's we can substitute the names of particular, specific events we have observed to occur. But a string of words containing a lot of blank spaces cannot be known to make a true assertion about the constitution of nature, because it cannot be known to make an assertion of any sort. The blank spaces guarantee that!

If as so interpreted, however, the principle is only a sentence schema, it makes no true assertion about the world. It makes no assertion at all. It is empty. Immediately we begin filling in the variable spaces with the names of specific events, on the other hand, the result is a statement of the very law whose induction we sought to guarantee by appealing to the principle of uniformity. That is, instead of the sentence schema we call the principle of uniformity, namely, "Similar conditions are always and everywhere attended by similar effects," we fill in the blanks as follows: "Every condition wherein equal masses collide with equal and opposite velocities is attended always and everywhere by the effect that they rebound with equal and reversed velocities." But this latter is just the law whose ultimate justification we sought in the principle of uniformity.

Hence the unsatisfactory conclusion that the principle of uniformity, by itself, is only a sentence schema; it makes no assertion and is therefore neither true nor false, i.e., it is empty. And once we try to give the principle an assertive interpretation, it just turns into one of those laws whose validity we hoped to establish by appealing to the principle. Hence on this interpretation the principle is either empty or useless, or both.

If on the other hand we try to strong-arm the principle and just *clamp* it into the procedures and findings of science, it turns out to be patently false. For if it were true in this simple, fact-stating way, I should have to conclude from having encountered five or six bespectacled Princeton undergraduates that all Princeton undergraduates wear glasses.

Of course, an inference may be formally valid even if one or more of its premises are false. But according to customary conceptions of the principle of uniformity it must not only lend *deductive* force to inductive reasoning but, moreover, endow the inductive conclusions with a *probability* they would not possess were the principle in fact false.

Would it help if we qualified the statement of the principle with the words "probable" and "probably"?

No, I do not think so. For the word "probable" is actually a very vague predicate; in many ways it is like the word "bald" in this respect: There are extreme cases where the term applies completely and definitely, but there is also a middle region that is decidedly indeterminate. Other formulations of the principle, using, e.g., the word "sufficient," suffer from the same vagueness. So too with expressions like "same circumstances" and "similar conditions." Obviously, circumstances are never *exactly* the same, conditions are never *precisely* similar. Since (presumably) no state of the universe is ever repeated exactly, no two states could be required to be similar in all respects (and, indeed, even if such a state were exactly repeated, how could this possibly be known?). What inductive evidence could ever be provided for the principle?

Of course, one could argue that the principle requires *no* inductive evidence since it is logically demonstrable in the following way:

Suppose that a state α of the universe is followed by another state β. If state α' is to resemble α exactly, then it also must have the property of being followed by a state β' which resembles β exactly. Hence it follows from the assumption that α resembles α' exactly that the consequent state β' resembles β exactly.

This is like proving that Utopia exists because by definition Utopia is perfect, which it would not be if it did not exist; i.e., one of the constituents of perfection is existence, hence Utopia exists! Logically the two arguments have much in common.

Once again, however, this argument reduces the principle to an empty tautology. And no matter how many such tautologies we add to the empirical premises of an inductive argument, we will not thereby increase the argument's probability one jot.

What if we asked only for similarity of conditions *in some respects?* If this were all the principle required, it would be so indefinite that it could justify the most unlikely and implausible predictions. Surely there are respects in which a flash of lightning and the striking of a match are similar. But who expects thunder every time a cigarette is lit? If on the other hand we *specify* the respects in which the similarity must hold, the result is a straightforward law of nature, of either the colliding bodies types, where we assert that one particular event will always follow another particular event, or the acid-litmus paper type, where we say of a substance that it will always behave in such and such a way under specified circumstances.

To avoid, then, the extremes of vague indefiniteness on the one hand, where the principle asserts either nothing or something too slippery to grasp, and thoroughgoing specialization and specification on the other, where the principle becomes a redundant statement of some natural law, we might try the following compromise:

If the conditions are similar in all *relevant* respects, then the same kind of effect will follow.

This formulation, clearly, is designed to relieve entomologists of the responsibility of taking the disposition of Jupiter's moons into account when formulating laws about the ways in which the angle separating a fly's antennae is related to the insect's speed.

But, so far from solving old difficulties, this formulation only raises new ones. What is the criterion of relevance? *Why* should we dismiss as irrelevant to an antennae-airspeed experiment the behavior of Jupiter's moons? Surely only because, on the basis of past experience, the same effect (namely, an alteration in its airspeed) is known to result from altering the angle between an insect's antennae, regardless of the disposition of Jupiter's moons. So it turns out that we distinguish between relevant and irrelevant conditions solely on the basis of similarity or dissimilarity in the observed effects. If we observe that alteration in its airspeed *is* affected by fixing the insect's antennae at a given angle, but not by the disposition of Jupiter's moons, then we say that the former condition is relevant while the latter is not. And once again the principle of uniformity when framed in terms of relevance has become a useless tautology, certainly incapable of *justifying* inductive inference. For now all the principle asserts is this: Whenever all the necessary relevant conditions for an effect are present, the effect will occur. A miracle of the obvious.

By now it will have begun to dawn on us that the problem of

justifying induction is a puzzling one perhaps just because we are not at all clear about what kind of justification we are asking for.

If to justify an inductive inference means to show that the conclusion is deductively *entailed* by the evidence, then, of course, it is analytic or true by definition to say that an inductive inference cannot be justified, just as it is analytic or true by definition to say that a round quadrangle cannot be constructed. A quadrangle is by definition not round, so few would find in *this* a problem to be solved. But (also by definition) inductive inference is inductive, not deductive. Yet so many find this a methodological torture chamber: How to turn induction into deduction? Is this really any less absurd than asking how to turn a quadrangle into a rounded figure?

Hume claimed boldly that *no* logical justification of induction is possible: He said that our beliefs about the future are produced, not by reasoning, but by mere habit. Humeans would have it that we have no *good reason* for expecting the sun up tomorrow, we are just creatures of habit and habitually expect the future to resemble the past. In this case we expect that the sun will rise tomorrow as it has done in the past. Thus some Humeans would lament their inability to give a logical justification for our beliefs about the future. They provide instead a psychological account of how our beliefs about the future are built up. But what exactly is being lamented? What would it be *like* to justify an inductive inference in this manner?

If the contention is that inductive beliefs cannot be justified in the sense that no *a priori, logical, deductive* demonstration is possible, the contention is as trivial as it can be. By definition an empirical proposition just *is* a proposition which cannot be demonstrated from logical principles alone. The lament in these terms comes to no more than the fact that one cannot demonstrate what it is logically impossible to demonstrate. This is hardly the sort of situation to grieve over.

Perhaps, however, the quest for a justification of induction is the quest for a demonstration that evidence of a certain kind makes a conclusion of a certain kind *probable*.

In order to see what such a demonstration might be like, a comparison with justifications of deductive arguments might be helpful.

For example, how do we justify that, say, "All whales are warm-blooded" follows from the premises "All whales are mammals" and "All mammals are warm-blooded"? To do this, one refers to a *principle of deductive inference,* a rule of logic. The rule which justifies this particular inference is the so-called "principle of syllogism": If all α's are β's, and all β's are γ's, then all α's are γ's. Reading *whales* for α, *mammals* for β, and *warm-blooded* for γ, we have just the inference we began with—an inference which is justified because this substitution is possible.

But suppose, then, that it were asked why we should accept this principle of deductive inference as itself true. Clearly, inasmuch as the principle (as so stated) contains the unspecified variables α, β, and γ, it makes no assertion whatever, and hence cannot be said to make a *true* or *false* assertion, i.e., be true or false.

What are we doing, then, when we claim to be *justifying* a specific deduction in terms of a rule of logic? What we are doing is this: We are showing that the specific deduction is a member of a class of arguments all of which are valid merely in virtue of their form. We can bring a doubting Thomas around to our side by calling to his attention familiar arguments of ordinary discourse which are also members of this class and of whose validity there would be no question, e.g., "If there is a tax on all musical instruments, and all violins are musical instruments, then there is a tax on violins," "If all dogs have incisor and canine teeth, and all spaniels are dogs, then all spaniels have incisor and canine teeth." Will not our doubting Thomas see the validity of these arguments? (If not, perhaps he is a candidate for the Bellevue Hospital.) But if he *does* see them as valid, will he not see also that it is their structure that makes them so, and that the argument "All whales are mammals, all mammals are warm-blooded, hence all whales are warm-blooded" has this same structure?

And will he not see also that the structure *itself* is completely set out in the principle: if all α's are β's, and all β's are γ's, then all α's are γ's?

With this example of a kind of justification of deduction as our guiding analogy, we might construe the task of justifying inductive inferences as being just the task of finding rules, implicitly exemplified in common-sense reasoning and in the predictions made by scientists, that might help to elucidate the expression "The evidence makes it *probable* that such and such is the case, or will be the case." This then would be the proper and genuine province of inductive logic.

Now, in accordance with this account, we may see what *genuine* problems of inductive logic are like—as opposed to this spurious "problem" of justifying induction. Genuine problems of inductive logic would consist of considering inferences made by scientists, inferences whose validity is not obvious, and comparing and contrasting them with inferences whose validity is very much more obvious. Another genuine problem of inductive logic may be this one: Why is it that on the basis of thousands of confirming and supporting instances we will attribute a smaller probability to the statement "All crows are black" than we would to the statement "At standard pressure phosphorus melts at 44°C," when in this latter case we rely on but one or two carefully controlled and observed experiments? In our many reflections on the nature of laws and generalizations, on theories, and on the calculus of probability, we have attempted to meet this problem. In general, we would try to answer this genuine problem of induction in some such way as follows: We should refer to and distinguish the ways in which statements of a higher order—laws and principles—are confirmed and disconfirmed, contrasting these with the manner in which we should go about confirming or disconfirming statements of a much lower order—generalizations and statements of statistical regularities. And in doing so, we, as inductive logicians, would have "justified" the scientist's intuitive conviction that the probability of a law depends, not only or even primarily on the number of confirming instances, but also on the nature of the properties which the law asserts to be invariably connected in the nature of things.

So here runs our justification of induction on exact analogy to our earlier justification of deduction. Scientists as a matter of fact distinguish statements like "All crows are black" from statements like "Phosphorus melts at 44°C." Their making of this distinction is shown to be justifiable when the inductive logician makes explicit the factors that are operative in the scientist's thinking at the time the scientist does so distinguish these types of proposition. For we show that the one sort of proposition falls into a class of statements resulting from arguments based on evidence that is merely cumulative, qualitatively similar, and thoroughly statistical—while the other sort of proposition falls into a class of statements resulting from arguments based on evidence that is largely non-statistical, non-cumulative, and qualitatively dissimilar, arguments which are in fact supported as much by deductive, systematic considerations as by any other. Pointing this out is, so far, showing the justice in (i.e., "justifying") the scientist's natural inclination to treat inductions leading to laws of nature in a manner thoroughly different from the way he would treat inductions leading to generalizations or statistical regularities.

Hence, to ask for a justification of the "rules" of inductive inference is to ask that the place of these rules within the cluster of concepts having to do with *probability, evidence, theory,* and *hypothesis* be brought out clearly. The analogy with our earlier justification of deductive inferences should be obvious.

But does *this* treatment of Hume's scepticism really do the trick? Let us reconsider for a moment what the sceptical position amounts to.

Remember our example in which we considered a bag containing black and white buttons. We make ten independent drawings of which nine are black buttons and one is white. If we were asked, "What kind of button will be withdrawn next time?" we must reply that we do not know, but that it is very probable that it will be a black one. Suppose, then, that we are challenged to *justify* our expectation. Shall we say that, since black buttons have been withdrawn more frequently than white ones, we therefore expect a black one again? But then our sceptical friend will come back at us, "Can you justify your belief that what has happened will happen again, that the future, in other words, will resemble the past? And, indeed, can you justify your belief that an event will happen in the future with the same relative frequency with which it happened in the past?"

Should we then reply that the uniformity we have observed in past events has, by sheer force of habit, led us to expect uniformity in the future, our sceptic will merely smile, indicating that we have just trapped ourselves rather badly. He will observe, "Either you just meant to explain what *causes* your *anticipation* of a black button (in which case you have not complied with my request that you offer a *logical justification* for your belief) *or,* if you did intend to give such a justification, you have only ended up in a vicious circle. For you propose to prove that nature must be uniform by reference to the fact that in the past it has always been observed to be uniform. But this is itself an example of an inductive argument—and since you were asked to prove the validity of this type of argument, but just assumed its validity in offering it, your argument is circular."

So speaks the sceptic, and he speaks rightly. For (as we saw before) if to justify inductive arguments is only to offer an inductive argument which would prove *all* inductive arguments of a certain form to be valid, then it is clearly impossible to justify inductive arguments, for the argument offered as proof, in order to be convincing, would have to be of the very form required to make an inductive argument valid.

But why should this rather trivial observation produce such a profound scepticism? Perhaps it is just because the triviality is not as obvious to the sceptic as we have made out. Perhaps he has many half-understood perplexities before him at once, and the dissolution of but one of these leaves behind an unsettled and incomplete feeling well known to us all. In fact, that is my diagnosis: The sceptic is wonderfully confused (puzzled) because he is wonderfully confused (does not see the nature of his difficulty).

Let me try one final time, then, to show that the traditional problem of the justification of induction—unlike the specific problems of inductive inference to which we have been addressed—is a pseudo-problem, easily exposed for what it is by analysis. In this final attempt let me again enlist our earlier comparison with deductive logic.

What precisely do we expect of a man whom we have asked for a demonstration of some proposition he has asserted? Let q stand for the proposition to be demonstrated. Then to demonstrate q is just to show that q follows from a set of premises, which we shall call p, all of which are accepted as true.

In demonstrating propositions, therefore, we are guided by the rule "If p is true and p entails q, then q is true." This rule, however, is not itself one of those propositions that might be substituted for q, and of which a demonstration could reasonably be demanded. For this rule ("If p is true and p entails q, then q is true") is nothing other than a symbolic rendition of the common logical form of all those deductive arguments that we regularly recognize as *constituting* a demonstration. In other words, the rule is just that logical instrument with the help of which we demonstrate the truth of propositions, just as our legs are, so to speak, the anatomical instruments with which we walk. The attempt to *demonstrate* this rule in the same sense as we demonstrate specific propositions in terms of the rule is bound to be as frustrating as would be the attempt to walk over one's legs in the manner in which *with* our legs we walk over fallen bicycles and sleeping cats.

Is there, then, no logical justification possible for the logical principle which is presupposed in all proofs of mathematics and science?

Yes, in fact, a justification is possible (though it will not be set out here), only it cannot consist merely in a demonstration of the very principle it must presuppose if it is to be acceptable as a demonstration.

Similarly, again by analogy, in one sense of the word, we *justify* a specific inductive argument like "Since a fair sample of the metals we have observed are heat conductors, it is probable that all metals are heat conductors" by reference to a rule of inductive inference. In this case the rule is as follows: "If M/n instances of a fair sample of α's are β's, then it is probable that M/n α's are β's." This rule, of course, is nothing other than an explicit formulation (with the help of variables) of the logical form of certain inductive arguments usually regarded by scientists as valid. But an everlastingly nagging puzzle will be generated if one asks for a justification of *this* rule in the *same* sense of "justification," as if the rule itself were an empirical proposition like "Since a fair sample of the metals we have observed are heat conductors, it is probable that all metals are heat conductors," as if the rule were itself a specific inductive argument of the sort we ordinarily justify *in terms* of the rule. "Justifying" induction in this sense of *induction* thus constitutes a task that is as absurd as it is impossible, or is absurd because it is impossible.

Still, the philosopher of science can justify these rules in a different sense. The rules of inductive inference are quite obviously not themselves empirical propositions. Nevertheless they are valid, if valid at all, in some one of the ordinary meanings of the word "probable." The justification consists of the construction of concepts of probability from which the validity of these inductive rules would be seen to follow, a task to which some of the most eminent philosophers, logicians, mathematicians, and scientists of the last fifty years have addressed themselves (e.g., Lord Keynes, Lord Russell, Professor Sir Harold Jeffreys, Professor Sir Ronald Fisher, Professor von Mises, and Professor Carnap).

It has been my intention to survey some of the contributions these men have made to our understanding of inductive inference, the theory of probability, and the uses of statistics. *They* have "justified" these procedures by making explicit, with careful, painstaking analysis, the *rationale* which underlies the varieties of reasoning which are fundamental in natural science.

EDITOR'S EPILOGUE

Editor's Epilogue

CHAPTER 1 began with some cautions about what the philosophy of science is *not*. It should now be possible to say somewhat more precisely what philosophy of science *is* and what sorts of questions it deals with.

Take the principle of uniformity in nature, for example. First in the context of clarifying the notions of law and principle and later in the context of considering elementary statistics and inductive inference, Hanson argues that the principle of uniformity is either (a) meaningless by virtue of vagueness and obscurity or (b), if meaningful, then a mere truism about scientific thinking. Neither way does the principle come out to be a grand metaphysical truth about the world. It is, at most, a common-sense rule governing the way in which scientists (and intelligent men generally) weigh evidence.

This analysis, however, applies more broadly. It can be carried through with varying degrees of difficulty for all of the metaphysical principles which are alleged to inhere in scientists' thinking as uncriticized presuppositions: The law of causality, belief in the mathematical character of nature, the assumption that purposes (final causes) do not exist, belief in the "reality" of only a "world of appearances"—all these and more can be shown to be either systematically unclear or, if clear, mere platitudes about the procedures of scientific inquiry. None of them states any truth (or falsehood) about nature itself.

Philosophy of science, therefore, does not aim at providing "higher order" principles about nature than particular sciences yield. Neither does it seek to "justify" the rules of procedure scientific inquiry follows (since by and large these rules are the only paradigms we have of rules for finding out about nature). What philosophy of science *does* do is to seek to clarify and elucidate the process of inquiry by focusing attention on the concepts and rules which govern it. In short, philosophy of science is an attempt to understand science more completely.

In this book, measurement and counting, probable inference, and statistical analysis have been dissected with that end in view. The focus of attention throughout was on the elementary aspects of these subjects. An alert student will notice at once, however, that even in the more advanced stages of scientific work (e.g., quantum theory of measurement, sampling techniques in genetics, or population studies in animal ecology) the same kinds of analysis Hanson uses can be employed to get clear about the procedures and concepts involved. What has been presented here is a model for getting one's ideas straightened out at the frontiers of scientific research.

For professional philosophers of science, the most puzzling problems in recent years have perhaps been those discussed in Parts II and III: the problems involved in analyzing the notions of *observation, description, fact, hypothesis, law, theory, cause, principle,* and *explanation.* It is obvious that what one says about any one of these crucial subjects will affect what one *can* say about the others. Nowhere is this more clear than in the elementary quantum theory of modern physics: On account of their adoption of a particular view of what constitutes "an observation" or "a complete description" orthodox quantum theorists find themselves implicitly committed to certain ways of speaking about causality, explanation, laws of nature, and the like. What the philosopher of science has to offer here is a detailed study of the interconnections— and this we have had in Parts II and III.

By way of summary, let us run over Hanson's main contentions about the concepts listed above.

(1) *Observation.* Hanson argues that neither a sense-datum theory nor a crude retinal reaction model suffices to explain scientific observation. He therefore rejects implicitly the attempts of certain formalist philosophers of science to locate a unique "observation language" vocabulary within science. He opts for what has been called an "ordinary language" or "instrumental" account of the meaning of observational terms—an account which recognizes all observational terms to be "theory-laden" to some degree.

(2) *Description.* The idea of a *purely* descriptive statement is

rejected by Hanson as incompatible with the nature of the act of scientific seeing. We do not merely *see* things, we see them *as* things of a certain kind. We can, of course, by exercising a bit of control, bring ourselves to see in a bare, lustreless, phenomenal way. But that, Hanson argues, is not the kind of seeing that is relevant to scientific inquiry, nor is it the logically fundamental or logically prior way of seeing. Scientific description involves *seeing as* and, indeed, *seeing that*. Both of these are dependent upon the theories we hold or the hypotheses we entertain; hence, description in the pure, theory-independent sense does not figure importantly in the scientific enterprise.

(3) *Fact*. Science is built on a foundation of fact. But what facts are is not easy to say. Hanson, following P. F. Strawson, contends that facts are what we describe in "that-clauses." He maintains further that the link between observation and facts resides in the use of expressions such as "seeing that" which are, one might say, ambiguous with respect to the issue of "pure" perception versus "pure" conceptualization. "Seeing that" something is the case seems to involve a mixture of seeing and thinking. Accordingly, it is misleading to say simply that we "see facts." Only in virtue of the theories we hold and the hypotheses we entertain can we identify certain states of affairs as facts at all.

(4) *Hypothesis*. Newton asserted, *"Hypothesis non fingo."* This is difficult to translate and even more difficult to believe: "I do not fashion (feign? frame?) hypotheses." Hanson, following the more common view among scientists, claims that hypotheses are essential to the process of scientific inquiry and discovery. Their role is to mark out for us the *form* of the facts, to separate the relevant from the irrelevant. They do not literally *make* the facts for us (nature takes care of that!) but in a certain sense—aptly captured by the Latin *fingere*—they frame or fashion the facts. Thus, the title attached to one of Hanson's crucial chapters, *"Hypotheses facta fingunt"*: "Hypotheses fashion the facts."

(5) *Law*. Laws have a scope or range of application and they also embody a rule, according to Hanson.[1] The former component is a proposition and can be described as true or false, probable or improba-

---

[1] The view of laws expressed here should be qualified by reference to Hanson's argument in Chapter 5 of *Patterns of Discovery*. There he shows that *any* general discussion of laws must take into consideration wide variety of uses and forms of statement which scientists call "laws." In general, Hanson did not believe that any single, univocal analysis could apply to all laws.

ble, depending on what the facts are. The rule component, however, cannot be described in this way. It is a rule for representing or analyzing the facts, a rule for drawing inferences about nature if you will, but not a proposition *about* nature. In science, laws are not so much laws *of* nature as laws of *our methods of representing and reasoning about* nature. Especially they should not be construed as "well-established hypotheses," as many philosophers of science seek to do. For hypotheses and laws perform entirely different functions in scientific inquiry.

(6) *Theory*. Hanson recognizes and distinguishes two types of theory: the formal, axiomatic sort of theory which predominates in pure mathematics and the somewhat more "open" theories (e.g., geometric optics) which predominate in empirical science. In the former, the meanings of terms are implicitly defined by the axioms but otherwise left undefined. In research science, however, theoretical terms get actual, empirical meaning. Like the laws they contain, theories should not normally be categorized as true or false, probable or improbable, however. The theory of classical mechanics is applicable or inapplicable, depending on the dimensions of the phenomena under investigation, but it is, according to Hanson, not to be considered a "false theory" on that account.

(7) *Cause*. Causal language is theory-laden. The concept of causality is shot through with "charged" words and phrases. Two misconceptions about it abound, possibly as a result of its theory-laden character. One is the assumption that causal claims signal the existence of "necessary connections" in nature. The second is the causal-chain model. Hanson rejects both of these, locating "necessity" in the theoretical inference structure which backs up our causal assertions about the world and dismissing the causal-chain model as mythology. He further maintains that causality is probably less important to the working scientist than most laymen believe.

(8) *Principle*. A scientific principle is an indispensable component of a theory. Thus, the principle of rectilinear propagation cannot be withdrawn from geometrical optics without the entire edifice collapsing.

On the other hand, certain things which philosophers *call* scientific principles are entirely different sorts of entity. Thus, as we have seen, the principle of uniformity in nature is, on account of its vagueness and/or mere platitudinousness, quite different from principles which actually occur in scientific theories.

(9) *Explanation*. The process of explanation involves all of the elements discussed above. The word "explanation" is a synoptic title for the resultant accounts of the facts scientists arrive at after observing, describing, posing hypotheses and testing them, applying laws and determining the limits of their applicability, theorizing, investigating causes, and invoking principles.

The overall thrust of this point of view could well be summed up in the expression "Instrumentalist-Empiricist" (though the two approaches to philosophy of science this expression suggests are widely believed to be incompatible and Hanson was not unaware of the difficulties of reconciling them):

"Instrumentalist" because it regards laws and theories as *instruments* (essentially rules) for dealing with nature, rather than as statements about nature; also, because it characterizes the meaning of theoretical and observational terms by reference to their *use* in the setting of actual research;

"Empiricist" because it affirms the validity of the division of all knowledge into analytic and synthetic propositions, and because it insists upon the existence of hard, objective fact as a basis for science.

It is a measure of Hanson's achievement that he was so well able to bring together the main threads of these two dominant movements in Twentieth-Century Anglo-American philosophy and weave them into a coherent view of the nature of modern science.

INDEX

Index

Achinstein, Peter, 8
Ames, A., Jr., 154ff., 158
Ampere, 52, 223
analogies, 317–318
approximation, 323, 325
Arber, Agnes, 65
Archimedes, 54, 61
Aristarchus, 239
Aristophanes, 115
Aristotle, 29, 50, 66, 183, 185, 196, 261
Austin, J. L., 60, 172, 188
averages, ch. 24, (see also *mean, median, mode*)
   dispersion of, 393–394, 398–400
   properties of, 394
Avogadro's law, 339, 340
Avogadro's number, 16
Ayer, A. J., 60, 71

Bacon, Francis, 278–279
Bahnsen, P., 159
Barker, S. F., 362
Bartlett, Sir F. S. C., 165
Bentley, I. M., 162–163
binominal expansion, 384–385
Bohm, David, 244
Boring, E. G., 161
Born, Max, 319
Boyle, 16, 223, 339, 340, 345
Boys, 33
Brain, W. R., 65, 67, 75
Braithwaite, R. B., 34, 212, 267, 391, 362
Bridgman, P. W., 8, 35, 40
Broad, C. D., 60, 71
Brouwer, 43
Brown, G. B., 189
Brown, Spencer, 404

Campbell, Norman, 8, 200
Capablanca, 16
Carrol, Lewis, 180
cause, 13, 169, ch. 16–18, 343, 346ff., 423, 426
   Aristotle's theory of, 279ff.
   final, 278–279
   Hume's theory of, 287ff., 290–291
   laws, and theories, ch. 18
   necessary and sufficient, 273–274
   and necessary connection, 288–297, 309–310
   requirements of, ch. 16, 285–288
Cavendish, 33, 52
Caws, Peter, 200
Charles' law, 339, 340
Chesterton, G. K., 28
Churchman, C. West, 8

Cohen, M. R., 8, 348, 401
complementarity, 319–320ff.
Compton, A. H., 207–208, 210, 211, 213–219, 252, 254
consistency and truth, 20–21, 24, 363
Copernicus, 78, 239–240, 243, 319, 333
copying relation, (see *meaning, picture theory of*)
correlation coefficients, 400–402ff.
    Pearson's, 402
Coulomb, 52, 239, 245

D'Alembert, 379, 382
Darwin, 78, 109, 214, 237, 244, 296, 333
DaVinci, Leonardo, 66
Davy, Sir Humphry, 251–253
DeBroglie, Louis, 319, 339, 341
deduction, 17, 34, 337–339, 349ff., 365–366, 371, ch. 25
definition, 17, ch. 2, 191–192, 265ff.
    axiomatic, 33–34
    by analysis, 29–30, 31
    by synonym, 28–29, 31
    by synthesis, 30–31
    denotative, 32
    historical-authoritative, 28
    implicit, 31–32, 266–267, 342
    indirect, 34
    inductive, 37
    in mathematics, 265–266
    operational, 35–36, 44, 49ff.
    ostensive, 32
    regular, 33
    stipulative-prescriptive, 28, 37
    systematic, 30
Descartes, 74, 278–279
DeMorgan, Augustus, 236, 385–386
Dingle, Herbert, 341–342
Donne, John, 66
Dryden, John, 299
Duhem, Pierre, 103, 200

Eddington, Sir Arthur, 60, 149, 150, 261, 269, 322
Einstein, Albert, 236, 244, 245, 253, 269, 270, 316, 319, 333, 354
Ellis, Brain, 8
Euclid, 33–34, 37, 232, 262, 264, 269
experiment, 17, 149, ch.14, 263, 333, 346–351ff.
    and measurement, 49–56
    crucial, 201, ch. 14, 250–255, 256
explanation, 300ff., ch. 18, 308, 342, 343, 427

facts, 13, 17, 85, 169, ch. 10–11, 204, 209–210, 333, 343
    and events, 188–189
    and true statements, 189–190
Faraday, 16
Fehling's test, 19
Feingold and Kingsley, 165
Fisher, Sir Ronald, 164, 362, 404, 406, 419
Flugel, J. C., 159
Foucault, 252–253, 256
Fox, C., 165
Frege, 43
Fresnel, A., 203–204, 209, 211, 213
Freud, 110, 227

Galen, 66, 109, 168
Galileo, 78, 109, 152, 182, 223, 230–239, 333
Gauss, 254
geometry,
    euclidean vs. non-euclidean, 254–255, 256–258, 269–270
    Riemann's 269–270
Gibson, J. J., 161
Goethe, 105, 109, 110, 185
Graham, C. H., 159

Hall, A. R., 200
Halliburton, W. D., 30, 222
Hanover Institute, 154

Hardy, G. H., 260
Harvey, William, 168, 169, 200–237, 249
heat, theory of, 251–253
Heath, Sir T. L., 34
Heaviside, 317
Heisenberg, Werner, 319, 333
Hempel, C. G., 8, 200
Herodotus, 87
Hertz, Heinrich, 204–205
Hipparchus, 239
Hirst, R. J., 60
Hitler, 391
Holmes, Gordon and Horrax, Gilbert, 67, 136
Hospers, John, 8
Hume, David, 172, 200, 287ff., 290–291, 356, 414, 417
Huxley, Julian, 322
hypothesis, 13, 17, 85, 88, 169, 201, 210–211, 217, ch. 13, 227 ch. 20, 343, 358, 370–371, 417, 425
 false, 235–236
 simplicity of, 238–244

icons, 142–145
inconsistency and falsehood, 20–21, 24
induction, justification of, ch. 25, 11–12, 17, 348ff., 364–365ff., 406
Isherwood, 127

James, William, 106
Jammer, Max, 200
Jeffrey, R. C., 362
Jeffreys, Sir Harold, 419
Johnson, M. L., 109, 152, 158, 168
Johnson, Samuel, 41

Kant, Immanuel, 12, 74, 149, 177, 181, 196, 248, 261, 269
Kepler, 16, 235, 237, 340–341, 352, 355
Keynes, Lord, J. M., 362, 419
Kingsley, H. J., 161

Kirchhoff, 52, 317
Kneale, William, 362, 372, 391
knowing that, 126, ch. 8
 and knowing how, 176
Koehler, 92, 106
Körner, Stephan, 200
Kuhn, Thomas, 200
Kulpe, O., 161
Kyburg, Henry, 8, 362

Lactantius, 240–241
Lamarck, 235
Lambert, 325–326
Laplace, P. S., 299
laws, 16, ch. 19–20, 425–426
 as forms of regularity, 340, 352
 causal, 290, 293, 299, 308, 346ff.
 of nature, 323–326, ch. 20, 352ff.
 numerical, 55–56
 scope of, 330, 335, 340, 342, 352
Lear, Edward, 180
Leeper, R., 165
Leeuwenhoek, 229
Leibniz, 16, 182
Lewes, G. H., 166
Lewis, C. I., 60
light, 201ff.
 quantum theory of, 209ff.
 wave-theory of, 201–204ff., 252–253, 255, 256
Linnaeus, 29
Locke, John, 106, 151
logical reconstruction, 14, 141, 248
Lysenko, 102

Mach, Ernest, 36
Malinowski, Bronislaw, 181
Malpighi, 229
Mann, Ida, 60, 65, 106, 107
Markham, S. F., 404
Maxwell, Clerk, 244, 314–317
mean,
 arithmetic, 394–396
 weighted, 396

meaning (see *definition*)
  in mathematics, 267ff.
  picture theory of, 132ff.
  vs. evidence, 370
measurement, ch. 3, 342, 343, 393
  additive, 50–51, 398
  and counting, 43ff.
  derived, 51, 54–56
  and experiment, 49–52
  and extensive properties, 51, 57
  fundamental, 51, 54–56, 398
  and intensive properties, 51, 54–55, 57
  and numerical laws, 54–56
  ordinal scales of, 48
  rules of, 44, 45, 47, 52–53
median, 398
Mendel, 16, 164, 333, 336, 406
Mercator, 325–326
Mill, J. S., 172, 188, 200
Mills, F. C., 401
Milton, John, 64
mode, 396–398
Molotov, 391
Monte Carlo fallacy, 390, 402

Nagel, Ernest, 8, 200, 348, 362, 391, 401

observation, 12–13, 17, part II, 108, 169, 245–250, 263
  vs. sensation, 82–83, 85–88, 94, 120, 125, 126–127, 147, 149–151, 246
Oersted, 52
Ohm, 52
operationism, 35–36

Pap, Arthur, 8
Pasteur, 78
Pauli, 16
Pavlov, 244
Peirce, Charles Saunders, 142
phenomenalism, 70–85ff., 92–93, 95, 98, 102, 110, 111–112, 114–115, 124, 125, 131, 132, 139, 141, 163, 168–169, 247
photoelectricity, 204–207
Piaget, Jean, 278
Plato, 183, 187, 277
Popper, Sir Karl, 322, 362
presuppositions, (see also *principles*) 12, 242, 356
Price, H. H., 60, 71, 94, 125ff., 247
principle of indifference, 374, 386
principles, ch. 21, ch. 25, 426
  and hypotheses, 358ff.
  and laws, 335ff.
probability, 343, part IV
  and beliefs, 385–388, 389
  calculus of, 372–376, ch. 23
  conditional, 380–381
  of events, 373ff., 389–390, 391–392
  of hypotheses, 33, 335, 370–371, 388–389
  and inference, 18, 364ff., 368, 412ff.
  and relative frequency, 368–369, 371–372, 388, 391
proof, 22–23, 258ff., 364
propositions, ch. I, 310–311
  grammatical vs. logical forms of, 321–322
  necessary vs. contingent, 18ff., 332–335
Providence, 21
Ptolemy, Claudius, 21, 239–240, 243, 318, 363
purpose, ch. 16ff., 423
Pythagorean theory, 258–259, 262

quantum mechanics, 316, 318–321, 424

Reichenbach, Hans, 362
Rhine, J. P., 401
Robinson, Richard, 8
Russell, Lord Bertrand, 43, 71, 74–

75, 132, 141, 172, 188, 247, 299, 351, 419
Ryle, Gilbert, 200, 280

Salmon, Wesley, 362
Savage, Leonard, 362
Scheffler, Israel, 200
Schroedinger, E., 319
science,
   history of, 9ff.
   language of, 16–17, 22, 124, 127, 137, 186, 266, ch. 19, 337, 343, 355
seeing,
   aspect shifts, 91ff., 159–160
   facts, 172, 172ff., ch. 11, 209–213
   and interpreting, 69, 86, 93–94, 98, 101–102, 119, 130–131, 169
   and knowing, 113, 115, ch. 8, 166 (cf., *knowing that*)
   linguistic component of, 124ff., 131–132
   and "organization", 91, 95, 103, 104
   phenomenalist theory of, 70–85ff., 92–93, 95, 98, 102, 110, 111–112, 114–115, 124, 125, 131, 132, 137, 141, 150ff., 163, 168–169, 247
   picture theories of, 125, 127, 131, 132ff.
   and psychological "set", 109, 159–168, 183
   and retinal reactions, 63–69ff., 109, 111, 125
   the same thing, 62ff.
   and seeing as, 105, 107, 110, ch. 7, 152, 172
   and seeing that, 112ff., 132ff., 152, 172, ch. 11, 245ff.
   theory-loaded character of, 110, 149ff.
   vs. feeling, 93
   vs. thinking, 86, 93–94, 102, 123, 125, 128, 138

self-evidence, 260–261
sense-datum, (see *phenomenalism*)
Sherrington, 106
simplicity, ch. 14, esp. 243
Smart, J. J. C., 8
Snell's law, 16, 229ff., 232, 307, 335, 337, 338, 339, 340, 341, 345
Spinoza, 278
Steno, Nicholas, 33, 37, 41
Strawson, P. F., 60, 193, 391, 425
systems
   formal, 258–270, 364, 385
   hypothetico-deductive, 264ff., 337–338

that-clauses, 124, 184
theoretical terms, 34–36
theories, (see also *systems*) 13, 85, 110, ch. 18, 306–308, 310, 313, 343, 417, 426
Thomson, J. A., 172
Thomson, J. J., 205
Thucydides, 87
Toulmin, Stephen, 167, 200, 329–330ff.
Trommer's test, 19, 30

uniformity in nature, ch. 21, ch. 25, 423

Vernon, M. D., 60, 132, 136, 141, 172, 188
Vesalius, Andreas, 169
Vicker's hardness test, 48, 49
Vigier, 244
Von Mises, Richard, 362, 419

Walter, Grey, 75, 85
Wertheimer, M., 165–166
Whewell, William, 60, 66–67
Whitehead, Alfred North, 188
Wisdom, John, 60, 132, 136, 141, 172, 188